Ain't Misbehavin'

Ain't Misbehavin'

The Groundbreaking Program for Happy,
WELL-BEHAVED PETS and Their People

JOHN C. WRIGHT, PH.D.,
with Judi Wright Lashnits

RODALE

© 2001 by John C. Wright and Judi Wright Lashnits
Illustrations © by Christopher Mills
Cover Photographs © CMCD/PhotoDisc, Patricia Doyle/Stone

Printed in the United States of America
Rodale Inc. makes every effort to use acid-free ∞, recycled paper ♻.

Cover Designers: Tara Long and Christopher Rhoads
Interior Designer: Joanna Williams

Library of Congress Cataloging-in-Publication Data
    Wright, John C., date.
    Ain't misbehavin' : the groundbreaking program for happy, well-behaved pets and their people / John C. Wright with Judi Wright Lashnits.
        p.   cm.
    Includes bibliographical references and indexes.
    ISBN 1–57954–195–X   hardcover
    ISBN 1–57954–519–X   paperback
    1. Dogs—Behavior.   2. Cats—Behavior.   3. Dogs—Behavior therapy.
   4. Cats—Behavior therapy.   I. Lashnits, Judi Wright.   II. Title.
   SF433 .W74   2001
   636.7'089689—dc21   2001031936

Distributed to the book trade by St. Martin's Press
  4 6 8 10 9 7 5   hardcover
2 4 6 8 10 9 7 5 3 1   paperback

Visit us on the Web at www.rodalestore.com, or call us toll-free at (800) 848-4735.

RODALE

WE **INSPIRE** AND **ENABLE** PEOPLE TO IMPROVE
THEIR LIVES AND THE WORLD AROUND THEM

Dedication

To my mentors—Domino, Peanut, Roo-Roo, Charlie, and Lucy—the loving cat and dogs who are teaching me the secrets of cat- and dogdom.—JCW

To Tom, Erin, and John.—JWL

Acknowledgments

We thank our literary agent, Jim Frenkel, for his confidence and hand-holding; Angie and Tom for their patience; and our editor, Elly Phillips, for her great enthusiasm, vision, and top-notch editing.

I (Judi) offer kudos to my kids, Erin and John, for keeping the volume down while I worked, and kisses to Sassy and Scout, who live on in my memory.

I (John) cannot express the gratitude and warm fuzzies I've felt for my four-legged girls—Domino, Peanut, Roo-Roo, Charlie, and Lucy—who share my home and allow me to learn from them, and for just being themselves. My deepest thanks to my clients and their critters, whose love for each other provided the inspiration for this book, and to my wife, Angie, the only two-legged female in my life, for her continued love and understanding.

Contents

Part One

*Chapter
One*

How to See Your Pet
through a Behaviorist's Eyes

Hi, I'm Dr. John Wright. For the past 20 years, I've been living an adventure story with more twists and turns than a Hitchcock thriller. My "official" title, complete with Ph.D. after my name, is "certified applied animal behaviorist." I am one of fewer than 50 such behaviorists in North America.

But I've also come to think of myself as a real-life pet detective—one who looks for clues to solve sometimes mind-boggling behavior problems presented to frustrated owners, puzzled veterinarians, and baffled trainers by a wonderful menagerie of dogs and cats who are doing things their people can't cope with or correct on their own. I'm talking about tearing up the house every day while Mom and Dad are at work; hiding in the bathtub whenever there's a thunderstorm; peeing all over the brand-new carpeting; wailing or barking all night long; or biting the kid next door. And those are just the ordinary problems!

Save That Pet!

I heard a popular radio psychologist field an interesting call the other day. A distraught mother was calling in, wanting to know how to break the news to her 4-year-old son that the family cat was about to be "put to sleep."

"How old is the cat, and what's wrong with it?" asked the therapist.

Well, the cat was 6 years old, and it was spraying. The family vet hadn't been able to find a way to stop the spraying, and putting the cat outside wouldn't have been "fair" to the animal, the caller explained.

"So you are going to kill it because it's spraying?" asked the radio psychologist incredulously.

"Yes, it's spraying <u>inside the house</u>!" the caller replied as if that justified everything.

"I'm sorry, but I can't accept your premise," the radio personality said. "Find someone who knows something about cats and fix the problem," she advised. "And by the way, I sure hope your <u>kid</u> never misbehaves!"

I had to smile when I heard this exchange. Not so many years ago, that cat would have been unceremoniously dumped outside, left at a shelter, or otherwise disposed of for a transgression that today is rightly seen as correctable. It also explains, in a nutshell, what I mean when I say that I work with dogs and cats with life-threatening problems. We have seen that desperate people sometimes resort to desperate measures.

As the cat in the phone call was, I hope, spared from discovering, certain behavior issues can doom a pet to an early demise if the owner isn't aware that help is at hand. I don't want to see cats and dogs defeated by common behavior problems, or just "put up with" because they are doing annoying things that no one knows how to resolve. So in <u>Ain't Misbehavin'</u>, I want to tell dog and cat owners how to use the understanding and tools of an applied animal behaviorist at the turn of the 21st century to ensure that their own pets have long, happy, and problem-free lives.

Distraught owners call me when they are struggling with the tough behavior problems that seem to arise for nearly all pet owners sooner or later, even those who have been through puppy kindergarten or obedience training courses. And after thinking about it and trying a few things and having little or no success, they ask their veterinarian what to do, and the vet says to ask me.

My technique is a little bit different from that of many of my colleagues to whom you might bring your dog or cat for a consultation. That's because I take my bag of tricks—well, my briefcase and pen, anyway—into the home where the trouble is occurring. In their own environment, I can see the dog or cat and the owners, "warts and all." I think there is no substitute for understanding the animals' own turf and seeing the way they live. With any luck at all, I can uncover the clues that are hiding in the corners of the owner's mind or under Fluffy's or Fargo's monogrammed water bowl.

The trick is in using those clues—the history of the animal's behavior, what the owner has observed, and what I can discern with my own two eyes and sometimes other senses—to figure out what gives and to decide what to do about it. Here I draw upon my own education and research and that of my colleagues and come up with some ideas on how to fix things.

I know that all of us want the very best relationships we can forge with our beloved companion animals. We want to help them when problems arise, not punish them or abandon them. With the information in this book, you will never need to feel that giving up your wonderful pet is the only way out.

Don't Be in Denial

Before they get desperate enough to consider giving up their dogs and cats, the owners I meet in the course of everyday life are more likely to be in denial about the things their pets are doing that make living with them less than copacetic. The standard response to trouble is just to ignore it and hope it will go away.

Well, it usually won't go away, and turning a blind eye to certain problems can be dangerous. I have walked into more than one home where a dog is growling or snapping at a 2-year-old child. The parents, however, are sure that the dog won't bite because Brutus is such a wonderful pet. They just wish he wouldn't growl!

Or they adore their cat; she is usually so affectionate. It's just that she has this one itty-bitty problem. When the husband walks down the hall to the kitchen, Snowball latches onto his calf and goes along for the ride.

These clients are risking serious injury by misunderstanding the bad situation they are in because they don't want to think of their pets as bad or problematical. It's natural for people to be willing to overlook one or two instances of bad behavior almost as if it didn't happen. Or maybe it was an accident. Or maybe it's just that even pets have bad days once in a while. That is not unreasonable!

But if the bad behavior becomes a pattern that is impossible to ignore, it needs to be dealt with. Today there are a dizzying array of resources out there so that even the loudest cry for help can be swallowed up in the general clamor of pet products, advice columns, Internet chat rooms, and books of advice from self-styled experts. There is precious little good, humane, research-based advice that will resolve problems for animal lovers. So I'm not surprised when people come to me saying that they have tried "everything," and "nothing" has worked. Maybe their pet needs psychoanalysis!

Dogs and Cats on the Couch

Ten years ago, the phrase "animal psychologist" was still drawing snickers. People thought that the dog or cat "shrink" was a high-priced trainer going into the homes of Hollywood stars to tell people that little Foo Foo needed more love, attention, and hugs. When I would arrive at distraught clients' homes with my briefcase, I would be ushered in quickly, lest the neighbors catch wind of the "head doctor" coming to treat the crazy cat or dysfunctional dog.

Today, the applied animal behaviorist is becoming more accepted as part of a pet support team that can include veterinarian, groomer, walker, trainer, kennel owner, and sitter. But as they say, a little knowledge can be a dangerous thing. People who wouldn't think of swatting their dog with a newspaper to inspire good behavior are going to the opposite extreme without understanding what they are doing. They've chosen what they think of as a Freudian route of trying to psychoanalyze their dog in order to "cure" it.

This must be what those dog shrinks do—put Fido on the couch so that they can learn to understand him and solve the problem. We can do that ourselves! But the absurdity of this approach is matched only by its futility. In the end, I get the phone call anyway, if the family hasn't already given up on the pet for its "stubbornness" or "inability to learn."

It leads to explanations like this: "Well, doctor, we thought that if we looked at these incidents and really talked about them and then faced what Cheyenne is trying to tell us when he destroys the entire living room... Well, we've decided that he's awfully hurt when we go out to the movies. Next time, we'll explain to him that we really, really love him and that we will be back soon. We shouldn't have just closed the door and left."

But "explaining" the situation to the dog won't help. It may seem reasonable and thoughtful, but it betrays a lack of knowledge of the animal's cognitive capabilities, which I will explain in chapter 3. And in a bigger and longer, guilt-ridden goodbye, these hypothetical dog owners have chosen a seemingly logical solution that's doomed to failure. That's because the more fuss the "parents" decide to make over their dog prior to going out, the bigger mess they are likely to find when they get back, as you will learn in our discussion of separation anxiety in chapter 14. So now the owners are calling me in an even more alarmed state than when they began their homegrown shrink-style "therapy."

Leader of the Pack

At the other extreme is the pet owner who hopes that a great deal of control and discipline will result in proper behavior. After all, the dog "knows better" than to nip at the hand of the master. Doesn't she? "We've decided that Silky has to do exactly what we say, when we say it," a client with a biting dog tells me. In this "nothing-in-life-is-free" strategy, the dog has to sit or stay or lie down to earn the privilege of eating, going outside to relieve herself, or even chewing on a toy. Many trainers advocate the kind of relationship in which the animal must do something for you to earn whatever it is she wants.

It is possible that this strict regimen will discourage the dog from nipping. On the other hand, this approach may not work if its only consequence is to discourage the dog! In the interests of discipline, the main effect could be to create a sort of robot that may be punished for not following "orders." At the very least, the poor dog clearly understands that she's subservient to you.

But isn't that what it's all about? Don't we have to become the "alpha animal," making sure that the cat or dog knows who's boss? Shouldn't we take charge of the "pack," which the ordinary family magically morphed into sometime during the politically correct 1990s?

Well, I hate to admit this—since I was one of the handful of people trying to correctly understand and document real behavioral characteristics of dogs and cats in the 1970s—but we had it wrong.

The theory was that people should behave like members of the pack, and the animal needed to be controlled by its dominant member. So I was among those behaviorists gripping a young dog by the fur on its neck to hold the too assertive pup down in a moment of discipline, just like the mother wolf was supposed to do. This technique turns out to be infrequent even for mommy wolves, and it is clearly not an appropriate method of instruction for a dog or cat owner. You won't hear that from many gung-ho "alpha animal" trainers because the word hasn't really gotten out yet. But now it's time to move on to the

(continued on page 8)

DR. WRIGHT'S
C A S E B O O K

Bruno

The alpha male syndrome seemed to be playing itself out with a group of five housemates I visited recently to see what we could do to help their flat-coated retriever, Bruno, return to being the good-natured dog they had started out with some years before. The 5-year-old male had taken to growling at all the people in the house, which alarmed them to no end.

And I can't say that I blamed them. Bruno was now growling or snapping at the very friends he used to love playing with. "He even growls if we just say his name and go to pet him," Richard B. told me. "Why is he trying to intimidate us and take over the pack like this?" he asked, perplexed. "He never acted like this before!" Richard reached for the dog, who growled as if on cue. "See?" he said, withdrawing his hand.

I saw the body language and communicative behavior of a submissive animal, not an aggressive alpha animal. He certainly was not bent on taking over the family's top spot. Learning to read your pet's body language and other visual signals will help steer you toward an understanding of what's really going on in your relationship. By reading his signals, I could see Bruno giving me the first clue that I needed to solve this mystery and discard the "alpha" jargon.

In this case, it seemed to me, Bruno was reacting out of either fear or pain or both. The dog's tail was tucked, and he had backed away from the young man's hand. That is not exactly a "dominant" reaction!

"Has Bruno had a physical lately?" I asked the assembled housemates. Yes, he had been to the vet; there was some problem with his hips. He was on medication of some sort, but no one was too sure if it was working.

Bingo. There was the second clue, and now my hypothesis began to take on a logical shape.

I asked a few more questions, focusing on exercise and any kind of stress to the dog's hips. The dog and his human friends loved to play

The effects of vigorous exercise contributed to Bruno's reputation as an alpha male.

Frisbee in the backyard or the park. They had a really good time last week, on Thursday night, the friends told me. "And how was his mood on Friday?" I asked. One of them explained that Bruno had not let anyone near him.

Poor Bruno! He loved the attention, but he was protecting himself. He was in pain from his hip problems, which are prevalent in more than a hundred breeds, so he did the only thing he could to keep people away. His pain would have been especially acute after an evening of racing and jumping for the Frisbee. His behavior the next day was what he had learned would keep his friends away from him at any cost.

When the housemates understood that there was a physical root to the behavior problem, we were able to work together to solve the problem. I consulted with their veterinarian, who prescribed something for the dog's joints so that he would feel better physically. The owners were careful not to aggravate Bruno's condition with a lot of roughhousing.

next level of behavior training. I will show throughout *Ain't Misbehavin'* that our role is not to dominate but to lead and enable the pets in our household to fulfill their needs.

Over the last decade, this wolf-envy among dog owners has become the "in" theory, tossed around by everyone who could tune in a daily talk show. This concept became so popular in recent years that the phrase "alpha male" came to replace the earlier "macho man" as the trendiest concept in the social fabric of our culture. Unfortunately, thinking of our male dogs—or cats—as alpha animals can stir up a lot more trouble than the original behavior problem.

11 Steps to a Well-Behaved Pet

I've met hundreds of cats and dogs during the last two decades. Most of them had owners who tried different approaches without success. That's because they didn't know what questions to ask themselves or what the answers meant in terms of behavior.

However, with a basic understanding of what their pet's needs are and how they can go about helping their dog or cat fulfill them, clients are nearly always able to help resolve the situation or improve it to the point where everyone is coping nicely again. We still have much to learn about what our domesticated animals are like; I learn something new almost weekly.

The important thing is that you know your pet better than any outside expert can. Together, we can make a great team for improving the daily lives of both your family and your dogs and cats.

So, if talking them out of their bad habits doesn't work, bullying them into shape is wrong, and treating the family like a wolf pack turns out to be misguided, how do we effectively deal with these sometimes naughty dogs and cats of ours? What tools, tactics, and procedures can a certified applied animal behaviorist like me use to help owners like you with their pets so that problems don't crop up? What resources can I give you to handle any problems that should arise? Here are 11 basic steps for you to take when a problem comes up. Within the chapters of *Ain't Misbehavin'*, you'll learn the rest of the story.

🐾 Step 1

Figure out what the dog or cat is doing wrong. This suggestion may seem obvious, but sometimes owners focus on what they think their pet is feeling, or why it is punishing them, instead of concentrating on the actual behavior. You need to think about what actually happens in terms that are objective and verifiable. "Spike chews up my shoes" is more tangible than "he gets upset." The dog chewing on the shoes is the behavior you will be documenting in Step 2. We'll get to his "feelings" later!

🐾 Step 2

Record the "bad" behavior. Put down the rolled-up newspaper or the bottle of migraine tablets, and get a piece of paper and a pencil or open up a file on the computer called "Muffy's Mistakes." Write down how many times the unwanted behavior has occurred in the past seven days, so that you will have a baseline from which to work. That's really the only legitimate way to see if the treatment program is reducing the problem. The numbers, week to week, will tell the tale!

🐾 Step 3

Note when "bad" things happen. You will need more tracking than just the number of times the dog or cat is misbehaving. The time of day things go wrong is another thing to note. With this information, gathered over the course of a week, you'll find out whether you need to work on changing things that are happening only from dinnertime to bedtime, from sunup to sundown, during every waking minute, or at random times 24 hours a day. (You have my sympathies if it's the last!)

Most behavior problems fall into a rather predictable pattern of time and place, so you'll be able to focus your energies accordingly. I remember one cat whose owner was going berserk because the feline was "constantly" spraying the walls. That rather general timeframe didn't really help me get to the root of the problem. I asked for his help in trying to pinpoint the time of day when the problem occurred.

After some thought and observation, he realized that the cat was actually spraying only within the first hour after everyone went to bed. It turned out that a rival tom was coming around outside at that same time each night! After we isolated the problematical time, both the cause of the misbehavior and the likely cures became quickly apparent.

🐾 Step 4

Jot down the circumstances. Going hand in hand with the second and third steps is the question, "In what situations does it occur?" Does the dog bite only when you walk toward his food bowl, or does he bite anytime he's holding a toy or other object? Is the cat missing the box all the time, or just when you're in the same room?

Perhaps she only scratches you when you try to get her out from under the bed. Or possibly the dog tears up the furniture every time you come home from work and then go out again but never touches it when you're at work. Follow these first four steps carefully; they'll give you the clues you need to solve your pet's problem as we begin our detective work.

🐾 Step 5

Find out if something's wrong. The next thing to do is to make sure that the cat or dog does not have a physical or medical problem. Have your veterinarian do a thorough examination. Frequently, something physical is causing or contributing to the problem behavior, as was the case with Bruno. You should also know the working status of all the dog's or cat's senses.

A dog who isn't able to hear a command is not likely to obey it. And a cat whose eyes are clouded by cataracts may just mistake your finger for a tasty treat. Isn't a physical explanation for your pet's problem more palatable than assuming that "she just hates me"? So you need to take care of any medical problems first and then observe if the behavior changes—hopefully for the better!—or remains the same.

🐾 Step 6

Stock your training toolbox. This step is a little trickier, but it's very important if your pet is going to change its misguided ways anytime soon! Try to become familiar enough with your dog or cat so that you're able to perceive what he or she is experiencing. In other words, look at the world from your dog's or cat's point of view. If you need to get down on all fours and follow her the next time she explores the house, so be it. Use your knowledge of her behavior and responses to compile a list of words and phrases that make her feel good. Try to put them in order, starting with the "best" and going to less best. Obviously, words that excite her the most will be motivating—"Go for a walk?" will usually rouse the most devoted couch potato, whereas "Where's your mouse?" might only rate one eye opening as the cat snoozes on the sofa.

Objects should also be included on this list. The sight of the leash or a dog treat has been known to induce paroxysms of joy in a dour Great Dane, and the sight and/or smell of dried catnip might drive an indolent Persian into a frenzy of enthusiastic purring and rubbing. These words and objects are the tools you'll use to change your pet's motivation and, therefore, un-wanted behavior.

🐾 Step 7

Form a hypothesis. The next thing you need to do is make an educated guess as to why the pet is misbehaving. (No, it's not because he's a "rotten apple," or she is "spiteful" or "spoiled." Name-calling will get us nowhere. Remember that you need to take the cat's- or dog's-eye view!) If you can get to the reason for the behavior problem, then you can usually remove the cause and be more successful than if you simply try to keep your pet from doing it.

At this stage, I'm usually with the clients in their home, and we create the hypothesis together. But the knowledge you have of your own pet—and what you'll learn as you go through this book—will help you become your own pet detective. You'll be able to use the insight you already possess regarding the companion animal that you know much better than I could ever hope to. So go ahead and offer a hypothesis about why Cooper is suddenly losing his house training or running from the toddler.

This may take more than just mental effort. You may have to walk out of the house and sneak back in half an hour later to see if Cooper has left a deposit on the Persian carpet. Or you may need to come home at noon to see if the deed has been done then.

In any case, you will be able to apply common sense and your knowledge of Cooper to come up with an idea or, more scientifically, a hypothesis. Try to figure out whether the behavior is fulfilling a psychological, biological, or social need, and what you would rather have your pet do to achieve the same end without driving you nuts. If he is peeing in the house (biological need), you simply want to divert him to pee outdoors or in his box instead. The need will still be met, but it will be accomplished in an acceptable way. I will elaborate on pets' needs in later chapters, particularly those in part 3.

🐾 Step 8

Interrupt the bad behavior. With an educated guess as to why the problem is happening, it's time to interrupt the dog's or cat's behavior just before its occurrence. Better yet, prevent it from recurring by giving your pet something more interesting to do. You may want to block access to the area where the misbehavior is occurring. This is when you should use that list of "happy" words and objects to put your dog or cat in a mood that will keep him from causing trouble. You need to draw on one of those emotional words that will get the dog's or cat's attention and let him essentially forget what he is about to do that you really wish he wouldn't do.

🐾 Step 9

Distract your pet. Teach the dog or cat to get rid of its energy by doing something that is appropriate or that is in the appropriate location or that fulfills the original need, but in an acceptable way. In other words, con him into doing the right thing. For example, Woodstock might be protecting his food bowl every night just before you sit down to your own dinner. So remove his bowl before he begins to protect it and, using a "good" phrase from the list, get him involved in some object play away from the table to stop the habit he has started. If he is happy and playing away from the table and his mouth is occupied by a play toy, he can't be growling and threatening family members as they sit down to dinner. And now you have given him something else to do that he can still get a lot of pleasure from or that fulfills the social need that may have driven him to the bad habit in the first place.

🐾 Step 10

Pour on the praise. Now that your dog or cat is doing something right, praise the heck out of her for her compliance! Praise can take the form of social play, verbal stroking, food, a scratch under the chin, or whatever your dog or cat responds to best. It's up to the two of you. Praise is a terrific motivator for continued good behavior.

🐾 Step 11

Add consistent reinforcement. Now, with the baseline established, be as consistent and persistent as you can with the treatment program. The number of incidents should lessen if you are doing the right thing. To make sure that you are on the right track, it's always helpful to compare a whole week of behaviors to the previous week's to see if you're making a difference. If it's clear that you are not making progress, you need to revisit your hypothesis and think about whether you've got the dog's or cat's motivation right. You may need to take another look at it and figure out a more logical approach. By the time you finish this book, you'll be more aware of your cat's or dog's motivation because I'll be leading you through every aspect of the pet's point of view. Don't be discouraged!

Finally, after your dog or cat has given you about 2 weeks' worth of problem-free behavior, you've got yourself a pattern. And patterns last as long as you continue to make them interesting and rewarding for your dog or cat.

Congratulations!

You've done the right thing, and you are now back on the road to a healthy, fulfilling relationship with your dog or cat. Give yourself a pat on the back! You now understand your pet better, and you're able to change his misbehavior into something healthy. You'll probably have an even stronger emotional and social bond with your pet than you had before, and you'll both understand each other better in terms of your expectations for each other's behavior.

Notice that nowhere on this staircase to good behavior is there anything involving physical punishment, yelling, or causing pain to your best friends. Most of the things you're doing will allow your companion animals to do the kinds of things they need to do, but on your terms—even if you make them think it's their idea. Remember that, most of the time, they "ain't misbehavin'"; they're just acting like cats and dogs. It's up to you to help them learn to fit naturally into your home and family. And I'm here to help you!

Part Two

*Chapter
Two*

The World According to Dogs and Cats

In this chapter, I'd like to tell you what we behaviorists have learned about dogs and cats—what they're really like and how they see the world. You can bet it's not the way we see it! By learning more about your pets' perceptions, you'll be able to figure out what they're up to and to "talk" to them in a way they understand. It's a great basis for successful training. It will also be the start of a better relationship with your favorite dog or cat. Because they're so different, I'll tell you about dogs first. If you can't wait to read more about cats, turn to "What about Cats?" on page 19, and you can start right in!

A Dog's Life

What do dogs really want, anyway?

Whenever I contemplate the nature of this wonderful animal, I think about a poster I saw a few years ago. It's a photograph of a forlorn little mutt with sad brown eyes, head on paws, a yellow tennis ball beside him. The type reads, "But Mom, I played with Freckles yesterday!"

Sure makes you want to go and hug your hound, doesn't it?

That poster, offered by a humane society, always forces me to think about what dogs are really like, and how we ought to relate to them. It also brings out some pretty uncomfortable feelings, since I think that many of us in the last few years have promoted an image of our canine companions that's really inaccurate.

Much has been written in the past decade or so that makes dogs out to be nothing but friendly wolves and that advises that we take on the role of less-than-friendly wolves in order to keep the upper hand in the relationship. Otherwise, the dogs will sense our weakness, seize upon their biological imperative to get the best of us, and we'll end up trampled under the dogs' triumphant leap up the evolutionary ladder to dominant wolfdom. Or so the theory goes.

Today every sophisticated dog owner around can expound on how to become the "alpha animal" in his or her family's pack. Never mind that the "pack" stands upright, goes off to work every day in an SUV, speaks a strange language, and smells funny. The dog can relate only if we pretend to be a not-so-distant cousin of his, the one without enough fur to even clog up the vacuum cleaner. But this theory has us barking up the wrong tree.

The fact is that although dogs were domesticated probably 12,000 years ago, our study of them, behaviorally, was in its infancy in the 1970s. And it still is, in the scheme of things. But animal behaviorists today are beginning to see the results of years of wolflike behavior by trainers and owners alike, such as pinning the dog to the ground to gain the upper hand. And it's not a pretty picture. Our companion animals do much better with leadership than with domination by "parents" in fake fur!

So today we are looking at dogness with a fresh perspective, and with a dozen more years of field research under our belts, gleaned by going into people's homes and seeing what is driving them and their animals apart. As a result, we have been able to see a more realistic, natural—even holistic—way of relating to our canine companions. It is based on dogs being dogs and people being people. Now that's a pretty revolutionary idea in this day and age!

❧ Not a Wolf After All

A dog isn't a wily coyote, a security system, mommy's little baby, or a job to teach little Johnny some responsibility. So what is he? What are his real needs as a dog in the household at the turn of the millennium? And what needs of our own can we reasonably expect him to fulfill?

To answer these questions, I need to go back to our old friend, the wolf, for just a minute. (Aha, you say, I thought dogs had nothing to do with wolves?) Well, dogs aren't wolves, but they *are* distant wolf relatives. Actually, the distance is greater than we previously thought, according to some recent molecular biological research. Our domestication of dogs may have begun 100,000 years ago! Much of what originally linked dogs and wolves has been bred out of our modern household companions, and that's the part that people need to realize and consider when they are training and living with their family dogs.

In general, dogs have been bred for traits of submissiveness and friendliness over the centuries so that man could end up with a best friend who wouldn't bite his hand off. The characteristics that allowed the wolf to survive in the wild—cunning, aggressiveness, willingness to fight to defend its territory or to get enough to eat, predation, fear of humans—have been characteristics that people did not particularly want to perpetuate when selectively breeding generations of dogs to be household pets.

We aimed for dogs who were tamer and who got along with other domestic animals in the area. These traits allowed them to scavenge for food but still warn us against intruders, being socialized to feel protective of us.

Along with tameness, or tamability, came dogs who, as they got older, remained puppylike in their appearance. (They had skulls that are unusually broad for their length, floppy ears, smiling faces, and large eyes, not to mention "toy" breeds that never got big.) They also remained puppylike in terms of their behavior—playfulness, whining, and submissiveness—all characteristics that wolves outgrow, but dogs do not. And so over time we got the appearance and characteristics we wanted.

Thus the "best" and most desirable animals for us today are those that are friendly but alert to threats, docile without being lazy, trainable but proud, and obedient without being fearful. Our ideal dog retains a trace of the wolf's bold character yet is docile enough to lay his head on your designer sheets at night and let the toddler grab him in a gleeful hug once in a while.

🐾 Dogs Aren't People, Either

Perhaps it will take some adjusting to the idea that dogs are not merely friendly wild animals who must be dominated by human sheep in wolves' clothing.

DR. WRIGHT'S INSIGHTS

Your dog needs to be able to protect himself from threats—thus the "fight-or-flight" response to fear. If he feels he is being threatened, his natural response may include biting or growling at a family member. It's a good indication that someone is doing something wrong.

Well, here's another shocker: A dog is not a person, either! We often treat her as one, with wildly unrealistic expectations of her understanding of us and our needs and demands. We misunderstand just what a dog can learn and absorb and remember and cope with, and what she really needs to have a good quality of life in our household.

We talk to our dogs as if they were one of us, then we get frustrated when they don't respond like a person. Yes, you can talk to your dog throughout the day—in fact, it's a wonderful idea. He becomes alert, his ears prick up, and perhaps a smile appears on his doggy face or he starts to pant a bit or wag his tail, even to the point where his rear end is swinging back and forth like crazy. That's great.

The problems start when you give your dog long, involved lectures. He doesn't understand what you say about how he should behave when you are gone or what he did last week that was just not acceptable. When you say that you will be back in an hour or two and remind him that he must drink only a little water right now so he'll have enough left until you get home, he might know the word "water," but that's about all he can glean from that lecture. So he goes over and drinks the rest of his bowl of water and wonders why you are giving him another lecture, this time in an exasperated voice. Hey, just put out a couple of extra bowls of water if it's a hot day. He can't understand you!

❧ What Dogs Really Need

Your dog doesn't have to live in the wild. He doesn't have to come up with his own food source, figure out how to keep warm in cold climates, avoid predators, or find members of the opposite sex to mate with. He *does* have to worry about things like, "Where am I gonna pee and poop? They don't seem to want me to go on this nice carpet." And he also has to be concerned with things like, "I'm hungry. I don't go hunting and I can't open a can. How am I gonna tell

DR. WRIGHT'S **INSIGHTS**

College kids are usually better off with a cat or two rather than tackling dog ownership along with their hectic schedules. Cats are the number one pet of choice for college kids, for good reason!

them it's time to feed me?" Then a few more subtle things occupy their attention: "That little room is too lonely. Wonder how to get outta there and be friendly so they'll pet me." That's something the wolf didn't worry too much about when he was roaming the woods avoiding hunters!

Like our own needs, every dog's needs fall into three categories—biological, psychological, and social. All three can come into play at once, of course, since the dog is a complex individual and not a collection of separate parts. For example, your dog Angel may be afraid of your son Johnny. Why? Well, there could be a biological reason for this emotional problem: Angel may be 12 years old, and when Johnny jumps on her back for a ride, it hurts her spine.

Or take the dog's evening walk. We know she needs to go out to eliminate, so that part is biologically driven. She needs to feel assured that when she goes to you or stands by the door, you will get up off the sofa and get her leash. That's a psychological need being met. She looks to you for leadership and companionship on the walk, so it has a strong social component.

The quality of your dog's relationship with you is partly based on how successfully you teach him—with humane and effective training procedures—to be a valued member of your family. For the main focus of domesticated dogs at the dawn of the 21st century is to form that strong, deep-seated emotional and social bond with their human counterparts. When we do things to block the formation of those bonds—like chaining dogs up, ignoring them, or not allowing them to be part of the family—we start to see serious behavior problems develop.

The most common of these serious problems are separation-related (due to separation anxiety), defensive aggression, and overly assertive, controlling behaviors. (To control their access to food, toys, and other possessions, these dogs will bite, especially if they can't get this access any other way.) These problems can all occur when a dog is not able to fulfill his role as a companion animal in the family.

What about Cats?

The needs of our feline friends today are generally the same as those for dogs, but without the vigorous exercise. Play is every bit as important, as are affection, kindness, and all the basic "living" requirements. We'll get into more specifics about each species as we discover the behavior problems they have in common, as well as those that are unique to each.

Even though breed and species differences give our pets different points of view to start with, each pet also has its own preferences. One cat will want to sleep in the kitty bed you have so thoughtfully provided, whereas another has to be between the covers with her head on your pillow. One dog likes to lie and look out the window, whereas another would prefer resting in a den-like area under a table. One pet will only sleep on a soft, pliable surface, whereas the next prefers a cool, hard tile floor. One pooch prefers to play Frisbee; another likes to amuse himself with a fleece toy.

One of the ways we can determine dogs' and cats' social and psychological needs is to look at what choices each pet makes if given an option. It doesn't mean that we always have to give the dog or cat what it wants or let it act as it likes. But if we can find out what our pet's preferences are by offering him choices and seeing how he responds, we can learn how to find ways to meet his needs in order to build a trusting relationship.

(continued on page 22)

Cats and Us

Cats have a much shorter history of life as pampered pets than our canine friends. The first domesticated cat was the African wildcat, which became a pet some 6,000 to 10,000 years ago. The Egyptians were quite successful at making pets of these proud beasts around 1600 B.C., as we can see from the art of the period. Cats probably hung around people to catch mice and other rodents that frequented human granaries and garbage areas. The people let them hang around because the cats helped preserve precious grain stores and keep down diseases. Eventually, the cats were even worshipped as gods because, by protecting the country's granaries from rats, they saved the Egyptians from famine. They weren't bred for as many different abilities as dogs were, but they were valued for their ornamental qualities. At some point, people noticed that they were soft and lovely to curl up with, too, and the rest is history!

DR. WRIGHT'S
C A S E B O O K

Pal

I like to think that all my clients love their dogs enough to meet their needs as best they can. But what happens if the dog's needs aren't met? That usually leads to a call to me. One family that didn't seem to have a clue about their canine's rights, needs, or place in the family comes to mind. I am thinking of Pal, the Lonesome Golden.

Pal lived in a semirural area of South Carolina. An accused hole-digger, Pal greeted me through the fence surrounding his backyard with an almost desperate enthusiasm. The 6-year-old golden retriever displayed the friendly, nonaggressive manner that has made his breed a top favorite among families with small children. And Pal's family, the Mitchells, did have two kids. But after they built a fenced enclosure, no one in the family had anything to do with Pal. They expected him to be happy in his large backyard.

"Go live there!" he was told. The idea went against everything we know about this friendly, devoted breed. Pal was not happy. Nor could he behave well living that way.

Pal was simply not allowed inside the house anymore. Period. The reason, Bob Mitchell explained to me, was that the dog had fleas. And they were the dog's problem, not his, the owner asserted. I have yet to find a dog who can get rid of his own fleas, but I wanted to hear the rest of what this dog owner had to say, so I refrained for the moment from pointing this out. Bob had called me because Pal was digging up the flower beds out back. He had also escaped from his fenced-in prison a couple of times and was in trouble with the neighbor, who wielded a broom at him after he destroyed her prize azaleas.

The Mitchells told me that Pal seemed to have developed a fear of thunderstorms and would huddle close against the side of the house (since he had no doghouse to protect him from the storm) until it was over. Bob said the dog would get over it because he had no other choice.

I could barely think of a need that Pal was having satisfied. His bowl of water was often empty. The weather was very hot in that South Carolina climate, yet there was no shade provided for the dog. The family had shaved him—to get rid of the fleas, I suppose—but the real effect was to rob his body of the insulating effect of the golden's lovely coat. So now his skin was exposed to the elements, and he was even more uncomfortable. His social need to be a companion to his family members was simply ignored as well.

Now if you were Pal and you escaped from that backyard prison, wouldn't you run as far and as fast as your legs would carry you and never look back? Well, I for one wouldn't blame you. But guess what Pal, the Lonesome Golden, did? After visiting the neighbor dog and being chased away with the broom, he simply went and sat down on his own front step. Now this was more like it! From here, he could see his owners coming home. And watching, after all, was a good job for a dog. Especially one who wanted—more than anything else, ever—just to be a part of a family.

In this case, I had to start with the basics. Getting rid of the fleas so that the dog could be allowed inside was the first step. Then I asked each member of the family to greet and pet the dog before they went to work and school, and again when they came home in the evening, to reestablish a social relationship with him.

"What about the digging? That must stop," Bob told me. I told him that Pal needed more variety in his daily life. If he wasn't trapped in the backyard for endless hours, waiting for something to do, he probably would not be amusing himself by digging holes and watching the roots shift in the soil.

This family had a lot of work to do, but fortunately, they were willing to try my suggestions. I was very pleased to be able to lead the Mitchells through some rethinking of their attitudes and toward a more rewarding and responsible relationship with their pet.

DR. WRIGHT'S INSIGHTS

One of the things behaviorists have noticed is that, when dogs become behaviorally and emotionally distraught, it's often because one of their needs is not being met, or the owners aren't permitting it to be met. By the way, "spoiled" dogs have not been shown to have more behavior problems than those who are treated with rigid disregard for their preferences.

Dogs can be made to stay in one room and sleep where you tell them to. But if you really want your dog to be part of the family, why not let her have access to the whole house? (Obviously, the dog can be taught to stay off the furniture if you prefer.) So if your dog has shown a strong preference to sleep or rest or play in a certain location, why not let him fulfill that need, unless it is absolutely out of the question for you?

Your Animal's Rights

Along with the dog's and cat's needs go his or her rights. The rights of animals have become an extremely controversial topic, and I don't want to add fuel to that fire by analyzing the merits of any particular set of rules. But I *do* think that some rights of the pet in the household are simply logical, humane, reasonable rules that any pet owner should follow if she truly cares about her companion.

As we teach our pets how to be part of the family, we must acknowledge that the animal has a right to receive kind, respectful, competent treatment. Here's my take on it; I call them Dr. Wright's Rules for Responsible Pet Owners.

🐾 Dr. Wright's Rules for Responsible Pet Owners

• During training and at all other times, we will strive to make sure that our dogs and cats don't feel fear, pain, or deprivation.

• We will protect our dogs and cats from excessive heat and cold and provide them with adequate shelter, nutritious food, and fresh water.

• We will do everything in our power to prevent injury and disease, including keeping our pets free of fleas and ticks. If we are unable to prevent a problem, we will provide a prompt veterinary analysis and consistent follow-up care.

• We will make sure our cats and dogs have adequate exercise, play, and the freedom to behave like normal cats and dogs.

Gone are the days when we think that our dogs and cats are objects, that our pets are going to be anything we want them to be, and that our dogs and cats don't have any needs other than the ones we think they should have. We can be thankful for this change in attitudes!

For dogs and cats to have a good quality of life with humans, we pet owners need to strive to improve and enrich the human-animal relationship. This means doing things to enhance our mutual affection, communication, and trust. These are all two-way streets. We learn how to be affectionate and receive affection, to communicate and receive communication, and to trust and receive another's trust. By the way, these are all great life lessons, even for people who don't have pets!

So it's important to allow a dog to be a dog and a cat to be a cat, to respect their preferences and understand their needs, and to enable them to fit into the family structure, be it a family of one or ten. And it's vitally important for us to live a lifestyle that will allow each family member—pet and human—to benefit from the other's presence. Of course, that lifestyle includes making sure that our pets are well-behaved. I will talk more about the difference between domination and firm, caring leadership in chapter 8.

The Senses—The Keys to Your Pet's View of the World

Like birds of prey, humans have "sighted" brains; in other words, we weight the sensory input we receive through our eyes most heavily. What we see usually makes the greatest impression on us, and we tend to downplay the data we receive from our other senses.

As you're about to discover, this isn't true for dogs and cats. In many respects, sight is their weakest sense. And, unlike most people, dogs and cats have an uncanny ability to put together an impression composed of the data from *all* their senses to create an amazingly "whole" picture out of the sum of its parts. This composite is truly a supersense that can tell them more about a person or situation than any single sense could ever hope to do. So, let's explore the wonderful world our pets experience every day.

🐾 The Better to Hear You with, My Dear!

"What was *that*?!" Something hit the house with a jarring impact. Hmmmmm. It sounded like a rock or a ball or perhaps a bird with dysfunctional radar. Better go investigate. We think that we can hear pretty well with our ears plastered against our skulls. Sure, we can hear if we turn our heads and point our noses

(continued on page 26)

DR. WRIGHT'S
CASEBOOK

Simba

If I had to pick an animal who had some very different individual needs, I would have to choose a cat I met a few years ago named Simba. Pal, the Lonesome Golden, wouldn't have understood Simba's point of view at all. And neither did her owner, Erin, who was one of the most patient and devoted caregivers I've ever run across. She even moved from an apartment to a house "to give Simba more room"! But the finest palace in the world wouldn't have made this feline content to sit on a windowsill. Erin had rescued Simba from the mean streets, and she never quite adjusted to life on the inside looking out.

Simba was about a year old when Erin found her eating out of a dumpster near her apartment building in the city. "I started feeding her outside my apartment door—she wouldn't come in," Erin explained as we sat together in her family room, the tiger-striped cat rubbing back and forth across the front door. "It took 3 months for her to come inside, and then she just ate and ran back out. By the end of the 3 months, I could pet her a little."

Erin told me that she had quickly found the cat's hiding place—inside the crawl space under the apartment building. She had the cat spayed and brought her indoors. But Simba was panicked by the idea of staying safely indoors for the rest of her life. In fact, the very night Erin moved to a new apartment, the cat escaped by breaking through the rubber around the air conditioner that holds the unit in. "Miraculously," Erin said, "she returned the next night!"

It was the beginning of a long pattern of seeking out street places to stay during the day. Now Simba hid in a large pipe under a sewer grating during the day, coming home at dusk or later. Attempts to keep her in the apartment were met with frantic cries at the door. There was another move, this one to a larger condo when Erin married. Simba then found hiding places in a gully near a construction site and under a house's porch.

Erin would go out every evening and lure Simba home for the night. She had to stay home with an infection for several days, Erin told me, and when the cat tore the carpet in a frantic attempt to escape, she let her out. It seemed that the cat was not actually going out to *do* anything. She spent the majority of her time hiding from birds and dogs who would try to attack or chase her. But this cat clearly felt she *had* to get out of the house.

Erin and her husband finally moved into a freestanding house, but Simba still wasn't happy. She was still outside all day and lived in fear of a neighborhood dog and cat. In desperation, Erin even slept on the couch with Simba many nights, trying to keep the cat calm. She was frantic for Simba to stay home now, before she was attacked or ran in front of a car. Simba wanted to stay home too, apparently, but she was so conflicted that she would cry to go out, then turn right around and cry to get back in. In short, she was in need of some help.

I worked with Erin for several weeks, giving her some ideas about keeping the house interesting for Simba. I recommended some toys and perches for the windows all over the house. I also had Erin fix a "cat room" in the basement, which was dark and dank, but probably reminded Simba enough of her outside hiding places to make her comfortable staying there.

Erin got the cat a cozy bed and established a routine of playing to-gether, and it wasn't long before Simba seemed at least somewhat content with her companion in the house. But old habits die hard, and Erin under-stood that she would never totally "possess" her cat. That stray from the streets was still somewhere inside Simba, despite her nice series of homes, and the individual preference to be "free" at least some of the time was something Erin needed to accept.

At the end of the 6-week program, Simba was still going out daily but now coming home before dark. So Erin enjoyed the time they spent together and stopped trying to change her cat completely. That's how I left them. It wasn't a perfect solution—the safety record of outdoor cats is not great—but it did respect the very individual needs of an interesting alley cat.

DR. WRIGHT'S INSIGHTS

Part of the formula for successful dog and cat ownership is for us to understand our dogs' and cats' evolutionary and domestic needs, to pay attention to their individual preferences, and finally to decide which options for fulfilling their needs are compatible with our family's routines and lifestyle. If we do this well, we enable our animals to fit into the family's social structure—whether we have a family of one or of ten—and into a lifestyle that allows each of us to benefit from the other's presence.

where the sound seems to be coming from. The sound waves must hit both of our ears with equal pressure. Only then can we announce that the strange sound outside came from (vague wave of the hand) "over there"—somewhere.

Meanwhile, Buster has perked up his ears to the same sound. And perked, peaked, or floppy, the dog's ears ought to have us humans green with envy. For Buster can, without turning his head, rotate his left ear to the rear to take in our "What was *that*?!" behind him, and at the very same time, rotate his right ear to the front to pinpoint the exact location of the suspicious sound, which has (unbeknownst to us) been followed by approaching footsteps. Buster, who has heard every step, will scoot through his doggie door, find and scoop up the neighbor kid's baseball in his mouth, and deposit it in the boy's hand before we can put down our newspaper and decide which side of the house to investigate first.

Even though they may not be as effective as eyes in the back of the head, the domestic dog has been nearly as well-served by these sensitive, flexible otoscopes. His ears have protected him from sneak attacks in the forest and alerted him to the opening of a new box of doggie doughnuts.

Dogs don't hear better just because their ears are bigger, more open, or more flexible than ours. In fact, dogs have it all over humans in two of the three components that make up sound as we know it. The first is the frequency of the sound waves that produce sound or tone. Most dogs can hear both lower and higher tones than we can. The secret lies within the dog's inner ear, where tiny hair cells in the cochlea respond to tones that are very high and those that are very low.

The second measure of a sound is loudness or amplitude, and here the dog has us beat as well. He simply can hear "better" than we can, from those footsteps on the walk to the distant rumble of thunder that we haven't noticed yet. In my own case, there have been many times when our four dogs perk up their ears and run to the front door barking at some unidentifiable sound, leaving my wife Angie and me looking at each other and wondering what the heck they're responding to!

The third measure of a sound is timbre or tonal quality. Timbre enables our canine companions to tell the different voices in the family apart, so that they'll know whether "Mom" or "Dad" is calling them.

🐾 How Did She Hear That Mouse?

In the category of best ears, cats win hands down over dogs or people. That's because they can hear higher tones—not to mention the fact that they can listen to two conversations at once. That's what she's doing when one ear is rotated toward you and the other is swiveling around, looking (it appears, anyway) for something more interesting to listen to at the same time. A cat can distinguish sounds up to 10.5 octaves, more than all mammals except the horse and the porpoise. (Man can hear about 9.3 octaves.) So those ultrasonic whistles that attract your dog's attention should work even better for calling a cat— if only she were more inclined to obediently come when called!

Cats don't really care much about whistles. Their hearing is designed to make them effective predators by tuning in to the ultrasonic vocalizations of mice and rats. Ultrasonic vocalizations in mice—about 35 kHz or 35,000 cycles per second—occur at a rate of about 5 or 6 per minute, and bouts of vocalizations last for about 4 to 5 minutes at a time. Man's range of hearing is 10 Hz to 20 kHz. Even rat vocalizations, which occur at about 22 kHz, are above the range of human hearing. The range of cats' hearing, on the other hand, is about 50 Hz to 60 kHz. They can hear extremely soft sounds—that is, sounds that are at a very low volume—in the range of 2,000 to 20,000 Hz. Some cats have been observed to detect sounds at volumes four to five times softer than an adult human can hear at these frequencies.

DR. WRIGHT'S INSIGHTS

If you can read your dog's ears, they can give him away. He may appear to be paying attention, fascinated with your lecture on why he should not rip up the garbage bag and eat all its contents. But watch him, and you'll see one ear starting to pivot around like a lighthouse beacon, searching for any little sound that's of greater magnitude in the doggie scheme of things.

DR. WRIGHT'S INSIGHTS

Both my sister, Judi, and I have dogs who have gone from extremely frightened and distraught in the face of thunder and lightning to completely oblivious. Although some animal behaviorists have recently theorized that dogs find storms disturbing because of the electrically charged atmosphere zinging them, I have a simpler explanation that may account for most of the storm reactions I see in my practice. Like Peanut and Scout, an elderly golden retriever who's blissfully unaware of the thunder crashing outside, many dogs give up their storm phobias at about the same time their hearing fails them.

Over the years, both our dogs have transformed from panting, whining, pacing stormophobics to calm, cool, and collected stormophiles—the result, I believe, of now being quite deaf. So even with sniffing the air for moisture, seeing the curtains move or the sky getting dark, being zapped by the low-pressure atmosphere, or however else a dog "predicts" rain with his sixth sense, my money's on the sense of hearing as the most important in this problem behavior.

If you're seeking a mouse moving around in the grass, the ability to hear blades of grass rustling underfoot could come in quite handy. Although man seems to be better at hearing sounds of very short duration, cats hear better than humans those sounds that are continued or repeated, such as rustlings or the ultrasonic distress vocalization of the poor mouse.

🐾 It's a Matter of Taste

Generally speaking, dogs like to eat the same things we do. But we don't feed them the things we feed ourselves because we seem to have more regard for their little bodies than we do for our own! Dog foods are carefully balanced to deliver the proper nutrition to a dog at the proper time, whether it's a young, adult, overweight, or older canine. We, on the other hand, often eat without regard for fat content, fiber, empty calories, additives, and other things that can shorten our lives or at least have an adverse impact on our health.

The Evil Twins—Fat and Chocolate

Dogs can't take a lot of fat. This point came home to me when I read a magazine article by an editor whose large dog—the breed escapes me—got up on the table and stole a whole ham and wolfed it down. Even a little fat can upset our domestic friend's stomach. Because of his inability to digest that fatty ham, he met an untimely death. That seems awfully drastic, doesn't it? Now, we probably couldn't eat an entire ham in one sitting, but even if we did, it probably wouldn't kill us. When we finished with the Alka-Seltzer, we'd just have to go into the back of the closet for the "fat clothes" most of us have lurking there.

Chocolate has the same type of effect on dogs. Some dogs become quite ill if they eat chocolate, especially if it contains a lot of cocoa. Our little Peanut, the mixed terrier, had a run-in with a birthday cake some years ago when my wife and I had just begun dating. Angie had made a devil's food cake with chocolate frosting, and, suitably impressed, I ate a piece, as did she. We left the rest on the dining room table when we went off shopping in Atlanta.

It never occurred to us that Peanut would take any interest it. But when we returned, laden with shopping bags, the first thing we noticed was a set of black, sticky footprints—er, paw prints—leading from the kitchen to the dining room. Oh-oh! There were more on the polished wood of the dining room table, where only a plate and a few crumbs bore witness to the evil crime perpetrated by the doggy vandal, who showed up a moment later looking as innocent as a dog covered in chocolate frosting is able.

"Well, what do you expect?" her innocent face seemed to reply to our anguished cries. "After all, the cake was designed in the shape of a dog!" That was the last time my superstitious wife decorated a cake with anything even faintly doggy! Luckily, there were no ill effects for Peanut, but another dog might not have been as lucky, for chocolate is one of those foods that can be toxic to some dogs.

What Dogs Taste

Why would a dog consider a piece of gooey, sweet cake such a prize anyway? (I know why I would!) Weren't dogs supposed to be carnivores, or meat eaters? That is true. However, going back to the wolves again, it appears that when a wolf couldn't get his fill of meat in the wild, he would make do with fruits and berries. The same was true of the wolf's relative, our furry friend. And over the millennia, there has been little change in the dog's taste buds, even though

DR. WRIGHT'S INSIGHTS

Two different taste tests by scientists have identified a clear hierarchy for the type of meat a dog generally likes. They are beef, pork, lamb, and chicken, in that order.

she gets her meat out of a can today. Unlike humans, whose taste buds recognize the four flavors of salt, sweet, sour, and bitter, dogs' taste buds primarily recognize only two: meat and sweet.

That a dog's taste buds are tuned to meat makes perfect sense. But why would dogs sense sweet? Perhaps because sweetness is found in their mother's milk and in those berries and other fruits their ancestors foraged in the wild. These days, sweet tastes may be present in the carbohydrates you might feed your dog, including dog treats and bread. But the taste for sweets also may draw dogs to candy and other sweet treats. So keep that cookie jar covered!

I guess the biggest difference between people and dogs, as far as their taste buds are concerned, is that we have no taste buds for meat and they have no taste buds for salt. This is probably due to the fact that the wolf's diet, overwhelmingly meat-oriented, was already "seasoned," so he never had a need to recognize a salty taste in nature. For other mammals—especially herbivores—the ability to detect salt is important because it is necessary for bodily function. Today, if given a salty pretzel as a treat, your dog probably won't "taste" the presence of salt.

Dogs can also taste bitter things, as we do. This may be nature's way of protecting them from toxic substances, such as might be found in plants in the wild. That's also why your dog gives the old "ptui!" to our designer greens when the vet suggests he should have a few vegetables, unless they are slathered in some sweet dressing.

Wolfing It Down or Being Picky

So a dog is "prepared" genetically to taste and enjoy meaty and sweet flavors, and we know he sneaks food whenever he can get it. Most of the dogs I've encountered spend a great deal of time waiting for or eating their daily meals. We tend to have a mental picture of the dog eating until he drops, given the chance. It's certainly not unusual for wolves to attack and gorge themselves on meat, snarfing the food down as quickly as possible. Hence, we have the expression "wolfing food down."

But most of our domesticated dogs don't wolf food down, although some of the pack hounds like beagles and foxhounds do because of the competition among them at feeding time. A more common problem is dogs who are picky eaters. After all, our dogs don't have to go out and kill anymore, and they get plenty to eat so they can afford to be picky. We've found that certain dogs—

toys and giants, mostly—are nibblers that sometimes have to be coaxed to eat with gourmet fare of one sort or another.

Sometimes, we create our own monsters in the picky eating department. We can turn a puppy into a picky eater if we provide only a few choices on the menu in her first months. Like children, pups are likely to grow up suspicious of new foods unless they are given a variety of things to taste when they are young. The willingness to eat a wide variety of foods made a lot of historical sense for both our species because, in case of a food shortage, eating a variety of foods could mean the difference between survival and death.

We know that the newborn puppy, suckled by his mother, is exposed to various flavors through her milk. What he tastes may influence what he prefers to eat later on. After that, the rest is more or less up to us, as the people in charge of what our pets eat.

To help determine whether early feeding habits matter today, an animal behaviorist did an experiment in which an entire litter of chow chows was hand-fed from birth. Some were given a soybean diet, some were fed a vegetarian diet, and others were given a mixed diet. The behaviorist found that the pups who were fed the soybean diet would eat no new and different food as adults. Pups reared on a mixed vegetarian diet would eat no animal protein. And pups reared on a mixed diet would eat any new food—except for foods with a bitter, sour, or stale taste.

Most dogs are prepared to try new tastes, and they'll stick with something new for at least a couple of days, until they finally realize that it is not nearly as good as their regular food. Vets now generally recommend sticking with one basic dog chow and a minimum of table scraps, although there are plenty of treats to provide a bit of variety for the canine palate. If you want to introduce a new brand of food, gradually mix it in with the old food, or your dog is likely to suffer for it with GI tract upsets. The modern dog's digestive system is quite different from the hardy stomach of the wild wolf, who ate pretty much whatever he could catch!

The Inside Scoop

So dogs like most of our kind of food, and we like some of their kind of food (meat, in particular). But some of their food choices are beyond our comprehension. Dogs are known to gobble up deer droppings, cat feces, and even other dogs' poop with equal relish. Just recently, I received a frantic call from a woman who told me she was being held "hostage" by one of her wirehaired

(continued on page 34)

DR. WRIGHT'S
C A S E B O O K

Peanut

Sometimes even my own dogs become subjects for the old casebook. This case involved our beloved mutt, Peanut.

I have to laugh sometimes when I consider the careful diet we give our dogs, contrasted with the diets that the dogs who start out as strays have gotten used to. This fact was brought home to Angie and me in an interesting way several years ago when we lived in a small neighborhood of homes built rather close together.

Angie had rescued Peanut from the railroad tracks several years earlier, and her previous history was pretty much a mystery. At any rate, every night, just before bed, we would let Peanut out the back door for a short ramble in the backyard. She didn't go far, being a female—yes, the theory is true—and we made no attempt to track her movements, knowing she was a very "good girl" and would ask with a bark to be let in.

But soon, I'll admit, we developed suspicions that something was awry. One Wednesday night, Peanut came home practically dragging a huge ham bone. Well, she could have filched that literally from the mouth of another dog, but we couldn't imagine such a thing happening. She just wasn't the grabby type. And she could have dug it up from somewhere on the outskirts of the yard, but that seemed unlikely too.

We didn't think too much of it. Until the next Wednesday night, when she came home with a big plastic leftover bag filled with precooked breaded pork cutlets, licking her chops. Now we grew alarmed. Was she hanging out with some other family? Was this some dognapper's way of gaining her confidence in order to snatch her away from us and then demand a king's ransom?

At any rate, Angie and I followed the nocturnal prowler the next Wednesday night. But Peanut quickly caught onto our surveillance, and she gave us the slip as we peered around the corner of the house. She had a valuable resource somewhere, and she wasn't about to let us find it and take it away from her!

So we returned helplessly home to wait. Sure enough, a half-hour later, here came Miss Peanut home with food. But this time it was all over her! Angie and I took one look and burst out laughing in spite of ourselves, all admonishments forgotten. For our little Peanut, the white mixed terrier, had a face full of red sauce and a few strands of spaghetti draped over her nose. She trotted happily past us through the open back door, heading for the water bowl. It seems it was Wednesday, pasta day, just like in the commercials on TV. We practically fell down laughing, and Peanut was re-named "Noodle" for a number of weeks following.

Soon after this we moved, and the weekly smorgasbord died a natural death. We finally realized that Peanut, the railroad tracks refugee, must have foraged for her meals during whatever length of time she was on her own. It became obvious that she found a way to escape from our fenced backyard and visit some neighbor's freshly stoked garbage can: Thursday was garbage collection day on our street. Mystery solved!

A caller recently complained to me that despite the fact that he daily served the finest dog food to his pampered pooch, the dog was interested only in digging in the garbage can. It sounded just like our little Peanut, who is mainly responsible in our house for the garbage being kept behind a closed door in the kitchen. "Was your dog by any chance a stray?" I asked the client, on a hunch. "How did you guess?" he laughed. The answer—for dogs as well as people—is that old habits die hard.

fox terriers, who would eat her second dog's feces before my client could race over with her scooper.

Hearing her tell this awful tale would have been comical, had the facts not been so disgusting and so disconcerting at the same time. The hostage part was actually voluntary; this client chose to stay home to nurse the culprit when he became ill and vomited violently after each episode. She found him so weak that she was afraid to leave him. Yet the next day, the same thing would happen all over again. Clearly, the dog learned nothing from his bad experience every day. So the owner had to be the one to change things.

I advised this woman to get a head halter for the poop-eating culprit and to walk the dog by himself. I suggested that they walk down the street in a new, more "attractive" area, and that she control his every movement outdoors until she could wean him of the habit of seeking out his littermate's defecations. This client had to rethink her routine: She used to just take the dogs out together on leashes and follow with the scooper, when it was already usually too late. But soon the dog went back to eating regulation treats, and she was in charge of her own work schedule again. It was a perfect outcome for everyone concerned!

🐾 Who Says Cats Are Finicky?

Cats use three gustatory cues in selecting their food: its smell, its temperature, and its texture. For feral cats and pet cats that are allowed outdoors, sight and hearing are more important for detecting prey, followed by smell. Although dogs depend on scavenging for nourishment in the wild and rarely reject a tasty "find" that they've come upon during a walk, cats prefer to capture and eat their small prey fresh. For indoors-only cats, smell and taste are equally important. Sight and hearing become almost irrelevant, since food bowls don't move!

The flavor of food consists of a combination of its smell and taste. Once eating begins, though, flavor becomes a mix of taste and texture. If it tastes and feels good to your cat, she'll like it! Cats also probably experience sweet

DR. WRIGHT'S INSIGHTS

People think cats are milk-mad, but they actually crave the sucrose (sugar) in the milk. In fact, cats prefer sugar in water over milk! Some cats have trouble digesting milk or cream (just as some people do), so consider this a word to the wise.

and bitter as we do, and like most people, they prefer sweet tastes and avoid bitter foods. Also, like most of us, they avoid sour flavors. But unlike us, cats are not very sensitive to salt. They don't taste it well on the tongue, but the research isn't clear on whether cats are completely unable to taste salty substances.

Normal cats prefer to eat many small meals throughout the day. They eat more during times of high activity (since they've been working up an appetite) and less during sedentary times. They are more likely to decrease or increase the amount of food eaten at one time to suit their appetites, rather than change their preferred number of meals, be it 12 or 20.

The plain truth is that cats aren't as finicky as we've been led to believe. They are quite ready to accept new tastes in foods. If you or their mother introduced them to a lot of different kinds of foods as kittens, they'll tend to be more likely to explore novel tastes as they mature. But some cats are neophobic—afraid to try anything unfamiliar. So if your cat refuses to eat a new canned food, either he finds it unpalatable—it just doesn't taste good—or he may have a thing against trying anything that tastes "different."

Chances are that given a choice between a new taste and the same old taste, your cat will prefer the new taste initially (the novelty effect). After several days, however, her "true" preference for one food over the other will be obvious and will be stable for weeks to come. Or she might eat about half of each food, just to confuse you—cats are such unique individuals!

🐾 The Cold, Wet Nose Knows

You and your dog are relaxing out on the deck. A slight breeze ruffles the pages of your newspaper. The dog lies motionless, nearly asleep. Suddenly his nose starts to twitch, and he lifts his head, eager to identify the scent. It could be the smell of a deer, a rabbit, or another dog, and he takes off quickly to investigate. How does he *do* that?? You can't smell anything!

Take a look at your dog's nose. Like his ears, this sense organ has our own schnozzles beat in terms of form and function. Whatever you think of its shape or size, the nose of a dog—wet, quivering slightly, nostrils flaring when some interesting odor floats by—is a masterpiece of design. Those big exposed nostrils are so efficient as to make us, by comparison, feel like a kid with a bad cold who won't eat his dinner because he can't smell it! Let's see how badly we stack up next to the dog in the sniffing competition.

In scientific terms, the olfactory epithelium (the location of sensory receptors in the top of the nose) tends to be a lot larger in dogs—about 75 square centimeters in the beagle, compared to about 3 square centimeters in humans. That's a huge difference in the availability of receptors to detect molecules in the air. It doesn't take many molecules of a substance, such as a bit of air freshener or cologne or the smell of urine, to activate those receptors in the dog.

Studies comparing the nose-power of people and dogs have shown that dogs are capable of sniffing out a variety of substances in concentrations ranging from about 1,000 to 100,000,000 times less than humans can perceive! This fact should end the mystery of how Boomer knows when you've opened a can of dog food and comes running from three rooms away. That doggie "sixth sense" people wonder about is probably just a combination of some of the dog's powerful five senses—led by his exquisite sense of smell—working in perfect harmony.

In addition to all those extra receptors in their noses, dogs have a narrow vomeronasal organ (VNO), which is accessible by way of a little duct on the front upper palate and leads directly up to the olfactory bulb of the brain. They may use this organ to "smell" pheromones (chemical signals), although it's not as important for detecting pheromones as a cat's VNO. Nonetheless, there is no doubt that dogs use their exquisite olfactory sense to smell pheromones in urine. If you have two or more dogs, you probably could not tell them apart just by smelling them. But dogs can! Using smell alone, they can tell apart all the people in their world and all the dogs in their world, including (by the time

A bloodhound's nose is so sensitive that it can locate a person buried in the rubble of a collapsed building just by smelling a scrap of the person's clothes.

they're 12 or 13 weeks old) a dozen littermates; they can also distinguish between puppies from their own litters and outsiders.

Scientists have checked out the scent-discrimination ability of dogs by taking urine samples from two puppies in one litter and a third puppy from another litter and mixing them up. They found that without fail, the dogs they tested were able to pick out the urine from each of the puppies. They can even detect the difference between genetically identical human twins—a feat that sometimes even escapes the twins' parents.

Who Goes There?

Dogs tend to urinate over the urine of other dogs, each dog picking the same spot as the dog before it. In nature, it looks like wolves do that so that they can mask the other wolves' marks and make their own the strongest and most recent smell. In this way, other wolves, or, in our case, dogs, can keep track of who was where, when.

Scratching the ground after urinating or defecating may provide a visual signal for some dogs, but it also leaves a distinctive scent to "mark the trail," so to speak, for other dogs. The scent is produced by either the interdigital glands, which are sweat glands on the footpads (the only place a dog sweats!) or by the sebaceous glands that lie under the dog's skin in the fur between his toes. So by sniffing the ground after another dog trots by—which always seems kind of silly to us—the sniffing dog is actually picking up some data. To be sure, it is a very satisfying spot check for the dog.

Many dogs don't pay nearly as much attention to visual communications as they do to the information gathered by smelling. In fact, breed development has played such a trick on the dog that he is not likely to be able to tell by sight what the heck he's looking at! If he is a German shepherd, he may wonder when he sees a beagle whether that animal is also a dog—look at those floppy ears. Maybe it's really a rabbit! Tails have a tale to tell, but on a Doberman it's just a stub—there's not much information there. And who can tell if that's a submissive grin on that pug or if his face always looks like that?

So rather than do all this guessing, he just tries to sniff out the information he needs. With his nose, he can tell for sure whether he's looking at a dog or a rabbit, how old it is, which sex, how friendly by their mutual nose twitching, and what it had for breakfast. With all that information coming in via that cold, wet nose, he doesn't need to have the most terrific vision in the world.

Those Sexy Pheromones

The most frequent reason dogs communicate with one another through scent is that they are drawn by pheromones—chemical signals in the urine. These are especially important for sexual encounters. Males can smell female urine over great distances, especially when the female is in estrus, or heat. This is one reason intact (unneutered) male dogs roam more: If they detect this odor in the air, they will take off after it, whereas neutered males aren't as likely to do so.

There are pheromones in people's urine, too, and this may explain why some dogs are prone to sniffing in the genital area, which we tend to consider to be quite rude. It is their way of meeting you and seeing who you are. For although we try to mask our natural odors with perfume, deodorant, soaps, and so forth, our dogs can see—or sniff—right through those cover-ups, at least until we push their noses away! They deal with other dogs in a much more thorough manner. Sniffing each other from head to toe, they gain information about whether they have met before, sexual data, and perhaps the contents of each other's last meal in case there's more of that to be had nearby.

Do I know you? Have I smelled you before? What's that delicious thing you've rubbed against? These are the questions dogs ask one another when they meet and sniff.

A Cat's Nose: Nothing to Sniff At!

Don't let Kitty's tiny nose fool you. The size of the sensory receptor area in a cat's nose is almost ten times the size of the comparable structure in humans, putting the cat on a par with her canine counterpart when it comes to sniffing around. Cats perceive smells when scent molecules land on tiny hair cells called cilia, the olfactory receptors, at the top of the nasal pathway.

A woman wrote me recently asking why her cat goes wild when Grandma comes over. "Saffron meows, rubs against her, and then bites," she reported. She wanted to know what was wrong with her usually sweet feline.

Well, applied animal behaviorists like me tend to make hypotheses based on hunches and then test their hypotheses one by one to determine what's going on. So if I had to guess, I would consider the rubbing that the woman mentioned to be a pretty good clue. Rubbing, in cats, is what we call an olfactorily mediated reaction. The nose knows!

I suggested that Saffron's owner try to figure out which of her mother's fra-

The apocrine glands, which secrete chemosignals or scents through a long duct to the hair follicle of a dog, tend to be more dense around the head, the anal region, and the upper surface of the base of the tail. Perhaps that's why dogs spend a lot of time sniffing in those areas. It also explains why they like to sniff our armpits and lick and sniff our dirty laundry lying by the washing machine. It helps them become chemically familiar with our taste and smell.

Although they'll put up with our colognes and powders and essential oils, dogs don't seem to like being any too clean and perfumed themselves. Have you ever noticed how your dog, fresh from the groomer or the bath, seems always to pick that time to roll in something smelly out in the yard? She's probably trying to rid herself of the "fake" smell you've coated her with so that she will smell more like nature. At one time, it may have been important for dogs to coat themselves with the smell of a fresh kill, in order to help lead the pack to this rapidly perishable food source.

Dogs who track lost children or people buried in collapsed buildings need only to have a bit of their scent on a piece of clothing. This is enough for them to identify the smell of sweat or individual amino acids and seek them out. The canine nose can also sniff out cocaine or marijuana through chemicals that produce telltale odors only they can smell.

grances created all the excitement. By using the same perfume, soap, shampoo, deodorant, and even lipstick (one at a time), she can see if the cat reacts more or less strongly to any one of them. Just about anything can set off a cat's olfactory paroxysms. The fun part, I suggested, might be telling Grandma that she has a choice when she visits: She can wear her favorite hand cream *and* a wild, biting cat, or she can choose to wear neither!

Have you ever seen a cat curl his upper lip and hold his mouth open a bit? Did you wonder what that was all about? The cat's vomeronasal organ, a specialized scent receptor, is highly sophisticated, even more so than a dog's. And

DR. WRIGHT'S INSIGHTS

Here's another wacky feline fact: Cats are put off by the smell of coconut oil and some other plant oils. So be prepared to be shunned by your cat when you're out sunning.

DR. WRIGHT'S INSIGHTS

Everyone is familiar with the sight of a cat licking his front paw. But did you know that cats also smell their paws? They do this because they can use their paws as a nose substitute. It's fairly common for cats to bat at a dead fly or something they perceive to be a play or prey item, and then sniff the bottom of their paw to identify the smell. It's their way of investigating things, and it's safer than risking a nose touch. (After all, committing your nose to touching an unidentified wasp could have dire consequences for such a sensitive body part!)

all cats raise their upper lip in a grimace and hold their mouth open for several seconds (scientists call this phenomenon *flehmen*) when they detect the presence of another cat's urine. It can be outside or on the carpet or a vertical object like the wall, a chair leg, or a dust ruffle.

Cats will actually smell and taste the urine spot. These encounters—mostly males sniffing female urine, but females will respond the same way if no male is present—are quite arousing. This activity may contribute to further marking or even aggression toward a person or another cat, depending on who moves first! Cats use their sense of smell to figure out what's available to eat, but they never use the vomeronasal organ in conjunction with feeding behavior.

Cats and dogs can see much more clearly at night than we do—at least eight times better (pets-eye view, <u>left</u>; our view, <u>right</u>). And they see many more shades of gray, which acts for them like another color.

🐾 The Big Brown, Green, or Blue Eyes Have It

Can Pooch rely on his eyes in the same way that he depends on his ears and nose? Not a chance—at least not for distance vision and clarity like human sight (or, at least, like human sight when we're wearing our glasses or contact lenses to make up for nature's flaws). However, in some respects, your dog's vision is much better than yours or mine.

Say you are now in bed, it's the middle of the night, and the deck where you and Sailor were relaxing earlier is now empty—or so you think. There is only a sliver of moonlight outdoors. All is quiet. Suddenly, Sailor starts barking like crazy. Because you can't see in the dark, you assume that he smelled something or maybe even heard something, but he's just as likely to have *seen* something—a raccoon, perhaps—out on the deck!

Dogs can see in low-light conditions much better than we can. No doubt, they developed this ability back when their prime hunting times were at dawn and dusk and good vision was necessary for survival. Dogs have the ability to see many more shades of gray than we do, which is a trade-off for their inability to distinguish colors other than (probably) blue and yellow.

Believe it or not, many dogs need glasses. Unless an object is between 12 and 20 inches from their faces, they can't see it very clearly. A dog's vision tends to be, in human terms, 75/25. In other words, he can see at 25 feet what the average person should be able to see at 75 feet. So the dog is nearsighted, and if born human, he would be wearing glasses or contact lenses. That's why he may bark at you if he sees you pruning the shrubbery at the back of the yard. To him, you look like an intruder hiding in the bushes!

Our eyes are built so that the visual field in each eye overlaps with the other, so we can see things in depth pretty well. Most dogs' eyes, set farther on the sides of their heads, can't perceive depth the way ours can. But before

DR. WRIGHT'S INSIGHTS

Because of dogs' excellent night vision, you can play Frisbee with your dog right up until dark. Even if you can't see that plastic disc too well, he'll be in his element. You won't need to worry about him. While you're out there with the Frisbee or a ball, you might want to throw it off to your dog's right or left side. If you throw the toy straight on, you run the risk of hitting your dog in the face because he can't accurately perceive objects coming straight at him.

DR. WRIGHT'S INSIGHTS

Keep this in mind before obsessing about the colors of Jezebel's cat toys: She probably doesn't have a favorite color! If you must choose colors that please her rather than you, try blue, green, and red toys. But if I were you, I'd buy the ones with colors I liked best— if you choose cat-friendly toys, your cat will like them in any color.

you feel too terribly sorry for this creature, who by all rights should go around squinting all day, realize that he doesn't have to drive, he doesn't have to read, and while things are a bit blurry from far away, he still has it all over us. That's because he can still rely on his other senses, which are much sharper than ours even on the worst of days.

Besides, dogs have a field of vision that's much wider than ours. Your dog can see, without moving his head, a scope of about 240 to 250 degrees. That's 60 to 70 degrees greater than we can see. So now you know why it's impossible to sneak out of the room and go for a walk by yourself. You thought the dog, ostensibly snoozing on the floor in front of you, would never notice. Ha! Bring him along; you'll both enjoy the company.

🐾 Those Wonderful Cats' Eyes

Like dogs, cats can see much better than we can in low to moderate light. They have a higher density of light-sensitive rods throughout their retinas than we do. This originally helped them find their prey at twilight and dawn. Now it explains why they can tiptoe around the curios atop the bookcase without knocking one off, whereas we can't even make it past the bedpost in our midnight bathroom trek without stubbing our toes! Every cat also has a receptor called the *tapetum lucidum* behind her retina, which functions somewhat like a mirror: It reflects light back onto the retina so that the cat has a better chance of seeing what's "out there." The tapetum is what you see when a cat's eyes "glow" in the dark. Dogs' eyes reflect the light, as well.

Cats may not be able to see in color, since they have fewer cones (the receptors in the eyes that perceive color) than people. Their cones are responsive to the wavelengths of light that we interpret as green and as blue and possibly red. But the inability to see the color of a mouse in the dusk is probably not too upsetting to a cat: It's more important to detect the mouse and see it move than to be able to tell if it's brown or gray!

Like dogs, cats are very good at detecting motion. And because their eyes are separated a bit more to either side of their head than ours are, they have a slightly larger range of peripheral vision than we do, so they can see someone—or something—sneaking up on them from behind pretty well.

🐾 The Touch of a Whisker

Who needs hands when you've got whiskers?

Wonder why Kitty has such long whiskers and eyebrows? And what's with those extra ones sticking out from the side of her face? Well, they're all called *vibrissae*, and their job, besides being so cute, is to give the cat a feel for how long she has been touching something and whether there is a new touch on top of the old. The whiskers are a handy tool to tell whether a passageway is wide enough to pass through: Forget size, weight, girth, whatever—those whiskers will tell Tuffie if he can squeeze through or had better back out gracefully before he gets stuck. My guess is that you've never seen a stuck cat? Now you know why!

🐾 Massage My Cat?! I Can't Even Pet Her!

Although many cats could lie on your lap for hours while you stroke their fur to the beat of a happy purr, a great number of cats don't want to be petted more than two, three, or four times. The number isn't important. What counts is that

In bright light   In dim light

When light shines in a cat's eyes, the ciliary muscles that open and close the iris make the pupil a vertical slit. These muscles are unlike our ciliary muscles, which make our pupils round. In bright light, the circle of our pupils simply becomes smaller.

DR. WRIGHT'S **INSIGHTS**

Cats' eyes, those marvelously deep and mysterious orbs, also get a helping hand from their nearby whiskers. With the aid of a camera that was implanted in a cat's body to record its eyesight from the inside out, we have been able to see just how blurry a cat's vision is "up close." So the whiskers take it upon themselves to help out at the crucial moment: Just when the cat is about to leap upon the mouse it has been in sharp-eyed pursuit of, things tend to start blurring. That's when the whiskers go into action, rotating forward and taking over as the preferred sense, lightly holding the mouse in place while the cat goes for the jugular. If your cat doesn't have access to a mouse now and then, she may attack her toy mouse in this way . . . or even, at times, the nearest hand!

if you don't remove your hand after that third stroke, you're almost certain to get nailed—and cat bites, especially those delivered when being petted, are notoriously painful. These cats are known as two-, three-, or four-stroke cats, depending on how long they can hold out before turning on the offending hand to give it a good nip.

The reason, I suspect, is not that your cat is an ungrateful wretch, but that she has very sensitive "slow-adapting" receptors in her skin. They can make anything more than light stroking too stimulating. For the same reason, cats have a hard time accepting a leash or harness. It's not that your cat's too haughty or proud to stoop to such confinement—it's just that she has more slow-adapting receptors than her doggy friends!

One of the most characteristic things that cats do is "knead," pushing their paws alternately against any invitingly soft area of their human friends' bodies like a baker kneading dough. Some cats will do this on a person's chest, whereas others like to give their human a hug around the neck and knead on a shoulder. Still other choose the belly, thighs, or neck of their favorite person to knead on, and I've seen cats knead the floor while they are being petted by their special friend. They probably knead because it feels warm and similar to the mother cat's belly—the ultimate feeling of love and security since kneading her warm belly brought them nourishment during early kittenhood.

🐾 Go Ahead, Pet That Pooch!

Dogs' preference for tactile stimulation is no accident, nor is it based merely on experience. From the time a puppy is born, touch is necessary for normal bio-psycho-social development. Newborns are licked by their moms in the anogenital area to stimulate elimination (otherwise, they would risk infection from the buildup of waste). And puppies who don't receive a lick-and-groom sequence from Mom—or handling and strokes from a human hand—tend to avoid social touch or other physical stimulation in later life.

The tactile receptors—the areas where we experience touch—are located and highly concentrated at the base of the hair follicles, although some areas of the body are more sensitive to touch than others. Both cats' and dogs' hair follicles are compound: There are many hairs per pore compared with human follicles, which have only one hair per pore.

Body parts not covered by hair, such as the foot pads, nipples, lips, and nose, may be selectively sensitive to various kinds of stimulation, like touch, pain, or intense heat or cold. Some breeds, like pit bulls and English bull terriers, seem much less sensitive to pain than other breeds, like shelties, although even "brave" German shepherds wince at the prick of a needle! Areas that are especially sensitive to touch in most dogs include the inside of the hind leg, around the muzzle, and the feet (which is one reason why trimming nails can be a trying experience for both dog and groomer).

But the true value of touch is that it causes pleasure, and each dog may have her preferred areas or touch zones for stroking and massaging. Most dogs love having their bellies rubbed. They also love being stroked behind the ears, tickled on the chest, and scratched on the rump. Touch and massage lead to a dog's returning the caregiving behavior by licking. Facial licking seems to be their preference, which may come from puppyhood, when they licked Mom to solicit more attention from her tongue. The two-way stimulation leads to a closer, more affectionate psychosocial bond between the dog and her caregiver. Dogs also love "kisses" (which they enthusiastically return) from their humans, if they're fortunate enough to have "parents" who enjoy that kind of bonding experience. But if you're not the type who enjoys having a dog lick your face, she'll still enjoy licking your hand or arm as a sign of caregiving.

The Sixth Sense

When we talk about a dog's or cat's senses, and we have gotten past the traditional five senses that we three species share, there always arises the question of the dog's or cat's "sixth sense." How does he know that quiet neighbor is approaching? Or that there's a perfectly good tennis ball hidden in the English ivy? Or that there will be a storm dumping rain on the garden 30 minutes after

(continued on page 48)

Massage Doggy-Style

Does your dog tend to get overexcited? Touch is just as important to dogs as it is to us, and it can set the tone of the interactions between you. When your dog gets too excited, try to combine a very soothing tone of voice with the soothing feeling provided by touch. I recommend that you provide your dog with soft places to lie down, which dogs have been shown to prefer, just like people. A dog's panting is a sign that she needs to slow down and quiet down. This is when you can use some "relaxation praises," such as (in a low, calm voice) "Go-o-o-o-o-od Gi-i-i-i-i-rrrrl."

When the dog is nice and calm, she'd rather plop down on your bed instead of on the hard floor, which is just good common sense! That thermotactile sense in dogs—their preference for warm, soft surfaces—is what makes it hard to remove them successfully from places like your $300 comforter once you allow them access to the bed. But that's up to you. They will love it if you care more for their comfort than for your designer bed dressing! I admit that Angie and I let our three dogs sleep on any surface in our house that they darn well choose. And of course, they choose the nicest sofas and most delicate duvets most of the time! In any case, the key to a calm dog—besides a soft, warm place to retreat to when it's time to relax—is a comforting, quiet manner of your own.

To follow the soothing voice, try stroking or massaging your dog, especially while your dog lies quietly on your bed or in her crate, to make it have good associations for her. As long as the dog will hold still, it's great to press and then release each major muscle group. That's what the classic massage is, and it's good for what ails dogs as well as people!

In this age of holistic medicine and an emphasis on natural remedies, massage and relaxation praises are a worthwhile alternative to the use of drugs such as Prozac to enhance a dog's ability to remain calm while learning new, better behaviors. At least they are worth trying first. Dogs love to roll over on their backs and want you to stroke or run your fingernails lightly up and down their stomach and chest. Patting is not much appreciated at this point, and many dogs get up and leave if you begin patting. That's not relaxing to them; rather they find it exciting.

You may want to end the stroking session with a pat, thus signaling to the pooch that you're finished. When he's on his feet, patting on the head tends to be a bit jarring. Stroking or brushing might feel better to him. But he's likely to take any kind of touching as a sign of affection and be quite happy about it, as long as it isn't a toddler pulling his tail! (Many dogs will tolerate even that indignity without a whimper, just as they will carrying small children on their backs.) Nevertheless, kids should be taught that the dog is not a toy and should be played with gently, for everyone's sake.

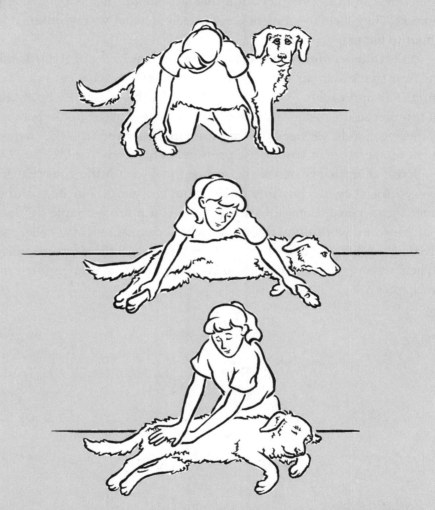

Here's a good way to massage your dog, borrowed from my colleague David Tuber, Ph.D. First, kneel behind your dog with his side to you. Then raise his legs, pushing them away from you so that he falls gently into your lap. Position him so that he's resting comfortably on his side, then begin a slow massage, pressing and releasing his muscles. Stop when he lets you know he's had enough!

you put away the hose? The answer is that dogs and cats use all their exquis-itely sensitive senses together to come up with the solutions to the problems facing them as they try to make sense of the world. In combination, these five senses can be so powerful that they create a whole sixth "sense." This sense is sometimes so uncannily accurate that we wonder if our pets have psychic powers. They seem to know how we're feeling, what we're thinking, or what's about to happen.

Take thunderstorms, for example. We humans have our arthritic knees and our scar tissue and our barometers (and who knows what else) that we use to predict an approaching storm. A dog or cat has his senses. He hears the sound of thunder in the distance and sees the darkening of the house because of the gathering clouds. He smells the moisture in the air and feels the change in am-bient temperature or barometric pressure. Who needs The Weather Channel?

With their marvelous senses primed to pick up the cues that tell them everything they can possibly know about the world, our dogs and cats can turn those perceptions into learnings. And after a couple of decades of watching and working with our household companions, my colleagues and I are finally beginning to understand how they use all these terrific built-in tools to help us teach them how to learn! We'll explore how our pets learn and think in chapter 3.

Chapter Three

A Behaviorist's Take on Pet Learning and Thinking

Do pets think?

This question always makes me recall the furor that surrounded a famous performing horse, Clever Hans. Hans astounded audiences a half-century ago by "thinking" his way through mathematical problems posed by volunteers. As his German trainer counted out loud, Hans would tap his hoof on the ground until he came to the correct answer. Hans rarely made a mistake, and his fame spread far and wide. Was this horse a genius? And did that mean other horses could be taught to add and subtract, even multiply and divide?

Well, as we say in science, there had to be a simpler answer. In science we say that we should apply the "law of parsimony" here. This law holds that a theory gains power when it can explain numerous results with only a few explanatory concepts. If two theories can explain the same number of results, then the one with fewer explanatory concepts is preferred. This means that we should not explain a behavior with a higher cognitive (thought) capacity if it can be explained or interpreted with a lower one.

What Hans was actually doing was paying attention to his trainer's non-verbal cues, his head and torso, which raised slightly (unintentionally, the trainer claimed) when the horse was supposed to stop "counting." The Clever Hans phenomenon was solved, then, to the extent that a reasonable explanation was provided for the horse's counting ability. It is reasonable to believe that Hans was responding to simple stop and go signals. By implication, it is unreasonable or less reasonable to believe that Hans either understood German or could do arithmetic.

That Hans was an observant horse is a parsimonious explanation of the Clever Hans phenomenon. Notice that we haven't proven anything; we have simply stated a preference for a theory you might consider when next marveling over your dog's or cat's thinking process or behavior (as in "How does Phantom know that?" or "There's no way Silky could have figured that out!").

Now, I am not saying that your dogs and cats aren't little geniuses in their own way. We certainly don't understand everything there is to know about the cognitive capabilities of canines and felines, and maybe we never will. But we know enough to help people deal with their pets in ways that will be the least confusing to them and most helpful in working toward good behavior and re-warding relationships.

Dogs and cats are not simply bundles of age-old instincts programmed to be blindly obeyed. Neither are they *Pinky and the Brain* cartoon-type creatures who sit around crafting intricate plots to take over the world, or at least to exact revenge upon you for going on that two-week vacation to Tahiti and leaving them in the kennel.

How Smart Are Dogs?

We know that domestic dogs use humans' and other animals' social cues to figure things out, just as Hans did when he looked at his trainer's face for a signal to stop counting. Dogs, we've confirmed via controlled conditions, have the same cognitive skills as some chimpanzees. These skills include gaze-fol-lowing, object permanence, and cognitive mapping. Let me explain each of these skills.

When they are puppies, canines learn by watching other dogs and their mother. But as they mature and become our household companions, they are

DR. WRIGHT'S INSIGHTS

Here's one difference between dogs and cats: Dogs will watch you point your finger at a ball in the grass and then go get the ball. Cats will just look at your finger. Many dogs will go fetch if you merely gaze at the ball lying out in the grass. If you try that with a cat, it will just wait for your admiring glance to return to it again. "What ball? Where?"

more likely to take their nonverbal social cues from humans than from other dogs. I find this quite exciting!

In one recent study, some dogs were able to watch either another dog or a person for cues as to where food was located. (The food was hidden and not "smellable.") The pups watched a dog look toward the food; then they went to find it. But the mature dogs watched the researcher as she looked toward the food or looked and pointed toward the food and then located it that way. They chose not to follow the other dogs' gazes.

Are Cats As Smart?

On the age-old question of who is smarter, dogs or cats . . . dogs are smart for dogs and cats are smart for cats. Cop-out, you wonder? I think not. By "smart" I mean the ability to figure out what survival behaviors are likely to work in a dog's or cat's unique ecological setting. Thus, cats have learned to survive in the wild in a manner different from dogs (who hunt socially and travel distances together, whereas cats do not). Thus, dogs and cats both became smart for being able to learn the special behaviors that increased the likelihood of their survival and reproduction, until we humans took over their breeding. Now, they no longer need to hunt or scavenge for food or fight to reproduce or roam to find shelter, because, as their caretakers, we satisfy all of their present survival needs. So it's largely unimportant to ask whether household cats are as smart as dogs in learning, say, silly pet tricks. As long as they continue to thrive, to satisfy our personal need for companionship, to increase our quality of life, they will have met all the survival skills they'll need to live to a ripe old age—and that seems pretty smart to me. Cats have a knack for capturing even the most elusive insects that find their way into your home. Who hasn't watched their cat's stealth attack on the unfortunate moth as it unwisely strays from the ceiling? She spots it, then stalks, rushes, leaps, and pounces on it to your delight! A fun-to-watch survival skill. Cats also have a knack for uncovering even the best camouflaged food treats, whether they are inside a jar on the countertop (how smart they are to leap up, knock off the lid, and consume the goody), behind the cupboard above the refrigerator (how silly we are to think that "up high" will deter a natural climber—and how useful a paw and triangular head are for wedging open the kitchen cabinet), or beneath the refrigerator. These are all good survival skills, developed in nature—not to traverse or reach under the refrigerator, but to climb a tree or reach into a felled tree trunk, and not to open a kitchen cabinet, but to access the nest of a well-hidden food source (mouse or bird). Smart Kitty? You bet!

A cool thing dogs can do is understand that when a squirrel runs into one end of a log, it hasn't just disappeared into thin air. And that's true of *all* dogs, not just hunting dogs. What's even neater is the fact that the dog will run to the other end of the log and wait patiently, "knowing" that the squirrel will shortly be making its appearance and will be eminently chaseable. (Cats do this too, but they wait for mice!)

Big deal, you say? Well, that's an understanding we humans don't have until somewhere between 7 and 9 months of age. If you hide a toy under a towel, Baby will simply stop looking for it, as though the rattle has vanished into thin air. It will take a few more months for her to discover that there is still a rattle on the table. But dogs know cognitively—thoughtfully, if you will—that even though they momentarily don't see some object they've seen before, it is still there somewhere.

A colleague, Dr. Steve Zawistowski, explained how experienced hunting dogs track a rabbit. Yes, the traditional nose-to-the-ground method is tried and true and might be the only method a younger dog would use to follow the trail. But an older dog will actually look ahead—across the street where the snow has melted—to see the tracks of the rabbit on the other side and make a beeline for them. The visual cue of the rabbit track tells the dog that it's likely that he will pick up a scent there. The dog knows that those tracks are associated with where he's going to pick up the trail; it is based on experience—it's not just an olfactory cue. You might say that these seasoned hunters are thinking ahead.

Getting the Lay of the Land

Our pets have it all over us when it comes to knowing where we are and how to get there from here. They use a technique called cognitive mapping. Cognitive

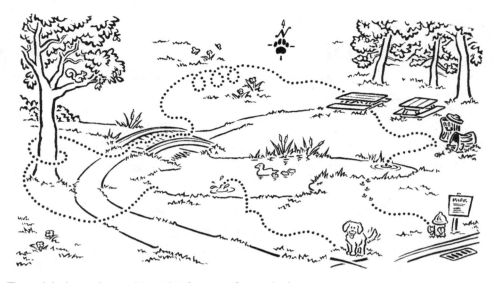

The park looks much more interesting from a pet's-eye view!

mapping is something we humans sometimes never perfect, regardless of our age, even though most mammals are quite expert at it.

If a dog and cat are romping in the woods, they do not need to return home on the same trail that they followed to get to the woods. That's because after they had gone the same route a number of times, they would form a cognitive map in their brain for knowing the topographic lay of the land. This means that instead of retracing their steps to go back home ("Where are those darn paw prints??"), they merely need to picture the map in their heads and trot home "as the crow flies," taking the most direct route back. It's *Homeward Bound* in the fast lane!

What about those long car rides to the best park around for Frisbee practice? If Daisy starts panting the minute you get in the car, how does she "know" she is going to the park? Well, you have probably given her a verbal cue. She can connect the sound of the word "park" with jumping in the car and ending up chasing the Frisbee with some doggie friends. But it doesn't necessarily mean she could lead the way to the park if you had to walk. Her body doesn't really know the way.

We share with our pets something called motor memory, where the muscles remember how it feels to do something. Consequently, simply visualizing an action—like a diver visualizing a perfect reverse two and a half—actually "fools" the body into believing that she is practicing the dive. But to establish

DR. WRIGHT'S
C A S E B O O K

Ranger

I once had a client who owned a mixed German shepherd named Ranger, who was truly terrified of going to the veterinarian and would bite any vet who had the duty of examining him. From the minute he was coaxed or dragged into the car until he arrived at the vet's, Ranger's radar apparently told him exactly where he was heading.

I made several suggestions to my client to "fool" Ranger into not anticipating the doom at the other end of the ride. His owner, Paul M., would take Ranger for a ride past the vet's and not go in, stopping instead at a pet store for a treat. Or he would take several different routes to get to the medical building. Then Ranger could no longer anticipate the horrors that lay in store: "When I reach the park and turn right, the next turn is the doctor!" Eventually, I asked Paul to bring Ranger to the vet's and have the doctor and the staff just pet him, give him a treat, and send him home.

All this helped to change what Ranger had learned about the car ride that ended in Hell. Now he had no reason to anticipate disaster, and he was finally able to arrive at an appointment with a smile on his face, looking for a doggy chew. We had to work on calming him during the examination, too, but it was a lot easier when he wasn't afraid just to walk in the door.

Plenty of dogs out there are like Ranger—vets who make house calls are having a field day. But with some understanding of the dog's thought process, the owner has a good chance of making things better.

DR. WRIGHT'S INSIGHTS

Researchers did a study of kittens to see if using their own four paws as they viewed something made any difference in what they could learn. A kitten was carried into a circular room with walls that were painted with vertical stripes and placed in a small merry-go-round located in the center of the room. The merry-go-round moved when a second kitten, harnessed to the opposite side of the merry-go-round, walked in a circle. Both the immobile kitten and the walking kitten saw the vertical stripes on the wall. Later, when both kittens walked into the room on their own, the "immobile kitten" clearly did not perceive the stripes on the wall. Only the cat who had seen the stripes while actively walking recognized the vertical stripes, showing that learning to perceive the world is an active, not a passive, process. (The phrase "soaking up knowledge like a sponge" clearly doesn't apply here. Ask yourself: Is it easier to instruct a friend on how to drive to your veterinarian's new office if you've actually driven the route, or if you've been a passenger in the car while someone else drove?)

the motor memory, first the diver must have actually learned the dive and practiced it repeatedly.

With dogs and cats, the same motor memory operates. But it's a little difficult to remember the way to the park if you have only ridden there in the car! Everyone knows the difference between riding in a van pool and being behind the wheel. The driver is likely to know the way to the destination without even "thinking" about it. The passenger is grateful to arrive on time for his appointment but couldn't tell you exactly how he got there.

Still, dogs are able to make good use of the perceptions their senses bring them on the way to the park, so that they "know" the way pretty well. In the same way, they use all their senses and past experience to figure out how to track the rabbit.

Planning Ahead

It's easy to demonstrate that dogs plan ahead: If you pull your arm in as if preparing to throw a Frisbee, you can bet your dog will be 15 yards downfield in a flash, looking wildly around for the disc you have just faked throwing. He takes off because his perception of what you are going to do next once the Frisbee is curling toward your body is based on his remembering the experience of running and meeting up with the Frisbee at the right moment.

(continued on page 58)

DR. WRIGHT'S
C A S E B O O K

Scout

It started out simply enough. Scout, my sister Judi's 6-year-old golden retriever, would stand by the back door a couple of times a day when he wanted to go out to do his doggy duty. When we went out alone, he was pretty businesslike. He would quickly look around the backyard for any encroaching deer, pee and/or poop, and trot briskly back to the kitchen door off the patio, perhaps giving a short bark to be let in if Judi had left the room. As soon as she opened the door, he would hurry over to "his" cabinet in the kitchen, where the "bones" were kept. Judi would take out one or two bones, Scout would sit, and then the treats were his to enjoy. This little routine went on for some time. But over the next couple of years, things changed. Scout went from a dog who wanted to be let out three times a day to a dog who would come in, complete his ritual, and then immediately go back and stand by the door!

"He just went out, and here he is again, begging to go out!" Judi complained about her wonderful dog.

"Sounds like a case of out-of-control chaining to me," I answered.

"Huh?"

Chaining is the way the dog makes sense of sequences of events that seem to be logically connected. Scout "knew" that when he stood by the door, the next link in the chain of events would be that he would be let out. He went out to do his business, looked around, and then trotted back and waited to be let in. The last but crucial link in the chain of events was to sit patiently while Judi scooped out a couple of bones from the biscuit box. Finally! To Scout's way of "thinking," sitting by the door was his meal ticket to Treat Heaven. And if he wasn't satisfied with one or two biscuits (and what self-respecting 85-pounder would be?) all he had to do was go out again to start the ritual that would end in his getting more of that treat.

He eventually figured out that he could simply skip the bodily functions and pretend to be sniffing around for a place to go, saunter casually back to the house, and oh, by the way, Judi, how about another biscuit for your adorable, obedient servant? And likely as not he would get the treat, just in time to start the chaining process all over again. Clearly, in Scout's mind, standing by the back door literally meant—after chaining all the other events together—getting a treat. With all these goodies, the dog was getting fat to boot!

Since this was driving Judi nuts, I suggested that she ignore him if he went to the door again immediately after going out. After a minute or two, she reported, sure enough, he would give up and go on his way. I also advised her to give Scout a treat at random times during the day—times that weren't connected to going outside—and to skip the bone some of the time when he did go out, replacing it with a pat or verbal stroking. This lowered the golden's expectations about a food treat, helped break the chaining habit, and left him without a "bone of contention!"

Scout had learned that the easiest way to get his favorite treats was to stand by the back door as if he needed to go out.

It's not just instinct—what we animal behaviorists now call prepared-ness—or habit. It's the same thing that happens to you when someone throws you a softball: You move under the ball because you have learned to do so from experience. You perceive the ball, its flight, and its speed, and you use that per-ceptual information to move your glove under it and squeeze at just the right time. This sensory-motor sequence is a combination of preparedness and ex-perience. If you see someone hit the ball way over your head, you may start to run to catch it but then stop because you know, based on your prior experi-ence, that you won't be able to catch it. Home run! We integrate all our per-ceptual and mental tools to solve problems. Dogs do the same.

The Truth about Treats

Treats certainly have their place as a reward for good behavior (going outside for elimination in Scout's case). The best way for a dog to learn to behave well, when he's not too emotionally aroused but is merely learning what is permitted and what is not, is for you to reward good behavior with a treat or a stroking or an enthusiastic "Go-o-o-o-od Bo-o-o-o-oy!" These simple reward procedures are part of what we behaviorists call operant conditioning—training in action.

Here's how operant conditioning works. Your dog does what you want. You reward him to reinforce the idea of what good behavior is so that he can remember the pleasant feeling he got from the reward, connect it to the im-mediately preceding activity, and do it again the next time to get that great feeling back.

Let me just get on my soapbox about treats for a minute. Used in this way, they work to reinforce good behavior. But remember that your high-energy pup is going to slow down as he gets older. And if you get in the habit of of-fering him *only* a lot of high-calorie treats, you're on the way to developing an-other kind of problem: an unhealthy, overweight pet. Remember that social rewards like praise and petting are calorie-free treats, and they'll motivate good behavior too! Use them interchangeably. If you just can't resist passing out treats, remember that most dogs love fresh veggies and fruits—apple and ba-nana slices, carrots, and sugar snap peas are big favorites around our house.

🐾 Time for a Time-Out

If your dog is acting too wild or out of control to pay attention to your wishes, a short time-out in a quiet location is the best bet. Pick a place where he can't have a bang-up time without you—someplace a bit dark where he can relax and settle down. You want to remove any potential rewards—like attention—that are possibly prolonging or maintaining the bad behavior. It's not important for him to understand your words. What *is* important is for him to be able to predict that too much activity results in a time-out.

If he's out of control, this may be the only way you can get him to stop—especially if he's still in the "wild" puppy stage before formal obedience training, or if his human friends have made a habit of yelling "NO" at the top of their lungs, to no avail. Sometimes a dog can't seem to control his own excitement, and he will actually welcome your putting him in a room by himself briefly (don't slam the door on him, please).

Only leave him for a minute or two; time-out isn't meant to be a draconian solitary confinement! Behaviorists have found that 30 seconds to 2 minutes is the most effective range of time for a dog to be confined. When he is calm, or acting out less and not barking and carrying on in the room or the crate—wherever you choose for the time-out area—give him another chance to be with you. Let him know that you are pleased with this behavior and that he may rejoin civilization, which is often reward enough; or offer a treat; or just give him an encouraging word or a pat.

Dogs that have a pretty good social motivation will learn this routine fairly quickly, especially if they are young—even as young as 8 months. Always reward his compliant behavior, whether it is sitting, lying down, or just being calm and avoiding emotional behaviors that he may do for attention, such as biting arms and hands and jumping on you. As with a child having a tantrum, the best thing you can do is to ignore his attention-seeking behavior or give a quick time-out.

Physical Punishment Doesn't Pay

Physical punishment for misdeeds is rarely productive and is often counterproductive. More to the point, it is not ever (*never*) the right thing to do. I have filled up my calendar with clients who thought physical punishment was the way to go and ended up calling me to undo the damage. And I do mean damage!

If you swat your dog every time she pees on the carpet, she may learn not to pee at all when you, the person who "doesn't want me to pee," is around, including when you take her outside. She is likely to hold it until you are both back indoors and you go away. Then she'll sneak in the back room and pee. That behavior is perfectly logical, from the dog's point of view! Instead of

(continued on page 62)

DR. WRIGHT'S
C A S E B O O K

Lulubelle

I remember one client who was quite surprised to learn what she was teaching her cat and how the punishment she was meting out only made things worse. She used the good old squirt gun, which has been recommended for years by all sorts of "experts." What the experts fail to get across to the thousands of frustrated pet owners who have failed with this technique is that the "squirt" must be administered within about 3 seconds of the "crime" for the animal to make any connection whatsoever between the two. But I am getting ahead of myself.

Mary S. was quite frustrated when I met her. Her cat, Lulubelle, had chosen to harass and bite Mary at least four times a week. The cat was aggressive and mean. It hissed at visitors, and it bit the heck out of Mary's leg every time she walked down the stairs in the morning. When I asked Mary what methods she had used to dissuade Lulu from this campaign of guerrilla attacks, she mentioned the squirt gun.

When I arrived at the suburban Atlanta home and met Mary and her husband, Austin, I was puzzled. The snowy white cat did bristle a bit at my entrance, but then she went calmly about her business, curling her fluffy tail around her as she found a spot on the sofa. Based on her owner's description, she was hardly the vicious feline I was expecting! (By the way, many cats will make a display of defending their turf when a stranger walks in; it's not just a "dog thing.")

But Mary's calves and ankles were indeed covered with scratches (she made sure I saw them), and when I turned to Austin and asked if he had matching wounds, I got another surprise. The cat never bothered him. In fact, they would play on his bed while Mary was getting dressed every morning for work. The cat would go under the covers and pounce on his arms (no claws), that kind of thing. Again, this was not the description of a generally aggressive cat.

By now it was clear to me that the cat's motivation in attacking Mary was, in a word, *play*. She had learned that it was enjoyable to go into the play sequence with Mary coming down the steps because she could leap out and surprise her. She got lots of attention when she latched onto Mary's calf! This antic was followed by Mary's chasing her fruitlessly around the house with the squirt gun.

To modify undesirable cat behavior, especially after it's ingrained, you must create situations where both you and the cat can win. Then you praise your cat's compliance and the desirable behavior. I told Mary to stop moving if Lulu attacked and not to challenge her. But getting the cat to be excited about something else was key. After Mary understood Lulubelle's strong need for play, she was able to distract the cat with plenty of toys tossed away from herself at strategic times (for example, when walking down the stairs).

Cats just want to have fun!

learning not to pee in the house, she learns, "I'm not going to pee when you're watching me—period!" We have such good intentions, but we don't always consider how the dog will read them. Owners commonly recite, "And we just took him outside, walked all over the place, and he didn't have to go!"

For cats, the same philosophy applies. The crime of punishment will come back to haunt you big-time. A good general rule for correcting feline misbehaviors is to interrupt the behavior just before he begins it. This rule applies whether you're interrupting a cat's staring at another pet cat before attacking her, looking up at the counter before leaping into the tuna casserole, or sniffing the clothes you have just packed into that inviting open suitcase on the bed before peeing all over them.

The remedy is to clap your hands quickly or to get his attention by the least intrusive, most effective means available. Then remove him from the situation or distract him with a toy or some petting or a treat. In any case, do no harm. And certainly, cause no pain. Cats become so emotionally distraught when they experience pain that it is unlikely that they'll learn anything at all about what *you* intend for them to learn. Instead, they'll associate the source of the pain with a fight-or-flight reaction in the future. When you're tempted to strike out, remind yourself: Cats simply do not understand punishment.

The idea is to keep the behavior from starting and then to provide Rocket with the opportunity to do the motivated behavior in an acceptable location and at the appropriate time. Creating the opportunity for correct behavior, and then praising your cat when she complies, is something she *can* understand!

Getting Through to Your Pet

Learning in cats and dogs doesn't depend on a huge vocabulary of "understood" directions, lectures, or commands. Dogs and cats are not robots. They process information, and they are pretty good at looking at us and predicting what we are about to do next. They try to respond appropriately by putting together the information they already know with information that comes next.

So, although there may be a disconnect between the words we use to convey our wishes to them and what they truly understand, our pets are much better at understanding the social cues they get from us—the things we do with our bodies and our faces. They pay attention to our eyes and our gestures, and they are much better at reading and understanding body language than they are verbal, spoken language. The implication for us, of course, is that you can forget the long lectures and explanations. You don't need to tell him when you'll be back and how he should be patient and not wreck the house or try to justify why she's going to have to go to the kennel because you couldn't find a pet sitter. It won't mean squat.

Your dog may understand in a limited way what you mean when you say, "Go to your crate." That's because you use the same phrase over and over again, and when she goes to her crate, you praise her. So she understands verbal communication to that extent, rather than the meaning of the words themselves. And when you point and look over to the corner, without even saying anything, he knows that you want him to lie down, and he will be praised for doing so. And if the cat is just looking at your finger, well, she's probably lying down already, so you can give her a stroking for that. Our pets really are doing the best they can to comprehend what we want and to please us. In chapter 4, we'll see how well we can understand how *they* communicate with us and with each other.

Chapter Four

What Your Pet Wants You to Know: Decoding Body Language, Barking, and Behavior

Picture this: A big dog—say, a German shepherd—and a large gray cat are side by side, coming toward you. Both animals spot you, and you notice that they are looking right into your eyes with a great deal of interest. Both have their ears pricked up and rotated forward a bit. They both trot forward with purpose. The most prominent feature on both of them, you notice as they draw closer, is their tails. Both have rather bushy, long tails, and as they approach you straight on, it looks as though their tails are growing out of the top of their heads. Both tails are straight up, moving back and forth.

Question: Based on their body language, do you think these animals are feeling friendly? Hostile? Or merely neutral?

Well, if you are anything like the group of college students I gave the same examples to a number of years ago, you would be all set to hold out your hand to that German shepherd, who is coming toward you in such a "friendly" manner. But if you did, you'd be risking a serious bite or worse. Most people don't realize that this dog is exhibiting classic communication signals designed to warn you that he is aroused and ready to act in an aggressive way at the slightest opportunity. My group of students—mostly freshmen—failed to recognize the danger signals. They were lucky that the quiz was only on paper!

What about the big gray cat? You could go ahead and pet him. He has the exact same body language, but in a different species, it has a different meaning. Go figure.

Both dogs and cats speak to us with their voices, tails, eyes, mouths, ears, and posture. When they "talk" to each other, they use scents we can't smell, subtler postures, looks, and, for all we know, ESP. Animal psychics claim to be able to communicate mind to mind with our canine and feline friends. That's great for them. But the vast majority of us must rely on receiving and correctly interpreting the unspoken messages our animals try to send to us and then try to make them understand us with whatever tools we have at our disposal.

Behaviorists have been working to decode cat and dog communication for only a few decades now, but we believe we're getting a handle on most of the everyday behaviors our pets exhibit to try to tell us things, as well as some of the more unusual ones. Let's start with dogs. (If you simply can't wait to read about cats, turn to "Cat Communication Tactics" on page 70 and dig in!)

The Canine Connection

First off, not all signals a dog sends out can be considered communication. It's like the old question about the tree falling in the forest. If no one is there to hear it, is there any sound? Canine—or feline—behavior is much the same. There must be a sender *and* a receiver for a message to count as communication. Behaviorists and owners alike spend a lot of time trying to understand what dogs are perceiving and what they are trying to tell us.

You can learn a lot from observing your dog. But before you look for meaningful information in your dog's activities 24 hours a day, you need to know that not everything he does is social behavior. Eating, self-grooming, exploring, eliminating, and other activities where no person (or animal) is present is what behaviorists call asocial behavior. That's not antisocial be-

The Vocal Life of Dogs: Whining

Vocal communications among dogs have not been thoroughly studied. You probably know what your dog's particular vocalizations mean better than I do. A little whimper that means nothing to someone outside the family may send a child racing to the bedroom for Spot's ball. ("Not <u>that</u> one, silly! She wants the yellow-and-blue one with the bumps on it! Didn't you hear her?")

We can recognize a few "classic whines" and what they mean. So here is Dr. Wright's Whining Dictionary!

Dr. Wright's Whining Dictionary

- A whine plus a paw held in the air: "Help! I hurt myself." This whine is meant to solicit caregiving to relieve pain or to emphasize his hope that you will dig the thorn out of his paw.
- A whine by a puppy prior to weaning may be a distress vocalization: "Please lick me, Mom, so I can find my way back to your warm tummy and food!"
- A whine by a dog shut outside the bedroom: "Please let me in there with you!" This time, the dog whines from a purely social motivation.
- A submissive whine may be your dog's way of trying to communicate his intention to be in a subordinate role to your other dog, or to you during scolding—usually in combination with a visual signal: "Please don't hurt me. I give up!"

havior; it's just behavior that can occur in a solo rather than a social context. So you don't need to read a whole lot into what your dog means when he eats the other dog's food as well as his own, as long as the other dog isn't around. However, what he does when the other dog is standing next to him, guarding his bowl, definitely counts as social behavior.

🐾 Subordinate Signals

With all the talk of alpha males, you're probably aware that dogs usually take dominant or subordinate roles in the family and with other animals. Now, I don't buy into the owner-as-alpha-male concept; I prefer instead to view owners as caring leaders in their relationships with their pets. However, I hope that a dog's submissive signals to its owner will not go unobserved and lead to a "failure to communicate." A dog may indicate her subordinate role by being either passive or active in her submissive displays. The more intimidated she is by your power, the more submissive the dog will be—unless she decides to challenge your role, which I'll talk about next.

Generally speaking, an extremely submissive dog will roll over on her back, lift her rear leg in the air, and expose her genital area. (Submissive males act the same way.) This display is sometimes brought on when an upset owner yells at his dog, who may even urinate a bit, making the owner even more angry! As the owner's anger level rises, the dog becomes even more likely to show its subordinate role by more of the same submissive behaviors—rolling over and urinating on herself.

This is the point at which we ought to take a cue from dog-dog communication. A dog occupying a dominant role generally will decrease his assertiveness or aggression toward a second dog (or person, for that matter) if the second dog's signals, postures, and behaviors indicate that it is assuming a subordinate role by displaying passive submission. Dominant people—including yelling owners—should recognize what their dog is trying to tell them ("I give up, you win!") and, like the dominant dog, back off.

Passive submission isn't really a way of life for a healthy, active dog in the family. If you are seeing these types of display regularly, you need to work on building up your pet's confidence with gentle guidance.

When she's not being yelled at, a subordinate dog is going to be more "active" in her submission, as you've probably observed in most dog-dog and dog-

If you see an alert-looking dog staring at you with his ears rotated forward, his mouth closed, and his tail up and swishing back and forth, he's giving you all the assertive signals that he's top dog (dominant role). But if he looks sheepish—he's not looking at you, his ears are laid back, and his tail is down and wagging (submissive signals)—he's telling you that he's the subordinate one, and you're the boss.

DR. WRIGHT'S INSIGHTS

A dog displays only one set of signals, whether the receiver is a dog or a person. Dogs already understand the signals because of their on-the-job training as puppies in the litter. Our job is to pay attention and understand what they're trying to tell us. We should always reinforce non-assertive, submissive behavior, since subordination is a good thing—unless you like the idea of the dog running the house! I am not suggesting that a dog is mere putty in your hands and can be changed from a cowering wimp to king of the hill if you don't let him know that following the leader is a good thing. But he should be able to understand how you want him to act through praise and encouragement when he behaves in a submissive, compliant manner.

person greetings. Actively submissive dogs will approach another dog or person while showing submissive signals like smiling, avoiding eye contact, rotating the ears back against the head, holding the tail down, and wagging. A dog may show more intense active submission by approaching with her body crouched down to the ground. Dogs frequently use these submissive behaviors to reaffirm an existing role relationship with another dog or person, to initiate a new relationship, or simply to try to keep their own "space" when greeting.

🐾 Defensive Behaviors

One of the most important signals a dog can give a person is fair warning that he may be aggressive if you don't allow him to remove himself, or that you should move away. These moves are the dog's way of trying to avoid a confrontation when he's experiencing discomfort due to your approach. Pay attention to them! If you don't, you may see the next step in the communicative sequence, the defensive component of aggression, which consists of growling, snapping, lunging, or biting, usually in conjunction with exaggerated submissive postures and signals directed toward the potential victim.

Dogs that are agitated sometimes run up to and then away from people, seemingly in conflict and unable to decide whether to bite or retreat. This approach/avoidance behavior is the same kind of quandary we find ourselves in when we get more and more nervous the closer we get to our play's opening night or develop all sorts of doubts as the wedding day approaches. Well, the play usually opens, you go through with the wedding, and the dog will probably bite you if you don't pay attention to the signals he's sending loud and clear.

Dogs who are feeling defensive also tend to attempt sneak attacks, approaching the victim quietly from behind rather than head-on. The main purpose of this form of aggression is to increase the distance between the victim and the dog.

If avoidance tactics and defensive threats in the form of a few nips don't work, overly assertive lunging behaviors may take their place. For example, intense staring is not good—on your part *or* his. A dog's displaying submissive signals, but staring or visually tracking a person or dog frequently precedes the lunging behavior that can end in an uninhibited bite.

The Vocal Life of Dogs: Barking

People and other dogs can tell the difference between different kinds of barking, aided, of course, by the circumstances a dog finds himself in. The communicative content of a bark from a dog left alone in a room for 12 hours is quite obviously different from that of a dog running to the front door after the doorbell rings. But they will sound different, too, and that sound may vary a bit according to breed. So let's concentrate on the various motivations for some of the barks you're likely to hear in the course of owning a dog. May I present Dr. Wright's Barking Dictionary!

Dr. Wright's Barking (with a Groan Thrown In) Dictionary

- The warning bark, while looking at the owner for help: "Someone rang the doorbell!"
- The aggressive bark, while looking over his shoulder at you because he really doesn't want the job: "Get out of my yard!"
- The aggressive bark, paired with a growl: "Get out of my yard, or I'll get you!"
- The continuous barking at the back door: "Let me in, dammit!"
- The nuisance bark: "There's nothing else to do out here alone in the dark."
- The freaked-out yelping bark (with tail tucked): "I gotta get off this chain 'cause something's coming to get me!"

Dogs also have a way of groaning, usually when they've had a big day, when they've given up trying to get you to feed them, or when you've finally decided to cuddle up with them at the end of the evening.

- The groan: "Oh, poor me, life is so tough" or "I'm worn out" or "At last, some creature comforts!"

I'll get into the specific types of aggression in chapter 15. Let me just tell you now that physical or verbal punishment with this type of dog will result in more than just the dog's piddling on himself—it will backfire on you. And you can't predict what will happen because a dog's emotional state and his propensity to attack can change abruptly from moment to moment. One last yell at your cowering canine may find him mad as hell and unwilling to take it anymore!

❧ How Dogs Use Dominance

Your family is not like a pack of wolves: Your dogs do not have to compete for food, shelter, or a mate. If physical force is used, establishing dominance over a dog is a risky proposition. It places both dog and human in a confrontational relationship where someone is going to lose. On the other hand, letting the dog establish dominance over you is an equally dismal—and potentially dangerous—proposition. Although dogs form role-role relationships with one another to decrease daily threats and aggression, some dogs haven't read the script.

Dogs who assume a dominant role may exhibit dominant signals (ears rotated forward, direct staring, mouth closed or lips puckered, tail up and swishing) and assertive, demanding, controlling behaviors, especially if the subordinate member of the relationship (not you, I hope!) fails to fulfill his role-appropriate behaviors with reference to the dominant dog. For example, if Thor (subordinate) fails to greet Mojo (dominant) on first meeting in the morning, Mojo may rightfully stop Thor from moving, or even pin Thor to the ground. This "demand" by Mojo for a full-blown submissive display has a purpose: to reaffirm overtly the dominant/subordinate relationship between the two dogs.

Most stable relationships, though, are reaffirmed either by the subordinate dog yielding to the presence of the dominant dog or by exchanging submissive and/or dominant communicative signals. For most dog duos, few instances of actual aggression occur from day to day.

Cat Communication Tactics

For all the millions of people who love and enjoy cats, a surprising number of people profess to "hate" them. "Cats are too cold, standoffish, inscrutable, and unaffectionate," they grumble. In addition, cats are not interested in people's opinion of them! I've always suspected that much of this dislike is based on fear—these people see cats as mysterious, sinister, and ultimately frightening. That's my theory, anyway. And it's a shame! These wonderful animals don't deserve to be so misunderstood!

It's quite true that most cats don't have the happy smile and wagging tail

DR. WRIGHT'S INSIGHTS

We animal behaviorists have actually identified several combinations of cat "personality characteristics" that regulate the degree of sociability cats exhibit. Toward people, these characteristics include friendliness versus unfriendliness/fear/hostility. We can also distinguish the degree of friendliness they display: initiative/friendly (taking initiative in being friendly) versus reserved/friendly (accepting others' initiatives) and play-contact/friendly (initiates friendly behavior only for play) versus petting/friendly (allows petting and stroking). These personality characteristics and behavioral styles influence the kinds of relationship some cats are able to form. They make it more or less likely that your cat will invite you to greet, play, include, and physically contact her.

of the people-adoring dog going for them. But cats do have a wide range of communicative behaviors that anyone can tune into, and they exhibit quite a variety of social styles. Some cats do nothing but hide under the couch, but others are as friendly as a puppy.

Some cats groom another cat as a means of caring for them; others groom another cat's head, licking it several times, which is a signal that precedes play-chasing and play-fighting. Cats lick our noses, knead on our arms and necks (which hurts if they get carried away with their claws!), and rub against us incessantly, indicating their pleasure with us. (Do you cringe when your friendly cat rubs fur all over you or against the corners of your couches or walls? Velcro-attaching combs help you remove the excess fur, making it easier to clean up.)

To refer to all domesticated cats as "social" or "solitary" misses the boat. Cat sociability depends on age, sex, early experience, recent history (experience with or without trauma and change in their daily living situation), medical health, and the availability of others for socialization. Communication allows cats to form and regulate interactions with others, which leads to relationships that differ in sociability and to the perception that a cat is either relatively social or solitary.

For example, cats differ in how social they want to be around the food bowl (some cats only eat alone, not with their littermate, person, or even the dog), during play (some prefer playing alone to social play and vice versa), and at bedtime (your cat may like to sleep alone, or draped over your head at night). Consequently, we may see one cat as a loner and assume that all cats are solitary or see another cat as the life of the party and generalize that all cats must be social.

Thus, to some extent, our view of the social life of cats is an "eye of the be-

(continued on page 74)

DR. WRIGHT'S
CASEBOOK

Laramie and Rush

Much of each cat's behavior is devoted to maintaining his social relationships with the humans and other cats (or dogs) who inhabit his world. Cats' relationships with each other can also vary widely, and multiple-cat households are virtual hotbeds of kitty communicative signals. I recently worked with two different clients with nearly identical problems: Their cats left messages by marking indoors. The Harts were cat lovers extraordinaire. They started out with Laramie, a 20-pound, 7-year-old male. Then they added Cinder, a 6-year-old male, who quickly assessed the situation and played a willing second banana to Laramie. He curled up on the floor, content to let the older cat command the best lookout posts.

But then the Harts adopted two kittens, Sugar and Honey. With the addition of these two sweeties, things began to turn sour for Laramie. Not only did the kittens command most of Mindy Hart's attention, but they also quickly claimed Laramie's resting places for their own. To add insult to injury, he had to start coping with a strange cat outside on the porch, which he had been unable to chase away because he was an indoor cat. It was all too much for Laramie to take.

Despite his gigantic size, Laramie was a cool cat, not one to throw his weight around. He simply would not start a fight with the kittens, and he would not stoop to the level of his subordinate pal Cinder, who was just lying around in the crummy spots the kittens had left empty. And he could not start a fight with the stranger. So Laramie used another means of communicating his displeasure: He went around spray-marking his territory for all he was worth—and unfortunately that territory was in the house!

By spraying his pheromones around, he was announcing to the other cats and Mindy, "I'm here! This is my house!" Although spray-marking is often mistaken for a litterbox problem, marking is as legitimate a communicative signal to other cats as meowing or purring.

The second case involves two cats, Rush and Roxy. Their frantic owner, Jillian, called me asking for advice. Both of her cats had "gone berserk." Jillian had captured two feral cats and had them spayed and wormed. Now she was feeding them daily on the back patio.

What she didn't realize was that while she was sitting out on the porch with these feral cats, doing her best to help them, and then bringing their scents in on her shoes and on the dishes, Rush was ready to drive these cats away. But Roxy was the only cat to whom Rush could display his aggressive signals. He would issue a warning growl to Roxy when he saw or smelled telltale signs of the feral cats out on the patio. Before he knew what was happening, poor innocent Roxy was being scratched and bitten.

In both cases, the cats gave ample evidence of what was bothering them—which was essentially the same thing—cats outside that they couldn't get to. But they responded differently—one by spraying and the other by attacking his pal. Both were using communicative signals that were misunderstood or overlooked by their owners.

In the first case, I introduced Mindy to a product called Feliway, which—in a neat sleight of hand with pheromones—turns cats' interest away from spraying and toward merely rubbing on the same spots where they once marked their belongings.

Remember that rubbing faces on items is a communicative tool for cats. The head and cheek areas contain glands used for scenting whatever and whomever they rub. And they love to lick and rub and roll around on interesting scents. The neighbors were persuaded to keep their outside cat inside, and the situation eventually improved dramatically.

In the case of Roxy and Rush, Jillian needed to wean herself away from the feral cats. She tried moving the feeding off the deck, but that didn't fool Rush. Finally, Jillian found a neighborhood teen several streets away who was more than willing to take over the feeding. The greater distance put the feral cats out of Rush's "smell distance."

Whether independently dreaming or curled in the nearest lap, there's nothing as contented as a sleeping cat.

holder" phenomenon. It's actually more interesting—and more relevant—to talk about how cats are social and how much sociability they display rather than to debate whether they are social or solitary.

Some cats are quite content sitting alone or lying in the window all day watching the goings-on outdoors. Others are more dependent and social, greeting their human when she comes home, eating dinner at the same time, sitting on the couch with her after dinner, playing with her before bedtime, and sharing a pillow at night.

🐾 Speaking Fluent Feline

Because cats aren't pack animals like dogs (don't get me wrong; I'm not perpetuating the myth that cats are loners), many of their communications serve the function of increasing or decreasing a cat's distance from another animal or person. The cat is trying to define her comfort zone—how much space she needs at any given time. If we can learn the signals and behaviors that indicate when our kitty is comfortable and when she's not, and we respond to her "ideal" level of sociability accordingly, we will build trust and a secure social emotional bond. She will also clearly communicate our success by purring!

Some cats want to decrease their proximity to other cats and do so by hissing, spitting, and running (to avoid them) or stalking, growling, and pouncing (to attack them). So how does a cat let you know whether to smother her with affection or to keep your distance? One way is to watch the signals that she uses repetitively.

Every night an assertive tiger cat named Jocko went through the same ritualized sequence of greeting his owner Beth when she sat down in the family room to watch TV. Beth told me that Jocko would look at her and establish eye contact, emit a short "Rrrrr," and move quickly toward her with his tail raised in a confident trot. Then he'd jump up on the couch next to her and proceed to rub her arm with the side of his face. Next, he would walk away a few steps and present his rear end, and finally head back to her for another rub. Sometimes he'd get more assertive and add a head-butt, and then go into another presentation.

Beth's typical response was to make a circle of her thumb and index finger and quickly run it up Jocko's tail. This made a "whip" sound, she told me, and seemed to encourage the cat to rub more. Finally, they would settle down to watch the evening news. This routine had gone on for as long as she could remember. It was not too different from the way Jocko might greet another cat. That meeting could include a nose check (a nose-to-nose greeting; I suspect that the idea is to allow the other cat to sniff, much as we say, "Hi, glad to see you!") followed by an anogenital sniff (a sequence we don't even care to think about, much less imitate).

Jocko used his eyes, his tail, his voice, his ears, and his whiskers to greet his friend Beth. All these signals and more play a part in the way cats communicate with one another and with their humans and dogs. Let's look at each of them more closely.

Tales Tails Tell

Cats' tails are wonderfully expressive! An upright tail when the cat is standing or walking lets you know he's alert, confident, and ready to greet you. A flick of the tail can mean that your cat has been disappointed in some way. Did you forget to toss a favorite toy in his direction like you did last night? Lashing or thumping indicates irritation or conflict. And if he is moving his tail from side-to side, watch out! That signal, or similarly a straight-up, fluffed-out tail, may mean an imminent attack.

The tail can also show you when your cat is not feeling so confident—in fact, he may be frightened. The typical Halloween cat looks pretty scary, doesn't he? He's always standing in profile, with his hair all puffed out and his body humped slightly—people are apt to think that he is a rough, tough cookie. But no, the fluffed-up fur happens automatically when the cat's sympathetic nervous system is in high gear, and he is scared. He stands side-ways to his potential attacker, which makes him look bigger. (Hey, it works for me, how about you?) He can also tuck his tail between his legs if the bold look isn't happening, which has the effect of making him look small and unobtru-sive, submissive. And an inverted U-shape can indicate that he's getting him-

The Vocal Life of Cats

Surprisingly, most cats have quite a lot to say for themselves. But what are they actually saying when they make all those funny noises? To find out, please refer to Dr. Wright's Cat Chat Decoder.

Dr. Wright's Cat Chat Decoder

Chattering. Anticipation of frustration: "I wish I could get to the bird on the deck."

Chirr. Mom's call to round up the kittens; a meow with a trill: "Get back here!"

Growl. Low pitch, aggressive: "I mean that!"

Grunt. Young kitten's comment: "Say what?"

Hiss. Defensive warning or aggression, with mouth open, lips pulled back, air forced out through arched tongue, often accompanied by a "Spitttt!" when sud-denly threatened: "Back off!!!"

Mew. Rather soft, short, high pitched—used by kittens to seek Mom: "Where are you?"

Meow. Greeting, announcement. "Here I am!"

Murmur (Mmmmm). During purring—pleasant social contact or arousal: "That feels so-o-o good!"

Wail. "Help!"

self in a snit, is sort of angry, and is definitely gearing up to strike back. (If he stalls long enough, maybe that tough-looking neighbor cat will go away.)

Catty Remarks

We tend to think of cats as quiet—compared to dogs, at least. Unless, that is, you have a very talky breed like a Siamese, or it's 3:00 A.M., and you're stifling the urge to heave a shoe through the window at the caterwauling toms outside the window.

Cats vocalize more than we give them credit for, using their various meows to express demands and cries. They hiss or spit when they're feeling defensive and growl when they're fighting. Beth knew that Jocko's little "Rrrrr" was his shorthand way of greeting her verbally.

Cats' vocalizations help them get closer to you or keep you away. In behaviorist talk, they may be care-seeking or care-soliciting, offensive-aggressive or investigatory. They can also be defensive (as in "Let me outta here!"). Vocalizations may also be a clue to kitty's mental state: a plaintive mew ("I'm lonely") or a wail ("I'm *really* lonely") or a pitch-modulated cry ("I'm afraid!").

A Cat's-Eye View

Remember how Jocko made eye contact with Beth? That's another sign of confidence. Often, though, direct stares may indicate a challenge or a threat, especially coming from a confident cat. Fearful cats will avoid eye contact like the plague.

When a cat is stimulated, surprised, or fearful, his pupils dilate. When he's focused or ready to attack, his pupils shrink. Don't be afraid to look your pet in the eye to judge his mood—but be prepared to make like a scaredy cat and glance away if you're worried that your communicative signals might trigger an attack.

Hearing What His Ears Are Telling You

Cats use their ears for more than just hearing. When Jocko jumped up on the couch, his ears were probably erect, showing that he was alert and focusing on a stimulus—Beth! Perked-up or slightly relaxed ears are great. But when a cat's ears are swiveled sideways and he's not listening for something, it may mean something else is afoot; he either is feeling aggressive or just wants to engage in some assertive activity. To complicate matters even more, cats also swivel their ears downward, which could mean that the cat is feeling defensive or submissive.

The Many Moods of Whiskers

A cat's whiskers, technically called *tactile vibrissae*, are positioned above and to the side of the eyes and on the upper lip. A cat uses them for figuring out whether she can squeeze through a small passage or hole without getting her head caught. They're sensitive enough to pick up changes in wind currents so that she can position herself to take advantage of smelling whatever morsel might be upwind. At night, she "sees" with her whiskers by flaring and projecting them outward and a bit forward. She greets others with them folded back along the sides of her head. She indicates fear by holding them tightly together and flattening them against her head. And if she would still need to capture her food to survive, her whiskers would help her be an efficient predator.

Eyes, ears, tail, voice, posture, fur, and whiskers can all tell us something if we want to really understand our cats and meet their social needs. That's a lot of information to be gained from our supposedly "inscrutable" feline friend, isn't it?

Which Cat Is Boss?

Whenever there's more than one cat in a household, the potential exists for competition over the top spot, which can lead to conflicts. Usually owners can figure out which cat rules the roost, but the higher the number of kitties, the tougher it can get to figure out who is lording it over whom. Here's a simple test a colleague with 14 cats, Dr. Penny Bernstein, shared with me when she wanted to identify the top cat in that menagerie. She calls it the Kitty Condo Time-Sharing Test. She simply took a small grocery store–size cardboard box, set it in the middle of her bedroom floor, and waited for somebody to notice it. And since no self-respecting cat can ignore a great place to jump in and curl up for a couple of minutes (or hours), she soon saw one of the cats doing just that. As the other cats wandered through the room, only one other cat approached the box.

My colleague sat a few feet from the box patiently waiting until the first cat jumped out. A second cat promptly jumped in—in effect, making the box into a "time-sharing" venture! None of the other 12 cats even attempted to enter the box the rest of the day, but the 2 who had shown interest in the box continued to swap places peacefully every so often. The 2 top cats had signaled the others that the box was their turf, and the others complied without incident. So it was clear to my colleague that 2 out of the 14 cats were sharing top billing in her household.

DR. WRIGHT'S INSIGHTS

Have you ever noticed how some cats have a very short fuse when it comes to stroking? One minute you're absently petting away, and Tigger seems to be enjoying herself. When you add another stroke, "Yeowch!!!!" Your loving companion has nailed you, biting the hand that strokes her! The stroking is just too arousing for some cats, and they "have to" get rid of that feeling by biting.

Well, help is at hand (so to speak). Next time you pet Tigger, watch her whiskers. Just as cats switch from the visual to a "whiskers alert" when they have a mouse between their paws, so they will telegraph you with their intent to assault. When cats are about to bite your hand during stroking, many will flare out their whiskers and rotate them forward in front of their nose, with your hand as the target. That's your signal to stop petting, wait until the whiskers relax, and then add another stroke. Thus, with a watchful eye, you can move your cat gradually from a "three-stroke cat" (the number of strokes you can get in before being bitten) to a "six-stroke cat."

You can use a similar box, or paper bag, or a "kitty condo" to try the same test with your gang. It will probably confirm the leadership qualities you already discern in your most assertive feline, or you might be surprised to find that that puss is sharing the spotlight with a friend. Try it and find out!

Here is one note of caution. In some multiple-cat households there is also a pariah cat, the one who receives the brunt of attacks from the others. These low-on-the-totem-pole cats would typically be chased off in the barnyard setting, but risk injury from repeated attacks by the other resident cats if not allowed their own space in the home.

On to Emotions!

Now we know a little bit more about what our dogs and cats are trying to tell us by observing what they do with their bodies, their voices, and their scents. Our next challenge is to peek into their hearts (well, actually, their brains) and try to fathom the range of emotions they are feeling. In the brief history of professionals studying animal behavior, the emotion "fear" is no newcomer; all other emotions are basically uncharted territory. Nonetheless, I believe that working with the emotions of dogs and cats is the key to transforming unwanted behavior into great behavior. And that's exciting stuff indeed! So join me in chapter 5 as we find out what our pets' emotions are all about.

Chapter Five

Managing Emotion—
The Secret of Good Behavior

People used to say that dogs and cats didn't have "feelings." Now we know better! The trouble is that we have gone from "Sure, I'm gonna tie up Bones in the backyard—I can put him wherever I want, he's only a dog" to "If Boopsie doesn't find a hand-knit sweater with her name on it under the Christmas tree, she is going to be totally crushed."

Both extremes show a misunderstanding of the emotional life of dogs, which we are just beginning to explore in a methodical way. No behavioral study can take the place of your own knowledge of your pet, which comes from years of devoted companionship. But in this chapter, I'll tell you what I think is the key to good behavior in many cases where pets are "acting out."

Most misbehavior is emotionally based. If you can learn to recognize the various emotions and know what to do to change them, good behavior is very likely to follow, as surely as Buster follows a trail of doughnut crumbs!

I don't make guarantees, but I am confident that if you take the time to understand this approach to behavior problems in pets, you will be able to nail what's going on with your dog or cat and help them over the rough spots much more easily.

The Anatomy of Emotion

The first thing I want to do is talk about emotional traits versus emotional states. Personality traits give us—and our pets—our stable behavioral tendencies. Some dogs are generally curious and explore everything within sight (or smell), whereas other dogs are generally disinterested and could not care less about what's new in the neighborhood. If these are persistent behaviors that don't change as the dog's situation changes, then you call them a personality trait. If the dog is disinterested at home but outgoing with strangers, or vice versa, then these behaviors are situation specific.

The same is true of emotions. Some dogs are always fearful, and because of this trait, they are said to have a temperament problem. Dogs that are hyperactive—and these are rare—are *always* turned on, whereas dogs with separation anxiety are calm most of the time—except when you leave. Most animals are on a fairly even keel until the beginning of some event or some perception that creates an aroused state.

What do I mean by arousal? Biologically, the sympathetic division of the autonomic nervous system gets animals—and us—ready for fight or flight, increasing the heart rate and so on. But there's more than just biology at work here—there is a whole spectrum of responses to arousal. For example, if the dog is in a situation that has made him happy in the past, then the whole context—the room you're in, the odors, the sight of the person coming with a treat—will stimulate that memory system to help interpret the arousal more or less unconsciously.

Even though the dog isn't really aware that this is going on, he feels aroused because you are holding up a dog biscuit. When you look at the dog's behavior, you see that he is smiling and wagging his tail or his entire rear end; it looks like he must be experiencing happiness or joy. Behaviorally, and in terms of physiological arousal, and in reference to those things that should create a happy response in the dog, we can reasonably assume that, to some extent at least, dogs and cats experience the emotion of happiness. We will never really know how similar their experience is to our perception of happiness. But we don't know if all humans feel the same emotions in the same way, either.

🐾 How Do Dogs and Cats React?

Individual dogs and cats react to stimulation differently. Some animals are aloof, laid back. Others notice the stimulus (person, food, squirrel, doorbell) and react by attending to it; still others get even more excited about whatever it is. And then there are those dogs and cats who just go nuts!! We behaviorists call the reaction to stimuli stimulus reactivity. Dogs and cats who show an interest in important changes and react calmly and confidently have a stable temperament (like the companion animal we all picture living in our family!). Those dogs and cats who have a personality trait that is "reactive" are quick to respond; their emotional response is unusually extreme—toward or away from the object or person—and it takes them a long time to de-arouse, or calm down.

Reactive animals are hard to live with because they're always "on." If Rocky hears the doorbell ring and starts his routine of barking, running to the door, zooming around the room, and jumping on the door (and no one is home to answer), and if it continues for 2 to 3 minutes after the doorbell rings, he can be considered to be very stimulus reactive. Now, imagine that Rocky's re-

DR. WRIGHT'S INSIGHTS

Too much arousal is not a good thing for cats. When they're all riled up, any additional stimulation produces more arousal and increases the likelihood that they'll respond by "fight or flight." If you try to console your cat, you may be rewarded with a "fight" response. It's better for you to just let the cat flee. If you have a scaredy-cat who has freaked out and is hiding under the bed, the most important thing to remember is not to reach for the cat!

Most cats stay aroused much longer than dogs do, so you'll need to wait for up to 2 solid hours for Whiskers' whiskers to stop twitching! Never try to grab a frightened cat. Wait until her fur unruffles itself, if possible, and try to entice her out from under the bed with some nice coaxing and a cat treat or two. If she steps toward you, curious about the treat, you've managed to affect her mood, and you're golden. Don't overwhelm her—in fact, don't even touch her for a while. If she gets scared again and runs back under the bed, give it 10 minutes; then go back to enticing. Cat scratches are prone to infection, so take it slow and easy!

sponse style is to react to *all* changes in his environment the same way: He just doesn't recover from stimulation easily. This type of trait is a big problem. It's an even bigger problem if he interprets the arousal he's feeling as anger and becomes aggressive. Dogs who bite and keep on biting continue to be aroused once their threshold for arousal is reached, and it may not take much more than staring at and bending over the dog to get him there.

Fortunately, many dogs who are easily aroused and who continue to be aroused after the stimulation is removed (we stop petting or playing, stop tossing the ball, stop offering treats, or come in from a walk) are friendly dogs who probably think of the arousal as "play time" or "happy time." With these dogs, we see play-bows, exaggerated movements with the front paws, hopping around, smiling faces, and so on during the aroused reactive state. Hyperactivity is different from this because there's no stimulus to set off the never-ending exuberance in the first place.

Some dogs, once they become fearful of thunder or a firecracker, or angry at another dog, continue to experience the emotional state long after the stimulus has been removed. These dogs do not respond as well to mood-altering words and phrases or to training by itself because they're just too wrapped up in arousal to pay attention. These stimulus-reactive dogs are often more likely to respond to a behavior-modification program if they're taking medication to calm them (not sedate them!) so that they can focus on good behavior without becoming overstimulated.

Cats' reactivity problems usually appear during a panic attack. If you have cats, you've probably seen this, simply because some cats' threshold for arousal is very close to their "resting state"! The slightest thing (a cupboard door closing) can set them off, but what is important is how slowly or quickly they come down. Many cats get into exaggerated, hyper-play bouts that allow them to expend a lot of energy. They may bounce off the walls, leap over the footstool—and a dog or two—zoom around, and then abruptly stop and start grooming themselves, with their backs facing toward us, of course—almost as if to say, "What hyper-play bout? Everything's cool!" And that *is* cool, as long as they don't leave a trail of smashed pottery in their wake.

The Range of Emotions

Do people all feel emotions the same way? That question is probably impossible to answer. The important thing, though, is that we understand the context of various reactions—like being scared and gasping when someone pulls out a gun, or feeling happy and cheering when we see a good report card. Our emotional responses and our behavior occur in acceptable, definable situations, so we are confident that our emotional experiences are the same, at least qualitatively (in kind) if not quantitatively (in amount or intensity). That's my best guess, anyway.

Our dogs and cats show the same consistency. Let's look at a situation that invokes fear. If you are glaring at your dog and snarling, "Get your butt over

When Your Dog Is <u>Too</u> Happy to See You

You come home after a long day in the office—a long, hard day. At least your dog appreciates you. You know that she'll come running when you open the door. And sure enough, there she is. The second you open the door she leaps up in the air, barking wildly with joy. "Hello baby! Hello-o-o-o-o!! How's my good girl?! Were you a good girl? Did you miss me?"

Encouraged by your enthusiasm, she runs around the dining room two or three times, screeches to a halt at your feet, and jumps up on you, panting with joy and covering your face with wet, sloppy licks. You pat her and squeeze her and squeal, "Yes, you're glad to see me, yes you are!" And this happens every night. It's the perfect picture of dog/owner bliss, right? It's bliss, if you're lucky. If you're not lucky, it's a prescription for disaster!

Your dog's response to your coming home from work is what behaviorists call <u>arrival elation</u>, and if prolonged, it is one of the surefire flags for identifying dogs with separation anxiety. With your encouragement right before you leave in the morning, this dog is likely to build up such a state of arousal that she may have to act upon it before you get home. And the destruction of your house is the most common way for Baby to release all that joyful energy—commonly in the first 30 minutes after you leave.

So both of you need to tone it down. Save your celebration for the backyard during a play session, but try to have comings (and for that matter, goings) evolve into more sedate affairs.

As grateful as you may be for all that exuberant acceptance, at least think about turning it down a notch. I've got dozens of clients with shredded armchairs and scratched doors who had to do the same. And their dogs still love them!

here!" the dog is likely to tuck his tail, drop his head, put his ears back, and shy away from you, avoiding eye contact. He is afraid and wants to avoid the situation. Fear is an emotion that our pets probably experience to the same extent that people and other animals experience it. People, dogs, and cats appear to personally experience and recognize in others the emotions of happiness and fear. In fact, both of these emotions are on the "short list" that various psychologists have theorized exists for humans: happiness, sadness, anger, fear, disgust, surprise, contempt, and shame. These emotions are the commonly shared ones recognized from culture to culture.

Love is an important emotion, too. I believe that we can infer that love exists in our pets from watching dogs' and cats' preferences for staying close to us, soliciting our care and attention (and returning it!), and choosing to be with us through thick and thin. (For dogs, this is true even if we're homeless.) But because there are so many different kinds of love (romantic, maternal, and compassionate, among others), and love is such a complex and difficult emotion to study in people, a firm answer to whether our animal companions experience love is "maybe." And that may also be why love is not included on all theorists' lists of basic human emotions.

After two decades of working with dogs and cats on their home turf, I can

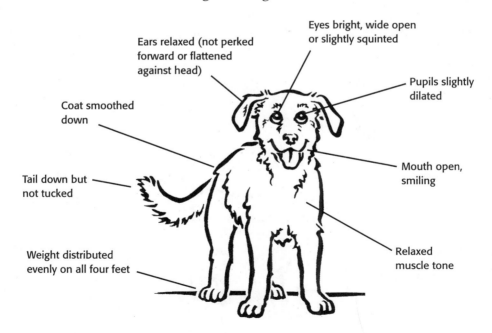

Eyes bright, wide open or slightly squinted

Ears relaxed (not perked forward or flattened against head)

Pupils slightly dilated

Coat smoothed down

Mouth open, smiling

Tail down but not tucked

Weight distributed evenly on all four feet

Relaxed muscle tone

These signs show that your dog is feeling great!

recognize with a fair amount of certainty (inferred from their behavior, since we can't really know directly what an animal is feeling) that dogs and cats very likely experience variations of happiness, jealousy, fear, depression, and anger.

🐾 Happiness Is a Warm Tongue

Let's start with happiness. It's really crucial that you understand what makes your dog or cat happy because you'll need this information to change his behavior when he gets off track. (We'll deal with that in more detail in a minute!) Dogs act as if they experience happiness. They smile and establish eye contact. They hop around with exaggerated movements, which makes them appear happy or joyful—an intense happiness. Charlie sure appears to be joyful when I announce that we're going for a ride in the morning to get (and share) a steak-and-egg biscuit! (Bad Daddy!)

Then there's love. If ever there was an emotion that we want dogs to share with us, this is it. Surely the many stories of dog heroism bear this out. And it's not just their unbridled joy at seeing us, playing with us, and hanging around us. Dogs engage in certain caretaking behaviors that we might think are motivated by love. Certainly we know that in a litter of newborns, if a puppy cries, the mother retrieves it, licks it, and feeds it. We don't know whether she is experiencing the emotion of love, but her behavior is not inconsistent with love. She's certainly "emotionally attached" to her offspring.

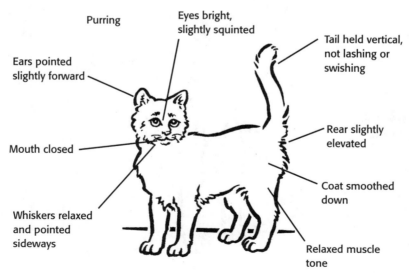

Purring

Eyes bright, slightly squinted

Tail held vertical, not lashing or swishing

Ears pointed slightly forward

Mouth closed

Rear slightly elevated

Whiskers relaxed and pointed sideways

Coat smoothed down

Relaxed muscle tone

These signs show that your cat is feeling great!

So when my dog wakes up in the morning next to my pillow and starts to lick my face, is it because she's happy, and she loves me? Some would answer no, it's because I have some tasty oil on my face. Others would say no, it's because the dog has learned that when she licks my face, I let her outside, which leads to relief of her biological need to eliminate, which feels good. When my wife, Angie, is awakened by our dog Peanut's licking, she is convinced that Peanut loves her and is taking care of her. Peanut is "Mom" at that moment. Our other dogs want to lie on us and be petted and stroked. Our Charlie brings her Nylabone up and tries to share it with us. Is that affection? I think it might be. We give her a rawhide, and before she will chew it, she'll bring it to us and try to put it in our mouths. Sometimes we will even let her, just to make her feel good. (Spoiled? Charlie?!) She then prances around wagging her tail, a very "happy" kind of thing.

Our dogs are certainly emotionally attached to us—they sometimes become distraught when we're not around, and they can be quite needy. I don't really know if the celebration the dogs carry on when we arrive home is due to devotion or relief. The question is whether Charlie's internal feeling is like mine when I feel love. I don't think a dog's experience is the same as ours because they don't have—or at least have yet to demonstrate—a concept of self, or a self-reflective self, which I believe is necessary for the experience of love.

Here's where I stand on the question: Because we have a hard enough time trying to explain love in people, let's just agree that the affection and attachment and excitement our dogs feel for us is real, it is emotional in nature, and it sure is great, whether you're on the giving or receiving end of the emotional bond.

🐾 What about Jealousy?

Pretty much anyone who has known love has also experienced the old green-eyed monster, jealousy. And cats and dogs are no different. Behaviorists are a little bit queasy about inferring a very human emotion like jealousy from a pooch's behavior with other dogs, cats, or humans. We would prefer to talk about competition for resources; it sounds so much more . . . well . . . knowledgeable! But it's natural to think of our pets as jealous when they try to get in between a hugging couple or bark wildly if they see us kissing. What's that all about?

In our house, we have been speculating endlessly about the behavior of our newest dog, little Lucy the cocker spaniel/Chihuahua mix (we call her a cockawawa). We adopted Lucy as a refugee from Hurricane Floyd. She has blended in beautifully with our other three dogs and cat, even choosing to sleep in the crate with Charlie, the Lab, a couple of hours a day as we try to ease her past the chew-up-everything-that-isn't-nailed-down stage.

But there's this one thing: Lucy is a love magnet—she can't get enough petting, holding, cuddling. That's fine; Angie and I can dish out as much affection as our pets can stand! But Lucy absolutely can't handle it when we attempt to pet, kiss, or get licked by any of the other animals. Luckily, the other dogs tolerate her when she blocks their approach or even puts her paw between their tongues and our cheeks, trying to stop them from licking us (that's what we think!). If that isn't jealousy, what is it?

I've noted that while she is trying to keep us from interacting with the other girls, Lucy shows very submissive signals to the other dogs, while still holding her ground. Her tail is wagging exuberantly, her ears are flat against her head, her front half is crouched down holding onto my lap for all she's worth, facing away from the intruder dog, while her rear end is raised in the air, and she may even be whimpering. This shows me that she doesn't want to try on such a dominant role regarding her newfound household pals. But she feels a strong enough emotion to be able to stand her ground anyway. Why?

My guess is that Lucy feels an insecure attachment to me and to Angie. Perhaps she is afraid that if we give the other dogs the time of day, she will somehow lose out. This is only an educated guess, but it's not too far off the mark, I believe, for a rescued dog who has been with us only a short time and is still feeling her way around the homestead. So, call it what you will—jealousy (an emotional state) or competition over resources (a behavior pattern)—the important thing is to realize that dogs like Lucy need a little extra love to fully bond with us and start to feel and behave more secure in their psychosocial relationships.

❧ That Fearful Feeling

Fear, the most scientifically studied of all the emotions in cats and dogs, wears many masks. It can be at the root of aggressive behavior that occurs when a dog feels under some kind of threat and strikes out to defend himself. I will talk about this kind of aggression in chapter 13. At the other end of the fear spectrum is the phobic reaction, which we will talk about in detail later. *Phobias*—the unreasonable fear of something that can't really be expected to hurt you—can be the stimulus for finding our pitiful canine companion quivering in the bathtub during thunderstorms, or refusing to walk out the back door because a bird might dive-bomb him.

Fear can even can take the form of whining and overattached, clingy behavior due to a personality trait or traumatic event that has made the dog react with an emotional response to things we would not find threatening, such as the inside of a car or a brush handle. Severe punishment is sometimes the culprit here. Here's another good reason never to touch your dog in an abusive manner. Physical punishment that causes pain is never justified, nor is it the path to problem-free behavior.

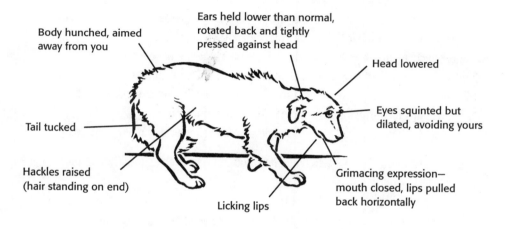

Body hunched, aimed away from you

Ears held lower than normal, rotated back and tightly pressed against head

Head lowered

Tail tucked

Eyes squinted but dilated, avoiding yours

Hackles raised (hair standing on end)

Grimacing expression— mouth closed, lips pulled back horizontally

Licking lips

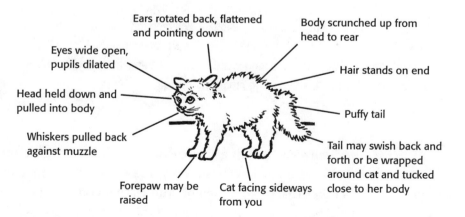

Ears rotated back, flattened and pointing down

Body scrunched up from head to rear

Eyes wide open, pupils dilated

Hair stands on end

Head held down and pulled into body

Puffy tail

Whiskers pulled back against muzzle

Tail may swish back and forth or be wrapped around cat and tucked close to her body

Forepaw may be raised

Cat facing sideways from you

If a dog or cat looks like this, it's a pretty sure bet he's more frightened than you are: tail tucked, ears smashed back against his head, body off to the side or pointing away from you. He may be looking back over his shoulder at you with hackles raised, staring at your eyes, but looking away now and then. Or he might look off to the side of your face, stiff-muscled and definitely not relaxed. It's your job to change his mood.

The Guilt Trip

Guilt in dogs—or what we call guilt—seems to be pretty obvious. The dog "knows" he has done something wrong: The minute you enter the room, he glances at you shamefaced, tucks his tail, and slinks away. "You should feel guilty," you scowl after him. "Look at this mess! Bad boy, Chuck!" He "knows" he shouldn't poop on the rug—it's obvious from the looks of him! Guess again. What would you say if I told you that his "guilty look" is more likely to be a combination of extreme submission and fear? He is probably reacting this way based on what happened the last time you saw poop on the rug, or the present look on your face and tone of your voice—or even the rolled-up newspaper in your hand. (I hope not!)

Dogs, unless caught in the act or confronted shortly thereafter, have no way of connecting your wrath with what they did, even assuming that they "know" it is wrong, which is questionable. They don't have the same moral sense we do. They don't take responsibility for their actions. They don't have a canon of ethics. They will react to the expectation of being treated well or badly, with submissive and fearful feelings or with happy ones, regardless of what they did.

If you don't believe me, try this experiment to see if your dog really feels guilty!

1. Purchase a fake plastic doggie doo-doo from a novelty store. (Ignore the clerk's smirk; this is for science!)
2. Place the doo-doo in the center of the pricey oriental rug you bought at auction. You know, the one you have to take your shoes off to walk on.
3. Call in the unsuspecting—yet totally innocent—dog.
4. Let the dog check out the fake doo-doo. She will probably want to go up and eyeball it and maybe even sniff it if she's never been punished "after the fact." But a punished dog will immediately cower and appear "guilty" even though he had nothing to do with producing the "gift."

People have described some breeds as more excitable and active than other breeds. However, because no behaviorists have really admitted that dogs experience emotions other than fear, no major studies that compare breeds or different emotions have been done. An exception is the "nervous" pointer dogs bred at a university lab to study the behavior genetics of fear. These dogs, after selective breeding from a stock of "normal" pointers, were significantly less likely to approach people or to allow petting.

5. Put on your meanest face and gear up to use your most irritated voice (the same ones you employ when that freaky bum comes to pick up your teenage daughter).

6. Furrow your brow, look your dog in the eye and snarl, "BA-A-A-AD PRECIOUS! NO POOP ON THE RUG!"

7. She didn't do it. You know she didn't do it, and she "knows" she didn't do it. Observe her tuck her tail and her ears, avoid your withering glance, and slink away. Looks pretty guilty, eh? Or is she reacting to your emotions and the present threat of impending doom? You're a bright person. You decide!

8. If you still don't believe me, wait a few hours, set the poop down again, and call Precious into the room. Let her look at the poop again. If she starts the same fearful responses without you saying anything or looking mean, she has had a "one-trial" learning experience, and she remembers the context in which she felt pretty uncomfortable the last time. If she's a bit of a slower learner, call her over to you.

9. Assume your most friendly, smiling face, and using your sweetest voice, say, "GO-O-O-O-OD Precious! What a go-o-o-o-od gi-i-i-i-rl to poop on the rug!!"

10. She still didn't do it, and you still both know it. But if I were a betting man, I would say that your dog is grinning from ear to ear, wagging her tail, and demurely saying, "It was nothing!" Give that dog a hug and a small stipend for participating in the experiment. (A couple of biscuits should do it!)

This is the power of Pavlovian conditioning and a small demonstration of how you hold the key to changing your pet's emotions and behaviors.

Sometimes dogs are a little bit more shy or fearful or prone to have reactivity problems because of what happened during early rearing rather than as the result of a breed characteristic. So my guess is that both nature and nurture—early rearing, nutrition, and health, as well as the genes pups share—are responsible for pushing littermates toward fearful behavior. (You'll find more on puppy temperament and what to look for when you choose a puppy in chapter 6.)

DR. WRIGHT'S
C A S E B O O K

Budge and LuLu

I remember one depressed 4-year-old cocker spaniel named Budge. His owner, Sally, had taken a second job three evenings a week after breaking up with her boyfriend, whom Budge was attached to, and having her room-mate move away with two cats in tow. The house was suddenly very empty, and Budge quickly sank into a hopeless lethargy. But the worst part was coping with Sally's new job. Budge was okay when left alone during the day, but after Sally came home briefly and then went out again, the dog couldn't take it and became so emotionally distraught that he peed all over the living room.

It was no wonder—Budge was so desperate for Sally's company that he would refuse to go out when she got home the first time. Or he would reluctantly go out the door while she was changing her clothes for the second job and just sit on the step until she opened the door. In any case, he was "holding it" to shorten the period they would be apart. Sally went off to her second job anyway, though, and Budge couldn't wait forever. The result was peeing on the living room carpet.

Another depressed dog I remember—LuLu the Chow Chow—used to spend her day wrecking the house. This dog would see the owner shaving his chin every morning and would just lie down on the spot and refuse to move as her owner tried to get her to follow him to the door and say goodbye. Perhaps she thought that if she didn't see him leave he would still be there. In any case, she felt bad enough to try to get rid of that emotion by ripping the carpeting to shreds.

These dogs' situations weren't unusual—leaving them alone while we work is the unfortunate plight of today's pets by the millions. But the owners and I were able to reduce the despair of these two by reducing the impacts of the goings and comings and concentrating on deliberately changing their moods when they did have time together.

🐾 Is Your Dog Depressed?

Your dog lies around all day. Every once in a while, he lets out a huge, dramatic sigh. He stares off into space and doesn't even seem to notice when someone speaks to him. His ears are droopy, and you rarely hear the thump of his tail when you walk into the room. He may have lost his appetite and his enthusiasm for playing.

Dogs who live in the house are not necessarily fulfilled socially if people ignore them throughout the day. Countless people have told me, "We are so busy, we just don't have time for him." I ask, "Well, does the dog come up to you and ask for attention?" Well, yes, they say, but we don't like it when he tries to lick us. So they put him outside (not take him outside) and when he comes back in and wants to be petted, they see that he is a little dirty, so they don't pet him. He goes and lies down. This kind of behavior on the part of the owner can eventually lead to depression in the dog, even though he's not tied up in the backyard. No matter what he tries, nothing he does gains your attention or praise.

Does this scenario sound familiar? If so, do something with your dog! Give him a job to do; ask him for some sit-stays. Give him a hug, stroke him, give him a massage, or put a little treat in a Kong toy and talk to him while he's playing. Employing any of these examples of social behavior will really strengthen the bond between you and your dog and can head off depression and loneliness.

Some dogs are just shy. They won't even ask for attention. They may lie in a corner, and because they are quiet you assume that they're content, but they may be miserable. Do them a favor; make them part of the family routine. Call Milo to you and tell him what a *great* boy he is. Shyness tends to be a personality or temperament characteristic, so you're going to have to keep enticing him to you. Fortunately, most dogs are a little shy rather than a lot shy. A little effort should lead to noticeable outgoing social responsiveness in slightly shy dogs. Don't give up, or he may slip into doing things by himself again, which can lead to loneliness and depression. Dogs just aren't bred to be alone.

🐾 Expressing Anger

When a dog or cat is biting, snarling, growling, attacking, or otherwise showing unbridled aggression, it's a little late to head off the animal's acting out of feelings of anger or fear. You should never intercede between fighting dogs because you are likely to be badly hurt when their aggression is transferred onto you. Of course, you would aid a child who was being bitten. But wouldn't it be far better to know how to head off likely attacks from dogs who have been aroused to the point where their anger takes over?

The average dog or cat can have an occasional ugly mood short-circuited if you know what you are doing and aren't afraid to try working with your pet yourself. After all, you are the one who really knows your own pet and what he likes and dislikes. You know the people who make him happy—and the ones who more often than not are met at the door with raised hackles and a low growl. Sometimes—if you can admit it to yourself—the dog takes to someone totally outside of those he lives with from day to day. You would be well advised to bring such a person into your circle of helpers if you have an angry or frustrated dog you need to help.

For example, Judi, my coauthor, had a lovable and very large old golden retriever. Scout was fond enough of Judi and her husband and two kids. But if the truth were told, he really came alive only once every 2 weeks. He would hear the distinctive sound of the car coming up the driveway, lift his head off his paws, leap (as well as an 85-pound golden can leap) to his feet, and run to the door. Judi would oblige by opening it, and Scout would run joyfully to the driveway to greet his best friend, Grace.

Never one to be without a bag of biscuits in her pocket, Grace would fuss over Scout and give him treats. During the next 2 or 3 hours, while Grace and her friend Claire cleaned the house, Scout would never leave her side. It didn't matter that she was bent over a tub scrubbing or running the vacuum (a machine Scout judiciously avoided at other times) back and forth, he would just lie on the cords and make a nuisance of himself. But Grace talked to him and gave him treats, and they had a wonderful time. Scout totally ignored Claire. And when the women left, he would sink to the floor and sadly put his head on his paws, waiting until the next time.

DR. WRIGHT'S INSIGHTS

A persistent, pervasive attitude of aggressiveness in a dog should be handled by a certified applied animal behaviorist or other qualified professional. It is never safe to try to discipline a dog like that, especially without a clear idea of what will make him even more angry, and what will calm him down. An overtly aggressive pet may need the intervention of drug therapy before he can begin to learn calm, socially appropriate behavior.

Now Scout didn't have any problem with anger, but if he did, I believe Judi would have done well to have Grace around to help change his mood. For when a dog is happy or content or amused or loving or entertained or whatever you want to call it, it's more difficut for him to feel that angry or fearful emotion that drives the aggressive behavior.

How Can You Tell What Your Pet's Feeling?

Trying to figure out if your dog or cat is experiencing anger, or love, or any other "human" emotion is a tough call, even for applied animal behaviorists like me. To date, no one has been able to provide convincing scientific evidence that we can distinguish one emotion from another by what our brains or hormones do. We obviously feel differently when we're in a loving versus an angry mood; however, what our brain does to influence us to feel those very different emotions is less clear. Nonetheless, we believe that our pets love us, they appear to be embarrassed when we dress them up in silly clothes, and they certainly look like they're feeling guilty after doing something they "know" is wrong. But *do* they know right from wrong, and when they do the wrong thing, do they *really* feel guilty?

Ask yourself this question: What would it take for you to be convinced that your dog or cat actually experiences a specific emotion? Is it possible that his appearance—the way he looks in his body language and behavior—leads you to confuse guilt with submissive, defensive behavior? Have you inferred that he's experiencing guilt from the way he looks or from the situation that seems to call for guilt (like I know *I* would feel guilty if I just tore up the couch)? If you were a behavioral scientist, what kind of evidence would you use to convince yourself and others that your dog or cat actually feels emotions such as guilt, love, shame, hope, pride, relief, regret, or revenge? It's an interesting question, and we behaviorists are still working on it. But let's look at it in terms of revenge.

Vengeance Is Mine!

So, some rude dude pushed in front of you to use the ATM? Or the obnoxious guy in the vet's waiting room refused to keep his slobbering Saint Bernard from terrorizing your cat? Do you want revenge?

When we think of getting revenge against someone who has "done us wrong," we think of doing something to get back at the person. We decide on the appropriate revenge by imagining how it would make the idiot feel to have such-and-such happen to him. If we think it would really make him feel rotten, and it would get back at him in an appropriate way, it makes us feel good—even if we just imagine it. We don't actually need to get revenge—we just need to imagine his reaction if he were to get his just desserts.

Your dog or cat probably doesn't have the ability to imagine how you would feel if he were to soil your bed as revenge for leaving him alone all weekend. And he probably wouldn't chew your favorite slippers as revenge for locking him in the bedroom and keeping him from enjoying that great catered party you had last night. The ability to look into another's psyche to imagine one's emotional response to a planned endeavor is what we behaviorists call revenge. Revenge requires a "theory of mind" that dogs and cats give no evidence of having.

Dogs and cats see the world from their perspective. That's why arranging their daily lives from their point of view works so well. Not only would it seem strange if my dog Charlie could understand why I won't give her a dog biscuit before dinner—it would spoil her appetite—but it would seem even more strange if she could be seen planning my undoing later that evening, to make me feel bad in some way for my stinginess. Yet we often find ourselves believing that this is exactly what our pet must have done when we discover that he's soiled our bed or chewed our best shoes. In reality, our pet's behavior is probably caused by a disruption of the pet's routine, an increased arousal or excitement, or a way to relieve discomfort or frustration.

Changing Your Pet's Emotions

Now that you understand your dog's and cat's emotions, you can go about changing them—manipulating them, if you will—to create a mood that is incompatible with the mood that drives his misbehavior. The concept is called the principle of competing motivations: A dog or cat cannot be angry, fearful, or depressed *and* happy or elated at the same time!

Here is a list of "happy" words or things that you can use to help change your dog's or cat's mood when you see him slipping away from a neutral or happy posture or facial expression into a fearful, depressed, frustrated, or angry one. Try them; they work!

❧ Ask Your Pet

In the most exuberant voice you can muster, with enthusiastic gestures to match, choose one question from the list to ask your pet.

Wanna go out?   Go for a ride?

Go for a walk?   Get some ice cream?

Go see Daddy (or Mommy)?   Go find Fluffy?

You'll also get a great response if you suggest the following games or actions to your pet:

Fetch your ball!   Go find Mommy (or Daddy)!

Where's your Kong?

Don't forget the positive distraction you can provide your pet by gathering his favorite things. Few things can distract a dog as fast as a box of treats. Other attention-getters for dogs and cats include:

Frisbee   Catnip

Kong toy   Tuna

Tennis ball   Fishing pole–type toy

Leash

You can think of plenty more. Use the toys and other things to grab your pet's attention. Talk to your dog or cat enthusiastically. Your goal is to get your pet out of his impending funk and into a good mood so that the unwanted behavior will be short-circuited and good behavior can take its place.

After you've distracted your pet and he's feeling happy, don't forget to let him calm down. After all, part of tameness in companion animals is our ability to calm them. So make sure that you give your pet an ample period to wind down after all the excitement of being happy, or you may have yet another type of misbehavior on your hands. Out-of-control dogs and cats are not what we're after here, even if they're deliriously happy ones! A sit-stay, a quiet place to lie down, some light stroking, a darkened room—all of these provide good ways to come down off that doggie or kitty high and back to neutral—which, after all, is where most pets are most of the time.

Moving Ahead

Now that you can look at your pet through his own eyes, you have a good idea of what motivates and moves him and how his senses shape his world. You know a bit more about how he learns and what turns him off. You have a realistic idea of what you can expect from him and what's pie in the sky. And you can interpret what he's feeling and what he wants more accurately, just by looking at him and listening to him. You've learned the powerful secret of changing his behavior by changing the emotion behind it.

Part Three

Chapter Six

Puppies, Kittens, and Older Adoptees

So you're going to add an animal to your family. That's great! But before that exciting trip to the breeder or animal shelter (skip the puppy mill–fed pet stores), you need to sit down and think about why you are getting a dog or cat, and what your new pet will mean to you. What do you want from him or her, and what are you prepared to give in return? How will you treat your new pup or kitty, and how will your pet treat you? What is a pet's role in today's frenetic world, anyway?

Because most people are looking for either a new dog or a new cat at any given time rather than one of each, I've split this chapter in half. If you want a puppy or adult dog, read the first half; if it's a kitty or cat you're after, start on page 114 with "Adopting a Cat or Kitten." (I'm starting with dogs because there's so much information—and misinformation—about how you should choose one.) When you're ready for another new addition, you can read the other half!

Dogs: The Behavior-Based Fit

In my experience, most people decide to get a dog for emotional reasons. They may not have thought much about adopting a dog when that "pitiful pair of eyes" attached to a body and bushy tail turns up on their doorstep. It probably hasn't even crossed their minds when they turn the corner at the mall and the cutest puppy they have ever seen sits looking out at them from the window of the pet store. Sometimes a holiday like Christmas or an occasion like a tenth birthday will spur someone to pick out a puppy, tie a red bow around her neck, and give her to a friend or child as the most adorable gift ever. Or maybe they're empty nesters or live by themselves, and are just so lonely they decide to go to the local shelter and adopt the first dog that doesn't bark at them.

All of these reasons are rather impulsive—if good-hearted—ways of deciding to choose a dog as a companion animal. They are not bad ways. Many perfect matches have been made when a pet was brought home on a wild impulse, but it's a little bit iffy.

These emotional decisions often carry with them a justifying rationale, such as the fact that it will surely be a good lesson in responsibility for little Josh and little Emily to have a dog to take care of. It will teach them discipline and reliability. This thinking is all well and good, as long as the puppy has stolen his way into the hearts of his owners and their kids before he quickly morphs into a dog and isn't so cute anymore. Someone in the house must be mature enough and responsible enough to take good care of the animal. Otherwise, the poor dog will end up neglected, likely to develop bad habits to compete with computer games and Nintendo.

Many "impulse" companion animals end up being sold or given away just as abruptly as they were brought home in the first place. Many more dogs are given up by owners who didn't pay for them or spend time training them than by people who spent at least some amount of money and time for social and emotional bonding. Certainly giving live animals as gifts is a rather risky proposition: Most people can't pick out a tie for someone else without its being stuck in a drawer as "not really me." Think of the plight of the poor "surprise" puppy!

🐾 Time Out for True Confessions

Now, as I said, I am not criticizing the emotion-based route to dog and cat ownership as entirely inappropriate or even wrong. In many cases, a dog or cat has had a wonderful life because she captured someone's fancy or someone's heart without prior planning or with a less-than-perfect setting.

In fact, I am not criticizing the impulsive move to pet ownership because I understand it in spite of my own experience and credentials—the big Ph.D. animal behaviorist who routinely has had to bail out countless animals whose owners didn't really think things through and were suffering the consequences of a poor fit. Despite being exposed to many tough behavior problems that began with an emotional choice of companion animal, my wife and I picked out all four of our dogs (only one was planned, three were "accidents") more or less by the same method. There, I've confessed!

The first one, Peanut, a sweet, now-elderly mixed terrier, was plucked from the railroad tracks by a soft-hearted Angie wielding some leftover fajitas. She brought the little mutt home to two parental units who were not given the choice to say no. Peanut happily came with Angie as a package deal when we married. The second dog, Roo-Roo, our regal black standard poodle, passed the sophisticated and scientific Dr. Wright's Lick Test at the breeder's. (Yes, she licked me—but only after she licked Angie and Peanut, and I felt left out!) And the third, Charlie, gave us both sad eyes and a half-hearted grin from a beautiful chocolate Lab face that we couldn't possibly back away from, although we had only gone into the PetsMart to buy dog food for Pea and Roo. ("We won't even look at the Humane Society adoption section. Well, maybe just one walk down that aisle won't do any harm.")

And talk about tugging at one's heartstrings! In spite of our resolve that "three is enough," my wife and I could not resist our latest girl, a tiny cocker spaniel–Chihuahua mix. She was initially rescued from drowning in Hurricane Floyd and brought home by a student at the university where I teach. Because puppies were not allowed in the dorms, she was reared for 2 months by a terrific woman who also fostered rescued greyhounds.

What a sight—a sweet little 3-month-old cocka-wawa making its way beneath five towering adult greyhounds. She needed to be adopted. After observing her with my "professionally trained eyes" and consulting with my wife for all of 20 seconds, I told the woman, yes, we would love to take the puppy! What was one more when there were already three dogs and a cat to feed, play with, and love?! Luckily, little Lucy has fit in beautifully, romping around hap-

DR. WRIGHT'S INSIGHTS

I am especially concerned about animals as gifts, especially to children as a means of teaching them something. If a little girl or boy is not up to taking on the responsibility of owning a dog or a puppy and understanding how to do things with it, then it may not be a good idea to let them talk you into getting this kind of pet. If parents and older brothers or sisters are willing to each take responsibility for the dog for a certain function or a sort of "time-sharing," then the adoption will probably work out fine. But getting a pet requires some planning and a number of commitments in advance—not when the dog is suddenly foisted upon a surprised family: "Hey, guys, this is Butch—isn't he cute? Who wants to paper-train him?"

Adopting a dog or puppy begins a commitment that—ideally—goes on for a long time. The lifespan of a dog is anywhere from 9 to 17 years (or longer), so the investment in time and energy is considerable. Don't give children a dog to teach them good citizenship or responsibility. Have them learn to take out the garbage or make their beds instead. Then the worst that can happen is a messy house, not a dog with hurt eyes and few prospects.

pily with the other animals. But our impulsive decision could just as easily have had disastrous consequences, had she not been the perfect, adorable sweetheart that she is!

Regardless of the scenario, emotion-based decisions make it less likely that the dog—and therefore the family—will turn out to fit the definition of "happy camper." If the right dog is put in the wrong environment, everything starts off on the wrong foot. And usually sooner rather than later, animal behaviorists like me end up answering SOS calls from baffled owners who don't get it. What the heck is wrong with that dog, anyway?

Get Out the Breed Book?

At the opposite end of the spectrum is the logical, rational, analytical method of deciding to get a dog. You pore over volume after volume of breed guidebooks, scientifically choosing the best pet for the family based on each breed's behavior and personality characteristics as they relate to your situation and lifestyle. Then you study all the relevant breed information dating from when Champion Whiskey Blue Regal-Biscuit first set its pricey paw on American soil.

There is nothing wrong with this approach. For the most part, it's a won-

Dr. Wright's Pre-Pet Quiz

For your peace of mind and your future pet's happiness and well-being, before you head to the breeder or the shelter, sit down with your family and ask yourselves the following questions. See what you come up with!

1. Does everyone in the family really want a pet? Is everyone willing to care for a pet? Is anyone allergic? Is anyone against the idea or extremely indifferent? If so, is avoidance, hostility, or neglect on the part of one or more family members a real possibility? Be honest.

2. Considering everyone's level of interest and ability to devote time to an animal, who will be responsible for what? Are there some jobs certain family members will refuse to do, such as changing litterboxes, cleaning up messes, or removing ticks? If so, is there someone else willing and able to take on these duties?

3. Is there enough time available to walk and exercise dogs and play with cats on a regular, predictable basis? Or are walks seen as a luxury if the dog could simply be let out? Does everyone agree on the need for these activities?

4. Are you able and willing to provide good veterinary care, food, and pet-sitting or boarding services when necessary? Have you spoken with a vet to determine the yearly costs of these services in your area?

5. Are children who volunteer to take care of a dog or cat at least 10 years old? It is generally unrealistic to expect a commitment of time and effort from children any younger than that. If they have ever owned a guinea pig or snake or hamster, you will have an idea of their pet-care credentials!

derful idea to do some research before you take the plunge. But, like the career-changer who takes a battery of tests to determine what he should do next, you may not like the logical results of your research. (My test results say that I would make a great architect, but I want to be a pastry chef!) What good is paying top dollar for a papillon because some quiz says it fits in best with your lifestyle, if you really have your heart set on a Westie and would have a hard time loving anything else? Motivation counts for a lot, even when making the best "logical" choice.

Also, if every dog and cat were pedigreed, *and* every pedigreed dog and

6. Is this really an appropriate time to be getting one or more animals? If you're about to go on an extended vacation or are expecting a baby any minute, you might want to rethink your timetable. How much time will the pet be spending alone, and how much do you think would be too much? How does everyone feel about crating?

7. Can you all agree in advance never to use physical punishment for misbehavior, but to work with the animal to correct the problem? It only takes one family member to set back a nice pet's chances to succeed in the family.

8. Does the entire family acknowledge that the pet is a long-term commitment, and is everyone comfortable with that? Do they agree to take their time, do their research, and obtain the animal from a good place? Cats routinely live to be 18 years old these days, and many dogs can reach 15. That's a long time to live with an impulse buy from a pet store.

9. What kind of cat or dog do you want? Would it make sense for you to get more than one? If so, what combination? Do you want a puppy or a kitten or an adult dog or an adult cat? Which sex?

10. Why are you doing this? What's in it for the dog or cat, and what's in it for each of you?

Decisions, decisions! I can't make them for you. But I can urge you very strongly to give each of these questions some serious thought before you open your home and your hearts to a companion animal.

cat were absolutely true to its breed characteristics, *and* what happened in puppy- or kittenhood and beyond meant nothing at all, then we would be able to say with certainty after all our research, "That's the kind of dog (or cat) I need!" But obviously, there are exceptions to every rule.

Because genetic diversity exists even within a breed, there will be a wide range of variation in each litter of puppies. Obviously, we must accept the notion that breed stereotypes (or breed standards, as they are called by the American Kennel Club) are more correct than incorrect. Thus, any Rottweiler will be more similar to the "average" Rottweiler than it will be to, for example, an "average" sheltie. And if Rotties are "by nature" (breed) more stubborn than shelties (or more assertive, or whatever), then different nurturing—which includes family lifestyles, settings, relationships, and nutrition, as well as training strategies—may be appropriate for one breed, but not as appropriate for another breed. So in dogs—and cats—nature and nurture definitely share equal billing when predicting a good outcome for the family pet.

🐾 Picking the Right Breed

Almost everyone agrees that dogs should be purchased or adopted with the idea in mind of selecting pets that will fit within particular households where they plan to hang their leash for the next dozen years or so. I am not suggesting that a nonshedding Portuguese water dog—very popular right now in some segments of American society where both allergies and income are sky-high—can find happiness only when found soaking wet with a bird in its mouth. Or that the various terriers keeping people company in condos and colonials will go to dog heaven unfulfilled if they have never run a rat aground or dug one out of hiding.

After all, dogs today are bred primarily to be companion animals, not to fulfill a dimly remembered biological destiny. However, you should give some serious thought to what kind of environment a new puppy or adopted dog is going to be facing when it joins your household. Applying logic will help you determine whether the adoption will be a good fit or present barriers to your mutual happiness that could have been avoided.

This analysis involves taking a good, hard look at what the dog's quality of life will be. You need to evaluate the space the dog will consider his domain; the people he will come to protect from all real and imagined dangers; the other four-footed creatures he will interact with daily; the activities and regular, predictable routines he will share with his family; the neighborhood he will be calling his own; and the turf he will defend with all his heart.

Deciding whether to adopt a cat deserves the same serious consideration, even though cats are seen as "easier" because they seem more self-sufficient. How do you know you wouldn't be better off with a baby, a bicycle, or a tank full of goldfish? It might be useful to sit down with everyone concerned and go over each of the questions in "Dr. Wright's Pre-Pet Quiz" on page 102. Use your answers as a guideline for doing the right thing for your family and your future pets.

DR. WRIGHT'S INSIGHTS

It may not be popular to say so, but some of the characteristics of champion show dogs, who are bred for top dollar and shape the way winners will look in the future, may be in some cases perpetuating undesirable behavioral tendencies. Dogs in the show world are bred for looks and conformation—the qualities that win dog shows— and because of that, they may happen to be carrying along some aggressive genes. The dogs who carry their heads a little bit more erect, keep their ears rotated forward, and stare a bit more easily than other dogs look better in the show ring. But these are also the signs of a dog who is likely to be a bit more assertive—not a good choice for most family situations.

Despite AKC guidelines, people's perceptions of a breed may bear little resemblance to the published breed characteristics. They may be based on limited exposure to a breed ("I lived next door to one for a year") or hearsay ("I understand that Dobermans are aggressive by nature") and may result in breedism—the assumption that certain breeds of dog (goldens) are inherently "better" than others (spaniels). Faulty generalizations about a breed lead to false stereotypes about breeds, which in turn may be confusing to unsuspecting owners of puppies who thought that what they "knew" was what they were going to get.

Even if owners have had previous experience with the breed, problems can still crop up if owners treat all puppies as if they were the same. If your puppy doesn't respond to the kind of training your other dog responded to, take time to consider other humane, effective, and sometimes creative options. You'll find plenty to choose from in part 4 of this book!

Some combination of rearing strategies will maximize all dogs' abilities to fit into the family, but don't expect all dogs to wind up like some imaginary prototype that may not really exist. Appreciate your puppy for who he is, and try to establish the best possible home setting for him to maximize his potential to become your best friend and companion—a setting in which both you and he can establish trust and find happiness.

What about a Good Ol' Mutt?

I don't mean to neglect mixed breeds here. (And that includes cats!) But when you introduce a mixed lineage, you automatically produce a crazy quilt of competing genotypes that hang a big question mark over that adorable "Heinz 57" head. Some behaviorists believe that the more mixed up, the better as far as canine behavior is concerned. Everyone knows of the horrors of temperament caused by the careless breeding of popular breeds. The good nature of the dog usually suffers as the "look" becomes all-important for consumer success. There may be something to the image of the healthy, fun-loving, uncomplicated, plain old mutt!

But on the other side of that well-worn coin is the possibility of genes from various breeds combining in not-so-fortunate ways, producing the worst of both worlds, if you will. And blending that dubious mixture into an environment where the owner doesn't have a clue what to do with the hand he's been dealt can produce yet another likely dog-and-family dysfunction—and another trip to the animal shelter.

Because 27 million cats and dogs end up in shelters each year (and only 20 to 30 percent of them are adopted), shelter personnel try to screen out the people who will be dumping the animal back on their doorstep a few months later. In return, today's animal shelters—most of which are usually in desperate need of more space, funding, supplies, and volunteers—offer the potential adopter a wide range of dogs who have been vet-checked and have received the necessary shots. And they'll also pass along whatever sketchy information they may have about each animal.

Most puppies will not have had time to develop bad habits, but there is no reason not to consider an older pet as well, if the shelter follows some reasonable behavior-based guidelines in choosing which dogs are offered for adoption. (And most do.) In a no-kill shelter, you may wish to ask how the animal's behavior has been evaluated before you fall in love with it.

🐾 Shelter Tests

Many large animal shelters now use puppy tests of some sort to rank the youngsters according to their desirability as potential adoptees. The trouble is that it is really difficult to say that a puppy's behavior at 6 weeks will be the same 2, 3, or 5 months later. Nor has anyone done follow-up work to show that the high-scoring puppies develop into "better" dogs than the low-testing ones. Tragically, but perhaps by necessity, the lowest scorers are usually put down before they have a chance to grow up. Or, in more than one case, potential owners have been told that the dog scored higher than he actually did to save the dog's life. Now there is a noble fib, but no one knows if it mattered one way or another. And, in a self-fulfilling prophesy, the "great" dogs will likely be treated as such. Don't *you* think they'll grow up to be great?

Here are some of the ways pups are evaluated: Call her (pick a name, any name) to see if she comes. Turn around and walk away and see if she follows. Clap your hands and say loudly, "No, no, no!" and see if she stops. Then say, "Come on!" and see if she follows you again. If the dog runs away and hides in the corner and doesn't recover from that, the idea is that maybe she is a bit too fearful.

On the other hand, if the first thing the dog does is run toward you, jump up on you, and then start to grab for your fingers and growl, she could be an overly assertive dog—a little *too* aggressive. Some people roll the dog over on her back, as though she were trying out a subordinate posture. Some puppies

Double Trouble, or Twice as Nice?

How about two puppies or kittens? The advantage of two, of course, is that they can grow up with each other, form a close relationship, and do things together during play. But you may get less individual time with each because they have each other, and that's already a stable relationship that they're forming. Or it could be that maybe one is a really social animal and one is more solitary. And now you feel sorry for the solitary kitty because every time he tries to get close to you, the other one bats at you and beats him up. So there are pros and cons to having two or more cats (or dogs) in a household.

The advantage of having only one kitty or puppy is that each person in the family loves the cat or dog and can pay attention to it. The kitten or puppy can get pretty much everything it wants socially, provided everyone knows that they are supposed to go along with its agenda. The down side, of course, is that he doesn't have anybody to play with when you are at work and the kids are at school and the puppy or kitty is all by himself. But how big a deal is this? Solitary animals just sleep instead of playing. (Unless they have separation anxiety! Then they keep busy destroying the house, and a companion animal doesn't really help them cope. In fact, an interesting study I'll tell you about when we discuss separation anxiety showed that dogs miss their people more than they do another dog when they are separated from each, and that may apply to cats as well.) You're just going to have to decide this one on your own!

DR. WRIGHT'S INSIGHTS

Does it matter when a puppy or kitten leaves the litter? You bet it does!

It's human nature to want to take that puppy or kitty home as soon as you see him in the litter or at the shelter. But if a pup or kitty is separated from his littermates and mother too soon, he will miss the valuable time when he should be learning how to communicate through displays and signals. And he won't learn one of the most valuable lessons of all: how to control the strength of his biting as he wrestles, nips, and tussles with his littermates during play-fighting. So which would you rather have, a puppy or kitten a little sooner, or one that is bound to be a safer family pet? Don't take home that puppy or kitty before he is at least 7 weeks old!

will do this fairly well at 4 or 5 weeks, but they are less ready to do this as they get older, especially with a strange person.

The idea is that if the dog will rotate onto her back calmly as you stare at her from above, then she's supposed to turn her head and be very prepared to be subordinate. There's probably a grain of truth in that, but if the dog tries to get up and you shake her and she starts to growl, that may be just a normal defense reaction from the dog. Some people will say this shows that the dog will not be subordinate easily, and she wants to be dominant. That, of course, can be utter nonsense. So whereas we used to think that these sorts of tests were fairly useful, now we're taking them with a grain of salt, until the appropriate behavioral studies can be done to assess their accuracy.

Dogs, like kids, can "test poorly." The best we can do is use these tests as assessments of what a puppy is like at one point in time in the shelter. She may act entirely differently in a roomful of rowdy kids or a quiet nursing home.

With these caveats, you can check out the puppy yourself in most kennels at shelters or breeders' facilities. Rather than depend on the kennel's tests, therefore, I challenge you to pick out the dog you like, take her home, and treat her as though she had aced the test!

Bringing Home Baby

When you bring your puppy home, you'll want to make him feel good while he's lying down or on his back. Give him a little stroke or an encouraging word. But don't overdo it. If you make the pup stay in that subordinate posture and he stays there, it teaches him a lesson both in submission and in dominance. That may seem okay. But the problem is that the dog also learns confrontation—not just with other dogs but also with people. He might just say to himself, "Maybe you are dominant to me and you slam me to the ground and I freeze, but boy, don't let someone else try that with me, because two can play that game!"

I believe that dogs can be taught to be compliant without using force and confrontation. That's why I don't recommend shaking the dog by the scruff of the neck and pinning him to the ground, even though that's what wolves occasionally do to establish dominance. As I've pointed out, no matter how many times you've been told that you are the alpha animal, the fact is that people-dog relationships are not like wolf-wolf relationships. Certainly dogs' teeth can inflict more damage than people's hands, so the wise thing to do here is not to start the confrontation.

Instead, start out with a companion-animal relationship where there's mutual respect for each other's roles (yours is to communicate direction, hers is to respond appropriately). Your dog will try to please you and be compliant, and you will praise her for doing so.

This is the kind of relationship you want to start as soon as you bring the puppy home, even before you take her to puppy kindergarten or hire a trainer to get her under control. At 8 to 10 weeks—the time a puppy is old enough to leave her mother and littermates—she is ready to start being herself.

Don't physically force your new pet to do things that she's not ready to do. Let her get used to one room at a time. Make sure that you keep track of when she is getting overly excited. This is your cue to say to the puppy, "Settle!" or "Outside!" The word or phrase you use is up to you. Try to go out the same door each time for the same activity. You will need to take her outside and stand there while she sniffs around and pees or poops, and you say, "Oh, good puppy!" And then you go indoors and play with her in a different location, so that she gets the idea that when we go to this one spot it's time to pee or poop, and when we go out a different door to another spot, that's where we play.

As you start to teach your dog good housetraining techniques, you will also want to put him on a regular pattern of eating, usually three times a day at first. Occasionally, a pup will not seem too interested in eating. Besides finding out what he was eating when you acquired him, and offering him tidbits of lamb and beef from your fingers to whet his appetite, I have found that puppies like a bit of company when they go to the food bowl. So if there is a littermate or a neighborhood puppy about the same age as yours who would like a dinner date, let them eat side by side a few times in the location you've chosen for daily feeding. There's nothing like a little social facilitation to help jump-start a sluggish behavior!

Many books are available to help you raise a puppy. (You'll find my favorites in Recommended Reading and Viewing on page 426.) The idea is to make her comfortable and get her into a routine of regular eating, sleeping, elimination, and walking. Make sure that you don't do unpleasant things with your hands. Don't let your dog start to chew or nibble on your fingers or hands. Even if the nibbling doesn't hurt now, it will hurt when the dog gets older and can lead to a bad habit that's difficult to break.

Very soon, you'll be getting to know your new pet very well. Enjoy him! What it boils down to is this: Nobody really "needs" a dog, cat, or any other

Is a Dog the Right Choice for You?

Over the years, I've been called in to mop up after a lot of owners who should never, ever have gotten a dog. How can you tell if you're one of them? Run through this checklist and see if you want a dog for good reasons or bad ones. And if you want a dog for a bad reason, resist! Maybe you're just a cat person at heart, or you would be better off taking up a new hobby.

10 Good Reasons to Get a Dog

1. You have a stable lifestyle and enough time and money to add another member to your family.
2. You've always had a dog and would feel your life was empty without one.
3. You wouldn't let drooling, shedding, barking, or pulling off ticks bother you a bit.
4. You want to share unconditional love with a companion animal.
5. You love throwing balls, sticks, and Frisbees and riding around with a dog's head out the window.
6. Your kids are ready to graduate from hamsters and snakes to a real pal.
7. You are a patient teacher who will be thrilled as your dog learns fun things.
8. You have room in your heart and home for an unwanted creature.

pet. But more than 50 million American households have decided to get one anyway. I applaud those who are able to use research, common sense, and self-awareness to decide what is right for them. Bite the bullet, take that leap of faith, and bring that puppy or dog or kitten or cat home from the breeder or shelter or railroad track. Let the adventure begin!

Going to School

Puppies and their owners have the option of attending puppy preschool before they get into more serious obedience training after the pup is about 6 months old. Puppy preschool—also referred to as puppy kindergarten or head start programs—helps puppies learn wonderful, adaptive social skills that get them off on the right paw in their new household. Although some preschools don't accept puppies until they're 12 weeks old (due to veterinary concerns like

(continued on page 114)

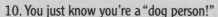
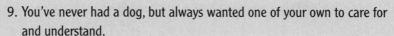

9. You've never had a dog, but always wanted one of your own to care for and understand.
10. You just know you're a "dog person!"

10 Bad Reasons to Get a Dog

1. Your wife loves to knit cute little booties.
2. Your husband wants something small and defenseless to dominate.
3. Your teenager has to have a dog just like the costar of this year's hottest TV sitcom.
4. Maybe the mutt you already own will stop following you around and whining if he has a friend.
5. You want to meet that attractive neighbor who is out walking a pooch every morning.
6. It's cheaper than a state-of-the-art security system.
7. It's the only way to get the kids to stop begging for a dog.
8. You've decided to try to like dogs if it kills you.
9. Your irresponsible child needs a character-building situation. Maybe he will even stop hitting.
10. You need something to practice on until you have a baby. By then, the dog can fend for itself.

DR. WRIGHT'S
C A S E B O O K

Scout

My coauthor and twin sister, Judi, wanted to adopt a dog, but she was a cat person at heart. So she went about finding a canine companion a little differently from most people. She called the various shelters and adoption organizations around the area where she lived and simply described the personality of the dog she wanted to find. It had to be as, well, catlike as possible!

Although the dog would be a birthday present for her 9-year-old daughter who had been begging for a year, Judi was a realist: She knew that most of the care of the animal would fall upon her shoulders. So it had to be the kind of dog she could live with and who would get along with the family cat. Sassy had ruled the roost for 16 years, and she wasn't about to yield pride of place to any uppity mutt.

"I don't care what breed it is, or how old, or what color or sex," Judi told each shelter. Her only criteria were that the animal be calm and quiet and lie at her feet while she sat at the computer. She didn't want a dog that was nervous or noisy or drooling. On about her fifth or sixth phone call, she hit the jackpot.

"I've got your dog," the woman responded immediately after listening to the requirements Judi rattled off. "He's been here 2 weeks—baby had allergies. His name's Scout. Come and get him."

"Are you sure you like dogs?" the shelter woman asked critically when Judi arrived. Judi assured her that she would like this one, as long as he didn't jump on her, slobber all over her, hump her leg, or drive her crazy with barking and racing through the house.

"He doesn't do any of those things," the woman confirmed. "He'll love to play with your kids, but then he'll lie around the rest of the day—at least until he loses some weight!"

Scout already knew how to sit, lie down, shake hands, and fetch, and he never pulled on the leash. He loved to find tennis balls in the bushes, his tail whirling around in a circle until he located his prey. He played with the kids, announced all visitors, and deferred to the cat, who took to drinking out of Scout's water bowl and licking his food while he stood by patiently, waiting his turn.

Scout quickly lost 20 pounds and turned out to be just what he looked like: a beautiful, good-natured pedigreed dog who gave his new family nearly 8 great years. And he would never have found a home if someone hadn't said the heck with the cute puppies, I'll take a chance on a big, fat, middle-aged shelter dog.

A puppy is like a baby. You're really starting from scratch. Older dogs are at least like teenagers; they've pretty much got all the basics down. They are often very well-trained, like Scout, and some people think that they even appreciate being rescued. In fact, Scout seemed so appreciative that he behaved like the perfect pet for at least the first 2 weeks, just so his new family would keep him. He showed no signs of missing the kennel or his old family. Judi wasn't too sure about the story in which the baby had allergies. That's what everyone says when they give up their dog to the shelter. Why would anyone give away a great purebred dog like Scout who had been in their family for 6 years? It would always be a mystery.

If Judi was waiting for a behavior problem to rear its ugly head, she began to breathe a little easier when Scout finally felt secure enough after a few weeks to dump over the garbage can, eat all the chicken bones, and scatter everything else all over the kitchen floor. Now that's what she expected a dog to act like! He didn't seem to like other dogs too much and would bark when he encountered one. So he just kept to his humans and his cat. Those were the worst problems Scout ever had.

If an older dog comes with behavior problems, I can attest to the fact that although it might take a little more time or effort to resolve them than with a puppy, the results will be just as satisfying. And there's a good possibility that he will already know how to shake your hand to thank you!

transmission of disease), others accept pups at a younger age. Both the pup and his owner benefit from the educational experience, which is shared with other people and their puppies in a 6- to 8-week program.

Among my favorite programs is one that takes place at St. Hubert's Animal Welfare Center in New Jersey (on the West Coast, there's a similar program at the San Francisco SPCA). At their wonderful facility in Madison, New Jersey, trainer Pia Sylvani and her colleagues help shape new, healthy relationships between "parents" and their pups by involving them in some pretty neat exercises.

Puppies (and their owners) learn basic behaviors and commands like Sit, Come, Wait (used for the food bowl and doors), Find It, Release or Give, and Settle (an informal "stay" or "chill out"). They learn the techniques of loose leash walking and how to sit politely for petting. And puppy parents learn how to cradle their pups and give them a massage (exercises used for restraint and relaxation purposes). They also work on bite-inhibition exercises, taking treats gently, and socialization with humans and dogs. In general, they create positive experiences in the environment for the puppy to build confidence in the presence of other people and puppies.

The trainers teach new caretakers how to help their puppies learn "house manners" and housetraining, the proper toys to consider for their pups (Safety comes first! Don't get toys, for example, that could get caught in a pup's throat), and how to play with a puppy. They also learn basic puppy psychology—how to get the pup to do something for them before getting what he wants. (For example, have the puppy sit before saying hello to another puppy or lie down before going to play with the puppy playgroup.)

Basically, the more time puppy and owner put in together now, the less time they'll have to spend mastering commands and basic behavior later on, or worse, trying to undo inappropriate socialization and bad habits that could have been prevented or minimized in puppy preschool. By successfully completing preschool, Brutus can more easily segue into other people-pet activities. He can become a candidate for formal obedience training, doggy day care, or a number of other options you may wish to consider for your well-socialized adolescent dog. I highly recommend puppy preschool for all puppy owners!

Adopting a Cat or Kitten

Generally speaking, adding a cat to the household requires a lot less of an adjustment than getting a dog. Cats usually take to a litterbox readily—you don't have to walk them outside several times a day. You can put out food and water and leave in the morning with confidence, knowing that if you opt for dinner and a movie with friends and don't get home until 10:30, the cat (and your house) will be fine.

Most cats appreciate a warm lap and some regular affection, but you don't have to worry about taking them for a run every day or going outside in the broiling sun or freezing snow to throw a ball for them to retrieve. Cats are low maintenance, and they still provide loving companionship. If you're not ready for a major responsibility, you just don't have the right living conditions, or you just love cats, then a cat or kitten is a much better option for you.

🐾 Long- or Short-Haired?

When you start thinking about bringing a cat home, one consideration is more important than age or breed: whether you want a long-haired cat or a short-haired cat. What's the big deal? If you get a long-haired cat, you'll have to brush it every day or every other day because it'll have a tendency to mat. If you get a short-haired breed, it won't have the long, gorgeous flowing coat, but it won't mat.

Some other maintenance hassles are associated with long-haired cats. Sometimes, even though they groom themselves, they get feces stuck in the fur of their bloomers, and you may have to wash them. And if the cat starts to lick quite a bit and she has long fur, she may be increasing the chance of developing hairballs, those disgusting mats of fur cats throw up all over the place, especially on the bed at around 2:00 A.M.

Do you want to sidestep the fur problem completely? A couple of breeds don't shed much at all; the Cornish rex and the Sphynx have almost no fur at all—sort of like the fuzz on a ripe Georgia peach. If you do choose a hairless cat, bear in mind that they're more fragile. They don't have the insulation that other cats have, so you're going to have to keep track of them and see if they are shivering and need a nice warm throw for their bed area. (This applies to Chihuahuas and other hairless dog breeds, too.) Of course that makes sense, but if you like to crank up the air conditioning in the summer, you probably don't want to own a hairless pet.

These two extremes leave a lot of room in the middle for the millions of cats with short hair. So take your pick from long, short, or bare!

🐾 Choosing Your Cat

My "rules" for buying or adopting a cat or kitten are much the same as for a dog or puppy—stay away from kitty mill–type pet stores. Use a reputable breeder where one or both parents are present. (Often, breeders use toms from other catteries for breeding, so both parents may not be present. But the mother should be resident, and there should be photos and a pedigree of the father and, preferably, access to some offspring of his past litters.) Or adopt from a humane organization that does behavioral and veterinary screenings before releasing adoptive animals. I consider all cats and kittens adorable. If you do too, and I have to assume that you do, how will you ever pick one out? Here are a couple of suggestions.

If you have little children and you're not sure whether you want a kitten or a cat, you might want to consider that kittens are a little more fragile than cats and can't escape as well as cats. However, if kittens get ticked off, they can't do as much damage as cats! Then again, kittens eventually become cats.

You may encounter a bunch of newspaper ads by "backyard breeders," people who have let their cats have kittens but are not breeding them as a profession. Be very careful. What kind of environment do the parents live in? Why has the backyard breeder let the cat have kittens? Why didn't they spay her? How many cats do they have? Are the cats well taken care of?

You shouldn't encourage irresponsible breeding when so many homeless cats are killed in shelters every day. Instead, rescue a cat or kitten from a shelter (or from a family that, because of allergies or another reason, must give up their pet), or buy a purebred from a responsible breeder. If you get the cat from a responsible breeder, you will have information about its background. (You should be able to get a pedigree for your cat, even if it's being sold as pet-

DR. WRIGHT'S INSIGHTS

What makes a cat standoffish? Besides his genes, a cat's early rearing in the litter and the human contact he's had can account for his living up to a reputation of being "too independent." Typically, cats are most open to accepting "others" (cats or people) when they're around 2 weeks to 2 months old. After that, they become more fearful of what they haven't been familiarized with, and the fearfulness seems to get worse as they age. So if you're bringing home a kitten and you'd prefer a lap-sitter to a ghost, invite everyone to lavish him with love (gently)—the sooner the better!

quality rather than show-quality, and information on its breed.) And any reputable breeder will take an unhealthy animal back, but you certainly don't want to have to cope with that, so look for bright, dry eyes, a clean behind, and no sniffles.

The Humane Society is a good option for finding a cat. One advantage of the shelter is that the cats are all readily available. They are not kept if there are any medical problems at all. Remember that if you adopt a kitten from a noisy shelter, initially the kitten might appear fairly timid or defensive because of the stressful environment. After he becomes acclimated to you and your family and his new home, he'll start coming out a little bit. But on the other hand, he may be in the shelter because of a behavior problem, or he might not have had the best care as a kitten. One down side of shelter adoptions is that you really know very little about the kitten's history. Try to get an idea of its age and when it was separated from the litter, if you can. Kittens who have been handled early and often will be more friendly and willing to accept the people in your household.

No matter where you get your cat, you should ask yourself these questions: How do the owners treat the cats? What kind of nutrition have the cats been getting? Have the kittens had their shots? As soon as you buy or adopt any animal, take it to the veterinarian you will be using over the years and make sure that it's healthy and has all the veterinary treatments necessary. The cat or kitten may need shots, tests for worming, a dental exam, and so on. And regardless of age, the cat should be neutered or spayed if it hasn't been when you get it.

❧ The Pick of the Litter

There are really very few guidelines on how to select a kitten from a litter. You certainly want to know if they are playing nicely among themselves. Ask if you can handle the kittens. Take a seat on the floor and see if they start to climb on you. If they're climbing on your legs, which are folded in front of you, do they look like they are interested in rubbing their heads or bodies against you? If so, that's a nice kitty greeting. Are they affectionate? You can probably pick out the loving ones pretty easily.

On the other hand, you probably don't want to take a kitten who is really hyperactive—she jets around constantly and can't control herself. Distinguish between bouts of energy and hyperactivity. If it looks like the cat is just really into some yarn or string she's playing with—tossing it up in the air, losing her

Greeting a Cat

The best way to approach and greet any cat—regardless of breed, temperament, and ownership (whether it's your pet, a stranger, or your next-door neighbor's)—is to give him a "nose check." Begin by stopping 3 to 4 feet in front of the cat, before he decides you're in his space and he would prefer to turn tail and run. Kneel down and hold your index finger out in front of you so it's pointed at the cat's nose, about an inch or two off the ground. Allow the cat to approach you—that's good cat manners.

If all goes well, the cat will touch the tip of your finger with his nose. This nose check is like a handshake. If the cat likes your style—and perhaps odor—he will proceed to rub your finger with his cheek or the corner of his mouth. Your next response should be to return the rub with your finger on the same locations. And that's the end of the greeting. If you need more fuss made over you, check out a dog!

Above all, don't let enthusiastic kids run up to the cat or kitten and grab him, or Snowball will likely be "greeting" everyone from under the sofa after you bring him home. If the prospective pet is too fearful or too wired to perform the nose check, I might be tempted to keep looking.

Cats use their noses to greet friends, family, and strangers. To cats, a nose rub is just like our handshake.

balance, having a good time—well, that's terrific! But pass on a cat who bounces off the walls or one who comes over and latches onto you claws-first if you try to stroke her. Cats who jump on you uncontrollably and clasp on, so that there is nothing you can do to control their behavior, are also bad news.

In some ways, picking the best kitten or cat for you is the same as looking at puppies or dogs that are up for adoption. You don't want the kitten who never approaches you, nor do you want the one who is just going nuts all over the place. You want the one who shows some interest in you right off the bat, is good at climbing on you, and has a good sense of balance. She looks at you with those incredible blue eyes, and she starts to purr when you begin to stroke her—she's probably the one you want. And by the way, kittens are also great if you hold them up on your chest. If they lick you on the chin, then they pass the "lick test" too!

You can't really subject a cat to a puppy "rollover" test, which I don't like much anyway, unless you're willing to risk going home with a few scratches and bites along with your new companion animal. Let's concentrate on "meeting and greeting " to help you find a nice friendly cat.

Introducing Your New Cat

The way you introduce a new cat or kitten to the household can mean the difference between a "fraidy cat" who hides under the bed for the first week—or longer—and a confident, happy feline who settles in with a satisfied purr! So it's worthwhile to pay attention to some do's and don'ts when you decide to add a new face to the family portrait.

Bringing a new kitten or cat into a household doesn't have to be a trying experience for either the kitten or the adoptive family. Cats are most likely to adapt to new living arrangements and homes—with or without resident pets—if the owner sets up a feline-friendly environment and thinks through some commonsense scenarios in advance.

No single plan will work equally well for all cats, of course, due to the wonderful variety of felines waiting to be brought home from the shelter or breeder. Each kitten will bring its own baggage: breed, personality, early experience, and so on. And the same kitten can be expected to react differently to various living situations (depending upon how small, large, quiet, noisy, busy, or calm the household is, among other things). But regardless of individual differences in nature and nurturing, the new kit on the block is most likely to settle in successfully—with less stress—if you can make the following arrangements before the cat first crosses your threshold.

🐾 Week One: Easy Does It!

The first week of your new cat's or kitten's residence in your home will go smoothly if you follow my 10-Step Plan to Make Your Cat Feel at Home. Try it!

Step 1. Pick a room. Select a room your kitten can use as its home base for the next several days. The room should have a door or some other way to make it private and to provide a safeguard for the kitten from the chaos of daily living (especially resident toddlers and other pets!).

Step 2. Locate the litter. About half a foot away from the wall opposite the entryway, place the kitten's litterbox, preferably with sandy, clumping-type litter. If the kitten has already established a preference for a different type of litter, use it.

Step 3. Fix food and water. Place fresh water and food along a separate wall, or as far away from the litterbox as possible. There should be a clear separation between eating and litterbox areas.

Step 4. Select snoozing space and toys. Select another corner (or area away from the litterbox) for Kitty to sleep, preferably on a worn t-shirt or towel that has her familiar siblings' odor or her own smell on it from the shelter. You will probably have to change this location to Kitty's self-proclaimed sleeping area later (how many kittens actually sleep where *you* want them to?). But that's okay. Add a few simple toys such as a balled-up bit of aluminum foil and a catnip mouse to the area, too. Most kittens are quite playful and will greet the toys with interest.

Step 5. Add the cat. Bring the kitten into the room with you and close the door. Allow the kitten to discover where everything is. Do not take the kitten to the various locations. DO NOT place the kitten in the litterbox! If you absolutely must place something in the litterbox, add a little soiled litter from the cat's or kitten's previous home, so she can tell where she's supposed to go. The nose knows!

Step 6. Let her look. Allow the kitten to discover the room's delights, such as window ledges to perch on and look outside; a comfortable couch, stuffed chair, or bed; tables to jump up on; and anything else in the room that might be used as an escape route or a hiding place when needed (from the kitten's point of view).

Step 7. Know when to go. You may wish to leave the kitten by herself when she appears ready to take a nap, or when she seems relaxed and confident in her new surroundings. If the new pet chooses to nap in your lap, consider yourself lucky. Of course, you can always try to place her in the sleeping area you've selected and then leave the room. If she cries, go back briefly if you want to. You won't spoil her!

Step 8. Open the door a crack. Your new companion animal will let you know when she's ready to take on the rest of the house. It is important for you to allow the kitten to do this at her own speed, which could range from an hour or two to a few days. A door that's cracked open just wide enough for Kitty to enter and exit from her room is a good starting place.

Step 9. Keep visitors away. This step is a tough one! Keep other pets and family members away from Kitty's room. Introductions to these individuals should be well controlled. Again, allow the kitten to approach and withdraw (back to her room if necessary) from people and other pets that remain passive. You'll need to control the dog—Kitty will eventually establish the house rules with other pets, but young felines will need help at first.

Step 10. Follow her lead. After Kitty has become familiar with her new home and playmates, you'll become aware of her preferences for playing, eating, sleeping, and other activities. At this point—again, anywhere from a few hours to a few days—you may begin to gradually change the location of the litterbox and food bowl to areas preferred by both you and Kitty. Hope that the new sites are one and the same!

For some of these steps, you need to judge the kitten's readiness for exploring spaces and establishing new relationships. This activity also provides a good time to watch for those communicative behaviors that show when a cat is being friendly and playful. Whiskers spread widely, eyes engaging yours, tail held proudly, and enthusiastic purrs and mews rather than crying should give you a clue about when she's feeling confident and ready to take on her new world.

Remember that most cats enjoy a friendly greeting in the morning and when their people arrive home from school or work. Although no cat will show up carrying slippers and the newspaper, she probably looks forward to this part of her daily routine. Don't ignore her ritualized greeting—the nose check mentioned in the discussion of cat selection. (See page 118 if you need a refresher on how to do it.)

Although coat length is not directly involved in sociability, the long-haired breeds, such as Persians and Himalayans, seem to be less excited about meeting and greeting than are many short-haired cats like Burmese and Abyssinians. So don't take it personally if your Angora won't twitch a whisker to say hello! Persians, for example, are likely to run beneath a piece of furniture if caught by surprise on the dreaded open space of the living room floor, whereas Abyssinians are more likely to say, "Oh boy! Moving feet to pounce on!" and proceed to get underfoot as part of their greeting. There's always an exception, though, and the hardy, outgoing Maine coon displays his gorgeous long coat proudly while tripping you up in his eagerness to say hello.

Settling In

When your new kitten is used to "her" room, you can certainly crack the door open so that the cat can feel free to start exploring throughout the house. Keep in mind that although this is not a baby and you are not going to have to kitten-proof your house, kittens are inquisitive and playful. So take some commonsense measures to protect the kitten *and* your possessions!

For example, move your most valuable things to locations where the cat can't knock them off a table, shelf, or counter. Objects are irresistible to cats and kittens because they're really interesting. Blown glass or things that don't weigh very much are especially appealing. The cat will think, "Oh, this looks like a great thing to bat around or latch onto," before you can teach her not to jump up on things.

If you're used to hanging your silk or wool clothes over the back of a chair where they dangle down, or you tend to leave your silk scarf or shawl with tassels lying around, change your habits and put your good clothes away. Kittens will find things that dangle, and when they do, they bat them and latch onto them. Pretty soon, you'll find your wardrobe in shreds.

It's important to start to think about keeping things like drugs, poisons, and frayed electrical cords out of your new kitten's reach. Make sure that you know the general area where your kitten is most of the time. Otherwise, you'll be better off keeping him confined to a safe area or room, especially if you are away. I'm not recommending that you shut him in a dark little bathroom for 10 hours a day. I am suggesting that if he has all those "dangerous" things available in your absence, he might, like a curious kid, get into trouble if you haven't thought of everything.

Besides doing some basic cat-proofing and keeping an eye on the kitten, you also need to teach your children some responsible, caring behavior around the kitten or cat. These rules are pretty simple: Don't terrorize him by jumping out or screaming at him. Don't play rough with him, pound on him, or pull his tail. Don't throw things at him—throw them *for* him. Be loving and gentle toward him. After you have made the house safe for your new cat and your family knows the rules, it's time to teach the cat what acceptable behavior is.

🐾 A Kitten's Daily Rituals

Your goal is to encourage your kitten to do the things that *you* want her to do. If she gets into a pattern that you set for her, it will satisfy her bio-psycho-social needs, as we psychologists say, which is to say, she'll be healthy, happy, and well-adjusted, and no trouble to you, to boot. She needs a good litterbox and some food and water to satisfy her biological needs and some stroking and playing to satisfy her social needs. If you have another kitten, they can play together. For the psychological needs, she needs to be able to trust you: She needs to feel assured that you are not going to surprise her or ambush her and make her do things she doesn't want to do. This feeling of security will give her a little bit more control over her environment and her life. Felines have a need to maintain control.

Kittens and cats like a predictable routine punctuated occasionally by an unexpected adventure—like a mouse in the house. Many kittens like to do today whatever they did yesterday and are likely to do tomorrow. So their "workday" may go something like this:

1. Wake up from their catnap (kittens don't sleep all night) with a nice stretch.

2. Go to the litterbox.

3. Jump on the bed and wake you up.

4. Encourage a short petting session while you think about getting out of bed.

5. Find a ball and engage in a little self-play while you dress.

6. Follow you around as you make coffee or make the beds.

7. Watch you eat and eat their own breakfast.

8. Jump on the windowsill and watch you go off to work.

9. Sleep, eat, play, eat, look, eat, switch windowsills, use litterbox, sleep . . . (set up a kittycam if you must know!).

10. Wait for you to come home.

When you do finally get home, the catnaps end and the action starts again! So whether you are home all day or go to work, pay attention to your kitten or kittens, because what they do is going to be based on what you do with them throughout the house and throughout the day and evening.

A kitten will set up his own routine and ritual. Your kitten will very likely greet you at the door as you come home, and she will probably start running around and showing you what a neat cat she is—racing to her scratching post and going to her toys. Make a little time to play with her now if you can. Then you can go about preparing dinner or making phone calls, and she will amuse herself (and you) by jumping up on things when she knows you can't reach her! Now it's dinnertime and you have to let her know it's *not* okay to jump up on the table while you're eating. Be persistent. Just say "AH AH AH" (no lecturing), and perhaps toss a treat in her food bowl on the floor before she jumps up. Or use one of the other strategies that we'll talk about in the second half of the book.

After dinner, you can enjoy some quiet time together on the sofa as you watch TV or read the paper. This can be a nice social time—petting time and talking time (yes, many owners talk to their pets, it's not just you!)—until finally, it's time to go to bed. Kitty will probably eat a little and then wash up before settling down with you in bed and purring herself to sleep. Or she may simply go to her resting area, wherever it may be. The day is done. And the next day, it's the same thing all over again.

Cats love these rituals, and you can enable your kitten to enjoy them by keeping things consistent from day to day. If this sounds too stiflingly predictable, don't worry, you'll never be bored with a kitten around!

🐾 Feed Me!

Your new cat's basic needs are simple: Besides your affection, he needs food, a litterbox, a place to scratch, and some quality playtime. Let's discuss these needs one by one. Food and water are basic needs for your cat. First, find out from the breeder or shelter what kind of food the kitten's been eating, and then check with your veterinarian to see if it's the kind of food he or she would recommend that the kitten stay on. There are numerous highly nutritional brands of food, and many of them have all the nutrients that your cat will need.

Your cat should have some cat food available (usually the hard stuff) most of the day. That's because cats will eat 7 to 12 times a day, even though they just nibble each time they eat. The problem with feeding only once or twice a day is that cats get hungry, and if there's no food available, they may start to chew on things.

DR. WRIGHT'S INSIGHTS

Most kittens are very easily trained to go in a box, but they have to be able to climb in and out of it easily and without fear. Put an inch or two of sandy, clumping-type litter in the bottom—my colleague Pete Borchelt has actually studied which kind of litter most cats prefer, and the sandy clumping type wins three to one. But if your cat's accustomed to a pelletized or clay-type litter and is using it without any problems, that's fine, too. Just don't use perfumed litter—it may smell good to you, but it's overpowering to your cat. Your goal is to make the litterbox inviting to your cat, not to you!

Make sure that the food and water are fresh, and that the bowls are clean (changed at least twice a day and washed regularly). Bacterial and gum infections are possible if you don't wash the food and water containers.

🐾 Life in the Litterbox

If you're going to keep your cat inside, which is your best bet if you want your feline around and healthy for many years to come, you need to become a litterbox expert. I would venture to say that your cat's life may depend on it! That's because the "litterbox problems" (ignoring the box, for the most part) are responsible for the great majority of families giving up on an otherwise well-behaved pet. We'll discuss the many reasons for litterbox problems in chapter 12. For now, let's focus on getting things set up in a way that will give you and the kitten the best chance of success.

Litterboxes come in all shapes and sizes. Some are covered; some are mechanized so that they empty themselves. Any of these can work just fine. But you need to start out with something a little smaller for a kitten and then switch to an adult-sized pan as he grows bigger.

When your kitten is ready to move out of that initial room, put his box in a location that is away from his food, but that's fairly private and doesn't have a lot of traffic. Choose someplace that's accessible to the cat 24-7 and has some nearby places to get behind or on top of—a spot with "escape potential." You will need one box per cat if you have a multiple-cat household, and that's true no matter how big your litterboxes are. Both cats need to be able to go at the same time! Some people like to provide one box per level if they live in a home with multiple floors.

No matter how many litterboxes you have, clean them every day, or soon your fastidious kitten will be looking for a more pristine bathroom. Sorry, folks, but unless you want to try to get your kitty to use the john (many have tried, but few have succeeded!), we're stuck with those darned boxes for the duration.

🐾 Cat Scratch Fever

Scratching is an innate behavior of cats, so one of the things you're going to have to teach a new kitten is how to use a scratching post. Although you could choose a carpet-covered post, posts made of sisal rope let the cat really sink her claws in, which seems to feel good. And there's no risk of inadvertently teaching the cat to claw your new wall-to-wall carpet! (After all, sisal rope feels much different than carpeting!) When you buy a post, make sure that it has a sturdy base. If an unbalanced post falls over on your cat, she's not likely to go near it again!

Probably the best place to put the scratching post is right in the middle of her room. Other good locations are areas where she might show a preference to start stretching initially, such as the side of the bed or the side of a couch. If you see her stretching against a piece of furniture or the wall, put a post there immediately. That's because the drill is: stretch up, sink in claws. If she

Make sure that your cat's scratching post is sturdy enough to support her weight and tall enough so that she can really stretch out on it.

DR. WRIGHT'S **INSIGHTS**

Do you want a happy, well-adjusted cat? Give kitty a condo! Cat condos have several different levels, so several cats can use them at a time, or a single kitten can move from level to level as she pleases. You can even keep one or two toys there for her to play with. In any case, you should have a place the kitten can jump up onto, be it a kitty condo or just a windowsill with a soft pillow on it. Your kitten really needs this so that she can look out and amuse herself by watching squirrels, people, falling leaves, and other fascinations while you're away. It's great to have a couple of window seats, especially if you can have one on the east side of your home and one on the west. In that way, your cat has the opportunity to have access to or avoid sun coming into the house. Frequently, cats will see dust particles coming through the window and jump up at them. So the condo or windowsill has a play function as well as being a resting place and entertainment site.

hasn't shown any preference, then put the post in a location where she can run over to it and it's clearly visible.

I recommend that you make more than one post available, especially if your house has an upstairs and a downstairs and the cat has access to both. Also, remember that cats want to stretch after they awaken from a nap. So if there happens to be a scratching post near or accessible to the spot where she's snoozing, she's going to go over and stretch out on that, rather than use her claws on your favorite curtains.

Some people have found that placing some catnip at the top of the post is helpful, whereas others have stapled a ping pong ball to an elastic band and stapled the other end of the elastic band to the top of the post. The battable ball dangles down a couple of inches and entices the cat to reach way up and bat at the ball, thus attracting him to the post. Anything that you can use to propel your cat toward the scratching post and to stretch out and use it will save you yards of fabric later on. You might even consider rubbing some of that catnip on the scratching post to attract him initially to go up and cheek-rub on the post, followed by scratching it (both behaviors are ways of marking it).

Some schools of thought suggest that you actually take the cat's paws and put them on the post and rub them up and down it. I never suggest that you force the cat to do anything like that. Cats don't force well! If you enable the cat to perform a desirable behavior by creating a situation that is attractive to

(continued on page 130)

Declawing: Pros and Cons

Some people worry about their cat or kitten jetting around the house and tearing up the furniture. Worse yet, they fear that the cat will wind up on their baby's tummy or their child's bare leg and then dig in and launch off to the next location, producing quite a scratch. And cat scratches can become infected. Moms and dads worry about that. So people have to weigh the threat of damage to their household and to their family against declawing—damage to the cat. And declawing should not be considered lightly.

If the kitten starts to use her claws with you during play, just take your hand away. Don't play with her, and don't stroke her. Essentially you should not do anything until the kitten simmers down; only then should you start in again. You want to get a kitten that's stayed with her litter until she's 6 or 7 weeks of age so that the kittens can get used to playing with one another and learn to inhibit their clawing and keep their claws retracted during play.

You can always transfer some of that clawing behavior from your hand or ankles onto something that the cat can get hold of, like a toy on the end of a fishing pole. As soon as you begin retraining her, she'll gradually learn that it is not okay to bite or claw Mom or Dad or the kids. Please give the kitten a chance before you decide to declaw your kitten. This decision cannot be undone.

I don't recommend declawing to any of my clients. As a consultant, I can tell them that there are many different solutions, including the use of little plastic sheaths (Soft Paws) that you put on the cat's nails and replace about once a month. That procedure is quite a hassle! But for people who are vehemently against declawing their cat, it is an option. Bear in mind that, for many cats, claws don't represent a threat to people or furniture. Well-chosen and well-placed scratching posts and early socialization keep the cat's claws where they belong.

Does declawing stop a cat from scratching? Cats will scratch themselves even if they don't have claws. They'll scratch in litter to cover feces, they'll stretch up and scratch on a scratching post, and they'll scratch other vertical locations to mark an area for other cats to see. ("I'm here!") They go through the same motions, extending and placing their digits in the material, whether they have claws or not. So whether your feline has claws or is declawed, she'll continue to scratch because it is an innate behavior that occurs with or without the presence of nails.

The actual declawing surgery is tantamount to removing the last digits or the last joints of your fingers. After the claws are amputated, the cat's paws are bandaged for the night and sometimes for the next day, and then she goes home. Cats probably experience quite a bit of pain when recovering from this surgery, and they probably continue to feel pain for several days after the operation. If a cat's not given any kind of pain reliever, it may be an even more difficult or lengthy recovery.

Typically, when you bring your cat home after declawing, and the cat is used to some kind of clay litter, you'll need to switch to a more paw-friendly substrate. Your vet will probably recommend that you get some kind of paper and tear it into strips or use pelletized newspaper litter, which actually absorbs urine a bit better and is more readily available; you can dump it out more easily, too.

Most cats seem to get over the declawing after about 10 days, so you can start to phase in whatever litter the cat had used before the surgery. Put a little bit of the old litter inside the box, so that the cat can get used to it over 2 to 3 days. In no time, you'll be reading your newspapers again instead of shredding them. If the cat looks like she doesn't like the hard clay litter any more, switch to a sandy clumping-type litter.

Has any research been done on declawing and resultant behavior problems? Three pretty large studies on declawing and behavior problems failed to show that declawed cats have any more behavior problems than intact cats. Now that's not to say that some cats don't have problems after declawing. But then again, some cats have problems when they're growing up and they haven't been declawed. Declawing has not been shown to be an important cause of feline behavior problems.

I do get calls from clients who say that their cat has recently been declawed and is not using his litterbox now. But is this behavior caused by the operation or the torn-up newspaper litter that the vet recommended? What seems to be a result of the declawing might be something else associated with it.

One potential drawback of declawing is that cats are not as likely to be able to escape rivals or predators by jumping up on things. If your cat needs to get away from another cat in the family or to latch onto the bedspread or other furniture to climb up high enough to avoid the 2-year-old chasing her, or, worse yet, the dog, she may not be able to get a grip on it. So when you bring the cat home, you are going to have to make sure that she can get up on things before you assume that she's going to be able to take care of herself again. Keep in mind that although she can no longer scratch, she can still bite if she is provoked or in pain. So watch out for yourself as well as for her.

her, she will discover how to do it by herself. And believe me, she will be much more likely to prefer it if you don't force her to try it.

Why Cats Scratch

Scratching may not only allow a cat to stretch her stiff, tired muscles but also serve as an emotional-release function. Sometimes when I walk into the house, Domino sees me and runs over to the scratching post and starts scratching away. She's just trying to get rid of that stored-up energy or relieve the excitement associated with my coming home and the impending petting. Sometimes cats scratch the scratching post if they can't get to the critter they see outside—a kind of displacement activity—because they can't actually go and chase or attack whatever is out there. At least they can get rid of their energy by scratching.

Some people suggest that cats use scratching posts to sharpen their claws or remove the sheaths that cover them. Although many cats scratch posts, they probably don't use them to sharpen their claws. Cats never use something like a scratching post or a tree to sharpen their back claws, even though these claws really inflict the most damage when cats are attacking or defending themselves. Scratching a post probably serves more of a marking or social function; Fluffy wants to show other cats that she has been there by depositing the odors from her footpads for them to smell. Cats usually stretch up and sink their claws into a post after they've rested, and it probably feels good to get those claws working again.

🐾 Rules of Play

Kittens learn that there are three different kinds of play, and only one involves people! Kittens learn how to play with one another and with humans, an activity called social play. Wrestling, running from others, and ambushing are all fun activities that kittens do quite readily. Next, they learn how to get up on furniture like book shelves, armoires, and even vacant wall sconces, to balance and even walk on these objects, and to jump from one object to another; that's called locomotor play. (Climbing up a person to balance on the shoulders on the way to the top of the person's head is a form of locomotor play.) Finally, they learn how to play by themselves, either with that tail that keeps sneaking up from behind or with objects and toys—a kind of self-play.

To start play, make sure that you're talking to or petting the kitten. Between 2 and 7 weeks of age is the single best time to stroke the kitten and play with her. Behaviorists consider this the sensitive period or critical period of primary socialization for kittens. (In puppies, that period is $3\frac{1}{2}$ to about 14 weeks of age.)

So if kittens are going to be as social with people as they can be, given their genotype and breed, you can help them by doing more things like playing with them, stroking them, talking to them, and including them in your activities in that period of time. If you don't, you'll get a solitary kitten who likes doing things by herself, torques around the house, and doesn't really want to do things with you. That's not to say that at 7 weeks of age you're done; during that period of time, however, a kitten starts getting used to how to play and how to do things like retract her claws when she plays with you. She also starts to prefer to be with people at that age because being with people feels good.

Your goal is to have your cat play with and tear up and bite things that you want her to. For this reason, you don't want to take her favorite little cloth

(continued on page 134)

A kitty condo will quickly become your cat's second-favorite place—right after your lap!

DR. WRIGHT'S
C A S E B O O K

Jazzy and Jessie

Sometimes people start out with the best intentions when setting up a home for their kittens, but bad advice or a lack of understanding about their pets' behavior leads them down the wrong path. That's what happened with Rachel and Josh Rothman, a compassionate couple who lived in rural Georgia and took pity on two kittens who were found inside a bag in a dumpster.

Rachel and Josh started out doing exactly the right things: rearing the kittens in a social environment with separate locations for them to satisfy different needs, having play sessions, toys, places for them to sleep, and all the things I talked about earlier in this chapter. The kittens, named Jazzy and Jessie, were shy and skittish, but they began to flourish in the rural Georgia home, and everything was fine until they were declawed at 6 months of age.

Unfortunately, nobody told the Rothmans about the fact that the clay litter they were using might hurt the kittens' newly tender paws, and that they should switch to something softer, like newspaper strips, newspaper pellets, or a sandy clumping-type litter. Well, I shouldn't say nobody told them—the kittens told them! They sent a clear message—by urinating and defecating on the carpeting right next to the two boxes.

"We even got two new boxes and put them upstairs," Rachel told me as we surveyed the scene at the colonial home. I didn't need my detective's hat for this one—the location of the "accidents" was not accidental at all. The kittens would have preferred to use the litter boxes, but they caused too much pain. So they went on the floor as close as possible to the boxes.

The new pets were trying to do the right thing, but they were shut in the garage for their trouble. The two new boxes sat upstairs, untouched, with urine stains beside them, Josh added. I explained that this confirmed that the boxes "stood for" the ones that represented pain. So the kittens wouldn't even try them.

The couple told me that they would get desperate and pick up the kittens and set them in the box. That decision was disastrous for these already skittish—and now hurting—kittens. The cardinal rule of cat training is that you can't succeed at trying to force a feline to do anything!

I explained to them that they probably should have tried the newspaper litter, and why veterinarians typically recommend that. It's kinder to tender paws after surgery. And it's a great idea to introduce the newspaper *before* the surgery, so it's not as strange to the cats when they come home from the vet. The unfamiliarity of new litter may lead kittens to avoid the litterbox after surgery as much as the operation itself.

To solve the problem, we had to make the litterbox look and smell different, and in fact feel different, so we went to the scoopable clumping litter. We also decided to put the litterbox in a place that was a little more private. Even though they were in corners, the boxes were now in a location next to the garage door, where there was a lot of traffic. There was a litterbox at the top of the steps, where you never knew if someone was about to come up or go down. After the Rothmans moved the litterboxes and changed the litter, the two kittens were well on the road to being part of the family again within a week.

Toys: From Homemade to High-Tech

Your kitten will enjoy many different kinds of toys. As Kitty grows up, a tennis ball might be a good toy. Kitty can get her claws into it and roll it around. (She may have to share this toy with the dog.) Ping-pong balls are fun, and they're light-weight enough for cats to bat around.

Fishing pole–type toys that you can use for interactive play are great, too. For a touch of realism, you can move the pole so that feather toys fly through the air or drag the mouse-type toys just above the ground. The beauty of these toys is that the pole and line keep your hands out of eager—and sometimes undiscrimi-nating—claws' reach, yet you can give your cat a great workout.

Not all cats love catnip, but for those who do, catnip mice or catnip-filled squares and balls can provide blissful entertainment. You can buy them or make your own. Pet stores sell a wide assortment, and you can even find organic catnip.

Consider buying a couple of nice little 2- by 3-foot area rugs. The cat will scrunch them up and attack them. When you flatten them, he will come over and scrunch them up again. (Just don't leave them around where harried adults, dis-tracted kids, or elderly relatives can trip over them when they're not expecting them.) Anything that is scrunchable, that cats can get their claws into, grasp hold of, and then get rid of, makes a wonderful cat toy.

You can even use homemade cat toys. The little circular plastic tops on milk bottles are great. Just make sure to throw them away when they get chewed up; the little pieces could get caught in the kitten's throat. You can also use a plastic straw for your kitten to bat and paper bags for him to climb in and out of. (Never

mouse and toss it up on the couch because she may jump up on the couch to attack the mouse, take a flying leap, and, in the process, leave some nice scratch marks on your couch. Or she may try to bat the little mouse around, get her claws caught in the material, and think, "Hmmmm—this feels like something I want to scratch." And, as luck would have it, she may start to prefer the fabric in the couch at $25 per yard to her scratching post.

To avoid this situation, play with your new cat in a location on a carpet or rug, in her play area, or in one corner of the den.

If you use a dangling toy (a stick or "fishing pole" with something like a feather or lightweight toy mouse or bird on the end of it) and move it around

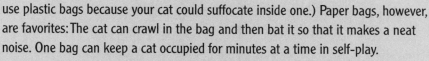

use plastic bags because your cat could suffocate inside one.) Paper bags, however, are favorites: The cat can crawl in the bag and then bat it so that it makes a neat noise. One bag can keep a cat occupied for minutes at a time in self-play.

You can also use wine corks or even larger corks as toys. They're great to roll around, and cats can grab them and bat them. If it looks like a cork is starting to get a little bit used, just throw it away and get another one. Rolled-up balls of aluminum foil, the old reliable ball of string (as long as it's in a ball and Kitty can't unwind it and choke on loose string), and balls of crumpled paper are all great toys for kittens.

Would you rather indulge in some conspicuous consumption? Some pretty high-tech toys are available for kittens today. One machine shoots out tiny plastic balls, similar to the machines for golfers who want balls shot back toward them so that they can stroke the balls toward the hole. This cat toy shoots plastic balls out across the floor for the cat to chase. You don't even have to be there. And of course, there are the kitty videos! My favorite is <u>Cyberpounce</u>, which is interactive.

Whether you make your own toys or buy them, make sure that they're cat-safe. They shouldn't have attachments like bells or plastic eyes that cats could pull off or choke on. The toys themselves should be too big to get caught in a cat's throat. Don't leave a cat unsupervised with a ball of string, an aluminum ball, or anything else that he could break apart and choke on. Put fishing pole–type toys away when you're not using them. Soft catnip toys are okay to leave out when your kitten is unsupervised. Otherwise, hoard your toys for playtime so that your cat will look forward to it even more!

the den, make sure that you don't move it on top of Dad's walnut desk or the shelf displaying Mom's hand-blown glass collection! Instead, pull it along the floor or over a little box on the floor that the cat can jump up on, into, or over to tackle the toy. Just remember to choose an area for play that's either scratch-proof or permissibly scratchable. If you do, you'll find that your frolicking feline enjoys family fun time and learns some important house rules in the process.

Mischievous Kitties

You get your cat home and you get everything set up, and now the cat looks like it's going to do something naughty. Do you get the squirt gun out? Grab the shaker can and throw it at the poor kitty? Do you yell at him, or grab him and flick his nose with your finger? No, you shouldn't do any of these things!

You're not going to stop bad behavior through any sort of physical punishment. Never shove your kitten's nose in the mistake he's made on the carpet. Never flick the cat's nose with your finger. He doesn't understand, and he has very sensitive nerves in his nose. And besides, what you want your hands to stand for in the kitten's mind are things that feel good, not things that the cat has to avoid or clasp onto and sink his teeth into because it feels bad.

A kitten needs to be communicated with in terms of what he should be doing, not what he shouldn't be doing. Now, obviously you're not going to reason with him. Instead, call the kitten's name, or make a noise, clucking, or a high-pitched "Kitty kitty kitty!" to draw his attention away from something. As soon as you've caught his attention, transfer the activity to a nice toy, or have him come to you, or walk over and present your finger to him so that he can initiate a greeting toward you.

If you're not getting his attention, go over and very gently pick the kitten up and then turn him around and face him in a different direction. Often, that's all you need to do to get the cat's mind off what he was about to do that you didn't want—like jumping up on the table—and on to something else. He no longer sees the table, he turns around, he now sees this nice little furry ball, and he goes and attacks that instead.

Again, all you have to do is interrupt the cat's undesirable behavior, change the cat's focus or what he's looking at, and direct his attention onto something else—anything else! Then, when he starts to play and do something that you want him to, make sure you praise him. Talk to him in a nice voice, and make sure that he knows that this is the right thing to do.

It's only too easy to rely on physical punishment to stop a behavior. As long as you think that way, you're going to use it, and the likelihood is that it's going to backfire. Instead, use the techniques I describe in Part Four on page 209, and you'll end up with a happy pet and a great relationship. But before we get there, let's look at your kitten's or puppy's awkward adolescence in the next chapter.

Chapter Seven

Adolescent Issues

Adolescent issues in dogs and cats are largely driven by training and/or hormones. So let's see what we can learn about this rather difficult period. Getting through it unscathed is an interesting challenge! Puberty marks the beginning of adolescence, and all kinds of mixed emotions begin to combine with dogs' and cats' desires to find their place in the family. I'll address some of the specific problems you might run into, but first I'd like to tell you about what is normal—though sometimes maddening—adolescent behavior.

We like our cats to come when we call their names and to stop scratching the armchair when we say "no!" A kitty treat and some verbal praise and strokes when they comply will help reinforce good behavior. But that's about as much "obedience" as we usually expect from our feline friends. Since we really don't connect cats with obedience training per se, this part of my discussion is aimed at the owners of adolescent dogs.

Dog Training: Just Do It!

One of the best gifts you could give yourself—and your adolescent dog—is formal obedience training. As a puppy, he may have learned some basic manners and how to get along with other puppies. In fact, you may already have taken him through a puppy preschool class. Sometimes puppy kindergartens or preschools start training puppies as early as 8 weeks of age. Early experiences and enrichment outside the confines of the home should enable your puppy to accept more of life's challenges later on in his life.

But even if you didn't start your pup in socialization classes at 8 weeks, it's never too late to get started. Some basic training classes for companion dogs (beginner classes) start before the pup is 6 months of age, and others allow you to register when your dog is 6 to 8 months old. So by the time your puppy is an adolescent, he is probably finishing or has finished a puppy kindergarten class, and you should enroll him in the formal obedience class.

Usually these classes are a lot of fun for the puppies. They typically start with a play session, when the dogs play together for about 10 minutes and get their energy out. Then the pups start hooking up with their owners, and that's when the serious learning begins. By the time the initial 6- or 8-week class is done, the pups will be allowed into the second class. Even if a dog starts when he's 1 year old, it's not too late. But he should be fairly aware of how to react to other dogs in a calm and playful manner, rather than growling at and biting them.

In a puppy training class, the pups should learn some behavior basics. They should be able to resist the urge to bolt when there is a distraction. They should be able to wait alone—to be by themselves and not panic—by the time this training is done. They should be able to sit down and allow someone to come up and pick up a paw or look into an ear, so that things will go much better when they go to the vet's or groomer's. They should be able to go for a walk without pulling on the leash.

By the end of the class, your pup will be able to approach strangers without too much trouble. He will be able to obey your command for sitting and the signal for staying. If you walk him outside and he comes to a curb, he ought to be able to sit and stay there just about automatically, until you're ready to walk him across the street. And he ought to be able to stay and wait in a position for a few minutes if you require him to. That way, you can talk to somebody informally and not worry about what your dog is doing behind your back.

DR. WRIGHT'S INSIGHTS

Head halters are really good for training your pup not to pull on his leash. A halter also allows you to move away from choke chains (gasp) and pinch collars, which are just unnecessary. I agree with my colleague Dr. Steve Zawistowski's instruction that if you start with a head halter, use it as a communication tool rather than a device to force your dog into compliance. Then he will be able to walk on a nice loose lead before too long.

🐾 Why Take Your Dog to Class?

As a behaviorist, I see a lot more benefits to training classes than simply teaching your dog commands. The classes give you some verbal control over your dog when she's still testing limits, your dog knows what to expect when she is in your presence, you build up your leadership qualities, and you learn to be in control. Besides all this, your dog will find that she wants to do things with you rather than compete with you because you make it such a pleasant experience for her to comply, by giving her attention, snuggles, play, and treats.

So from week to week, with some homework and training that you do with her from 10 to 30 minutes daily, your dog should turn out to be a really good companion. And when you have a situation that's dangerous for her, like when she goes outdoors (or even when she stays in and you just want her to stay off the couch), you can help keep her out of trouble by giving her a word or signal she understands.

🐾 Give Your Dog a Hand with Training

You want to teach dogs to be able to do things in a win-win situation, where both of you are gratified and you don't get involved in a confrontational situation. In training the puppy or adolescent dog to do something new, the first step is always to give her as much help as she needs to be able to perform whatever you have in mind.

Let's take the Sit command, for example. Once you have her attention, you can say "Sit," and at the same time use some kind of a hand gesture, with the palm of your hand facing the front of the dog. Or you can actually show her a food tidbit reward, passing it over the top of her head so that her head goes up and her rear end goes down. (Then make sure you give her the treat!)

Some pups do very well with a clicker, followed by a treat held over their heads—they sit, and you click the clicker and give them a treat. After about 20 click-treats, you'll only need the "click." For other dogs, you may need to actually kneel down in front of the dog and touch the treat to her nose; then, as she attempts to take it, move the treat back over her forehead. As she raises her head to get the treat and her rear end goes down, then you can click-treat and give big hugs. If she seems slow to complete the Sit, repeat moving the treat from her nose to just above her head until she sits down all the way. Don't hold the treat too far above her head, though, or she may inadvertently learn to jump up! Later in the training process, all those little treat-in-hand-prompting things you are doing can be dropped, or as behaviorists say, faded out. What you end up with is this sequence: You give a command, or use the palm-up hand gesture, the dog sits, and she gets her treat or a nice bit of praise. It works like magic.

If getting that first desired response—the sitting or the down or the coming to you—is taking too long, it is okay to prompt it even further by doing what we behaviorists call shaping the behavior. If, after many valiant attempts, your dog just doesn't get it, then begin with the first part of the sit, and when he lowers his rear end a little bit, praise that as if it was his idea! Then you say, "Sit," and try to influence the sit just a little bit more by holding a treat just over the top of his head. If his rear end goes down even farther, you say, "Good boy!" Eventually, he will get down low enough to sit.

Or maybe he'll sit, but he won't lie down. Begin by saying, "Farley, down," and then touch the treat to his nose and move it toward the floor just in front of him without offering the treat just yet. Then, when he attempts to eat the treat, slowly move it away from him on the floor, so he must stretch out his front legs on the floor to get it. As you see his legs begin to move forward toward the down position, give him the treat and say, "Go-o-o-od do-o-own!" as though it was the dog's idea all along. He'll learn that when you tell him to lie down and he does just that, he really feels good!

Most obedience training is done with guidance and with reinforcement training (like praise)—with things that both you and the dog can win at. You can also use play after a training session as a reward. After your dog finishes a serious half-hour training session, it's a good idea to allow him to get rid of all that energy—and some of the frustration of not being perfect. He has been concentrating, and now he needs to have a payoff, so don't forget about that. As a rule of thumb, it's a good idea to let him have a little playtime before he gets into the training and a little playtime after he gets through with the training.

🐾 Communication Tips for Training

Because animal behaviorists know something about canine communicative behavior, we can use it to facilitate training, to you and your dog's advantage. We
can especially use vocal communication as it relates to specific kinds of behaviors if we take advantage of some of the research applied animal behaviorists
have done.

Dr. Patricia McConnell looked at two different kinds of signals that are frequently used by trainers with lots of experience to see if they really had the
effect on dogs that the trainers wanted. Short, rapid repeated notes, such as
high-pitched whistles or words like "come-come-come-come," are most likely
to move dogs to action, preferably toward you.

Signals that inhibit or slow down behavior should probably be divided
into two classes. Frequently, you want the dog not to move or to remain exactly where he is. To keep him still, say, "Go-o-o-o-od bo-o-o-o-oy," so that
there is no frequency modulation, no change in frequency, up or down, which
is much more exciting to the dog than a low, steady tone. If you want to soothe
your dog, or slow him down after an exciting activity, say, "Slo-o-o-o-ow go-
o-o-o-od bo-o-o-o-oy."

DR. WRIGHT'S INSIGHTS

Do you want your adolescent dog to learn to behave quickly and painlessly? Use Dr.
Wright's Nearly Foolproof Formula for Canine Learning: Praise the good stuff, and ignore
the bad stuff. Unwanted behaviors that your adolescent dog seems to pick up in the
process of learning how to sit, for example, should just be ignored. You don't want to
scold your dog: That pays attention to the unfortunate things, and then she reacts against
you or overreacts trying to learn what she is supposed to do. As long as you ignore the
wrong things and don't make a big deal out of her mistakes, she will stop doing them;
she's trying to figure out what you want so that she can do the right thing. As soon as
she figures out that she doesn't get anything for misbehaving, she will try to do something else—the "just-right thing"—to get your attention and praise.

So if she starts to mouth or scratch the leash or jump up or complain (or do any of
those kinds of things) as an adolescent, you don't want to draw too much attention to them
so she won't fixate on them and make them a habit. It may be a little different if there is
snapping, but sometimes these kids will get into whining or complaining a little bit. Just
don't pay any attention to it. Your dog will quickly stop misbehaving if there is no payoff.

DR. WRIGHT'S INSIGHTS

When dogs communicate with other dogs, they use repeated whines to elicit an approach to one another. So it's okay if we say in a semipathetic, high-pitched, coaxing voice, "Come on! Come on! Come on!" As I am speaking this, my dog Lucy is looking at me from under the table, she is wagging her tail, and now she is up here wagging her tail, looking at me, and sitting down next to me.

On the other hand, if you want the dog to stop, then one short rapidly descending note is usually a good idea, according to Dr. McConnell. This is what I use with my dog. I say, "Wro-o-o-ng . . . Wro-o-o-ng," because with the word "wrong," the end sound, "ng," drops in tone just a little bit. And if I say "wrong" that way, my dog Lucy stops moving. A rule I learned from trainer Gary Wilkes is that it is easier to get her to stop when I say, "Wrong," than when I say, "No!" and try to get her to come to me. There is a mismatch between the sound of my voice and the behavior I am expecting.

It's much easier to get dogs to come to discrete high notes that are pitch-modulated ("come-come-come"). It is easier to get them to slow with kind of a continuous tone that doesn't modulate ("go-o-o-o-od bo-o-o-o-oy"). And it is easier to get them to stop with a tone that is really short but that drops off in a descending note at the end ("wro-o-o-ng"). When you get into obedience class, or you begin some obedience routines in your household with your puppy or adolescent dog, keep these vocalization signals in mind. Try them. You'll like them!

Want to Play? Dogs' Secret Signals

One of the things adolescent dogs like most is playing—as often and as enthusiastically as possible! And if yours is typical, she will invite you to play with her by using a "play-solicit" or an invitation to join her. This request generally takes the form of a play-bow: rump in the air, elbows on the ground, and an expectant smiling face mean that she is hoping you'll throw that stick or Frisbee. And if you don't, she will probably try again in a minute or two, because after all, dogs just want to have fun!

I knew a pair of dogs who had marvelous communicative signals. They were a neutered male Old English sheepdog about 3 years old and a 9-month-old German shepherd who already weighed about 65 pounds and was also a neutered male. Sometimes they would get into a scrap, but most of the time they were pretty good friends. Their play consisted of running around in the backyard, and the way they got into play was really interesting.

The German shepherd would be about 10 yards away from the Old English sheepdog in the backyard, facing him, and the shepherd would catch the sheepdog's eye. If the sheepdog didn't return that gaze, then they wouldn't play. But if the Old English looked back at the German shepherd, and they maintained their mutual gaze for a period of at least a second, then they would get into the game—one taking off and running, and the other chasing. First one would chase, and then the other would chase.

I noticed that they also had a ritual for ending the play session. Invariably, the Old English would signal that he was through playing by essentially "herding" the German shepherd into a pile of logs in the backyard and pinning him against the logs so that the shepherd couldn't move. Then the sheepdog would turn and walk away. If the shepherd didn't take the hint, he'd be pinned against the logs again. And if he still persisted, he would get a nip from the irritated sheepdog.

This is the way it works for a lot of dogs: The adults teach the adolescents, and the adolescents teach the puppies, "If you don't pay attention to my cutting off play behavior, then it is okay for me to nip you." In behaviorists' terms, that is the normal behavioral sequence of many dogs in stopping play.

Test Your Play Observation Skills

If you have a pair of dogs whom you can regularly observe, see if you can tell how they signal each other to start and stop playing. I have done this with my newest pet, the little "cockawawa" named Lucy. After being in our house for a couple of months, she was able to get any one of the three other dogs to chase her by first making eye contact with them. After she has their attention, her tail suddenly starts going a mile a minute, and she makes a quick huffing sound by breathing out. As soon as she does that, one of the other dogs takes off after her. She does the same thing with me and my wife; dogs have their own "personal" signals regardless of whether the recipient is of their own species or not!

Why Little Boys Get Bitten

It's a fact that 5- to 9-year-old boys are bitten by dogs more than any other group of human beings. And although each event is unique, I suspect that for many of the bites, the little boys persisted in bugging the dogs after the dogs indicated that they were through playing. We need to educate elementary-school kids better about some of the dog's natural behaviors that say, "I am through with play, leave me alone now." One is for the dog to just turn around and ignore you or to walk off and try to be by himself. If you keep bugging him, he is likely to swivel around and nip you because to him that is the normal way to say, "Stop!"

Adolescence in Cats

How do cats cope with adolescence? Left to themselves, as in the case of barn and feral cats, they have a pretty flexible social structure—a structure that begins to appear when the cats are only a month old. Typically, a mother cat will move her litter to a different nest site when the kittens are between 4 and 5 weeks of age, or when the birthing site becomes too soiled, or when an unfamiliar male becomes a nuisance. Weaning begins at about 4 weeks of age, at the same time free-ranging cats begin to bring their kittens prey, and household kittens accept moist "cat food." Kittens learn to prefer the food Mom eats, since they are exposed to the food's unique smell and taste through Mom's milk. Adolescent cats are social learners; they also learn to hunt particular kinds of prey from watching Mom, but they don't learn as well watching a sister or another adult hunt. Mom seems to be the preferred teacher. In some cases, after the mom cat moves her kids, the female kittens stay around and form a matriarchal group with the mother. Although it's rare, females may leave their home location, but the motivation is less competition for mates (as in males) and more likely to be related to competition over food resources (i.e., she only leaves her sisters behind when it becomes obvious that the catless neighbor has more and better food to offer). Females' home range—the area they commonly traverse—is much smaller than males'. These and other sex differences become apparent around puberty. Adolescent males may become peripheral members of one or more well-established groups, or they may leave the natal site and remain solitary throughout adulthood. It's usually these peripheral males, not the more well-established, older central males, who do a lot of marking (spraying). The testosterone in young male cats leads to the marking, in addition to lots of roaming around and fighting.

Although male kittens are tolerated by an "in group" and one or more central males, adolescent cats tend to disperse from their natal group with the beginning of puberty, and male-male competition over females starts at about 10 months of age. Thankfully, adolescent cats are quite adept at adapting to changes in social organization, and they learn to survive with changing circumstances. Their social groups may be disrupted when food and shelter from inclement weather become sparse, and the number of predators increases. But as long as the environment remains friendly, and if the breeding male and the female offspring from consecutive litters all stick with the mom, the group becomes pretty similar to what we would call an extended family.

Young, reproductively intact males may not prefer to remain indoors with older, experienced, grumpy toms. You can, however, make their life at home more palatable by spaying and neutering your cats. Don't become a cat hoarder (collector). There is no evidence that being "homebound" creates a problem for adolescent cats in a family of other cats, provided that all the family's cats have been neutered or spayed and there aren't an unreasonable number of cats in the home. Peace can be maintained as long as the cats disperse themselves throughout the house (don't forget vertical spacing!) and have control over their interactions with one another. The larger problem seems to be introducing a new cat into an existing family of cats.

Not enough research has been done in household situations to determine what the common social organization is among indoor cats, other than the one we described in chapter 4. But the age of the cats can determine the way they fit into your household. And, as we'll see, that fit can change as a new cat reaches adolescence.

We've discovered that cats don't become socially mature until they're at least 2 to 3 years old. So most cats mature at about $2\frac{1}{2}$ to $3\frac{1}{2}$ or even 4 years old, whereas sexual maturity (as we know!) can occur at about 6 months. So, as with dogs, it is helpful to distinguish between sexual maturity and social maturity—especially if you have a multiple-cat household. That's because most stable role relationships between two cats occur when they're not just sexually mature but also socially mature, and pretty comfortable with their roles in the household. Many dominant/subordinate relationships depend on one cat getting up and showing deference to the other one as she approaches. And that's okay for both critters. But not all cats establish role relationships with one another.

So, as a multiple-cat owner, you need to know the personality of your cats, whether they're likely to be sexually mature, and if they're confident enough in their roles. Signs that two cats have gotten past the uncertainty of adolescence include things like nose-to-nose approaches, cheek or head rubbing, and ear licking followed by flank rubbing. Then one cat will walk past the other one, presenting its rear end for sniffing, and then they do the whole routine all over again. That's a ritualistic kind of behavior that adult cats get into.

Cats certainly learn and become comfortable with what the options are as kittens and then as adolescents, and they finally firm things up as adults. So from the time they reach puberty until about 2½ years of age, they go through their adolescent period, during which they take on or try out different roles with one another—the roles that they will ultimately occupy as adults.

Some of the more popular roles cats take on include the Pariah ("I get picked on relentlessly by the other cats, even if I'm minding my own business"); the Athlete ("I jet around the house with all kinds of gymnastics and use your best furniture to practice contorting my body in strange ways during dismounts"); the Boss ("Don't even think of eating first, or doing anything before you pet me; now play with me; now feed me; now clean my litterbox—meow!"); the Player ("I'll bat at anything so watch out for my paw sticking out beneath the open door enticing you to play. Don't step on me!"); the Watch Cat ("No one had better darken our door, but if they do, I'll see them coming from the front window. Don't let them in!"); the Lover ("I love your lap, I love being petted, I love your gaze, and I love to purr whenever you're nearby"); the Macho Cat ("I strut around the house, daring you to touch me—or even look at me!"); and the Wimp ("Poor me!").

Does Sex Matter?

Males and females are different. There, I've said it! Sex differences matter in dogs and cats as well as in people. I know it's been more than "acceptable" for the last few decades to treat human girls and boys as though they were the same. That approach seems to be changing somewhat today, as differences between the sexes are becoming recognized once again as normal and healthy. We are beginning to recognize that each sex has strengths that can be built upon. Well, the same thing is pretty much true with dogs and cats.

When you think *male,* you might picture a sentry dog—alert, assertive, and bold. You also know that he will mark all over the house—dog or cat. Every time Fido lifts his leg on the dining room chair, or Buster sprays, you're reminded of their maleness. When you think *female,* attributes such as cute, affectionate, and nurturing to her owner may come to mind. You also undoubtedly know that she will draw all the males for miles around when she comes into heat, thanks to her female hormones and pheromones.

Cats and dogs show similar differences in *sexually dimorphic* behavior for "offensive" (overly assertive) aggression, home range, and marking (males outdo females in all three).

And then there are the neutered male and the spayed female dogs. Are they different animals altogether or just slightly "altered"? Bad pun aside, that's something that no one was asking before about 30 years ago, when dogs and cats were merely expected to fulfill their roles as pest eliminators without interference from humane societies or veterinarians with scalpels.

❦ Beware of Breedism and Sexism!

We've already talked about some of the misconceptions that can arise from breedism—the mistaken belief that all members of a given breed possess certain overriding traits. (For example, pit bulls are vicious, and Pekingese are nervous.) It's equally inaccurate to make generalizations about males and females. Despite opinions to the contrary, males and females are equally good at obedience and equally responsive to commands. Several trainers I greatly respect pass along the following advice: Don't let yourself use your male dog's sex as an excuse not to make him comply with your training. Just like breedism, sexism can and does creep into our assumptions about dogs and can lead to a self-fulfilling prophecy. ("Oh, he's just feeling his independence—boys will be boys!")

To some extent, we influence our pets to be more or less male-like or female-like based on our expectations of whatever we perceive their gender stereotype to be. Add to these stereotypes the beliefs people have about the way a particular breed should behave, and you have the same dilemma we applied animal behaviorists face when trying to figure out why Dusty or Daisy is behaving "that way."

So the next time you let your male Rottweiler get away with taking charge of a situation by being too independent or assertive, ask yourself what you would do if he were actually a female sheltie. Would you be more likely to "correct her" for the same indiscretion, confident that she should comply with your correction, and to scold her for being so bold? If so, perhaps some of the assertiveness, difficulty in obedience training, and other behaviors that seem to distinguish males and females of different breeds are based on the dogs' responses to *our* stereotypes of the way their breed/sex combination is "supposed" to behave!

Male and female dogs and cats are not different in any *real* behavioral way until they are postpubescent. All that "macho" play-fighting that people see

DR. WRIGHT'S INSIGHTS

Here's a fun little fact concerning the biological quality that makes a puppy or kitten female or male. Based on the activity of their hormones, all dogs and cats would become virtual females if testosterone was absent from their development prior to and shortly after birth—even XY chromosomal males. And even if the pups or kittens were hormonally normal in the womb, most males aren't able to begin being "all that they can be" sexually and aggressively until they hit puberty. Very simply, the hormone testosterone facilitates organizing your critter's male brain to influence its male life. Then, during puberty, the hormones activate again and really make the behaviors change. Ask any parent of a sixth-grader; it's the same idea.

little male dogs doing, and the "maternal" behavior of the female kitten licking your hand, stem pretty much from our own ideas about what is male and female. And as for which sex is "better," please don't ask me. I've had dogs and cats of both sexes, and I am sure my own stereotypes and preconceptions of sex-linked behavior came into play in molding how "male" and "female" they became.

🐾 Male, Female, or Neuter?

What are the differences between adolescent male and female dogs, anyway? There are many myths and stereotypes surrounding this question, and animal behaviorists are constantly striving to sift through the sexist baloney and arrive at some reasonable generalities. This question is central to all thought about dogs' gender-based behavior, and, of course, we are still struggling with the answers that fit males, females, and those neutered beings in between.

As with breeds, there is a wide variety of behaviors in dogs of both sexes. So don't hold me to what I'm about to say if your dog doesn't fit the mold, please! But here are some general guidelines that are backed up by enough research to make them ring true.

Ladies First

Let's start with the girls. Female dogs tend to stay closer to home and to have fewer episodes of aggression than male dogs. And they do like to lick. In part, this stems from licking their litters to stimulate defecation and to clean the puppies, but females are pretty good people-lickers, too. Many females like to care for their owners in a maternal sort of way. So if your idea of heaven is wet kisses from a devoted pooch, chances are you can expect more of them from a female dog.

In the absence of studies by animal behaviorists, a widely held stereotype persists that females are generally better than males in ease of obedience training and housebreaking and in how much affection they solicit or demand from their families (there's that lick again). Take your pick.

We do know, however, that the beginning of the heat or estrus cycle is designed to advertise the availability of a female to breed, and her part in attracting a male is to deposit pheromones—substances found in the urine—over as large an area as possible. So, females urinate more frequently (and as a result, in more locations) than when they're not in heat, but only for 2 or 3 weeks at a time, about twice a year. The cycle includes a bloody discharge that lasts about 9 days, followed by a clear discharge for another 9 days.

Boys Will Be Boys

What about the guys? We can pretty safely generalize about male dogs, too. They rank higher than females in terms of more reported bites to people, especially to young boys, who are bitten almost twice as often as little girls the

To Spay or Not to Spay?

The advantages of spaying a female include things like not having to put up with periodic bleeding—although there are now doggy diapers to address this problem—and the likelihood of accidents in the home during heat. A spayed dog or cat is likely to live a longer, healthier life. That's because in addition to reducing the stress and potential complications associated with pregnancy, giving birth, and rearing puppies and kittens, spayed females reduce or eliminate their risk of uterine, ovarian, and mammary cancer. And neutered males have a reduced risk of prostate problems, including cancer in later life.

Behaviorally, an intact (unspayed) female in heat means that any male dogs nearby can be counted on to go ballistic. And along with that comes a lot of urine-marking on your bushes and flower beds. If you keep your female dog inside during this time—as well you might, unless you're interested in adding to the zillions of litters of cute mutts languishing in the pens of the local humane organization—she will be more solicitous of your attention. In fact, she'll make quite a nuisance of herself—to say nothing of the unwelcome rubbing! Unless you have registered, show-quality purebred pets and are planning to both show and breed, you have no reason not to spay your female dog or cat—and plenty of good reasons to do it.

same age. This may be due to the nature of little boys, who are known for more risk-taking behaviors in general. (See "Why Little Boys Get Bitten" on page 144 for more on this.) Haven't we always associated little boys with snips and snails and pulling puppy dog tails?

Male dogs will get bigger, they will roam more, and they tend to roam farther afield than females. Some years back, I had direct evidence of this with a brother-and-sister duo who lived in the neighborhood with me and my 7-year-old male German shepherd, Zuk. One boxer, the male, used to trot past our house each morning, "doing house rounds." His sister would invariably follow only as far as our house and then meander back to her yard at the end of our street. These siblings who came from the same house had different roaming range—based, I believe, on sex. Males roam farther than females.

Sometimes I hear people use the terms *territory* and *range* interchangeably. But to animal behaviorists, they are two distinct areas. A territory is generally defined as that part of a dog's home range that the dog is willing to defend; the home range is potentially as far as a dog can roam but is usually some area traversed by a dog on a regular basis.

I once videotaped a friendly female cocker spaniel mixed breed being approached by a "walker" (my colleague) on a quiet side street. When the walker got within 20 feet of the dog, the cocker happily turned 180 degrees and trotted down the street in front of the guy for another 15 feet or so. When she reached "her" driveway, she turned abruptly around and growled and snapped at the walker, daring him to come closer. This sort of scenario is not news to letter carriers and meter readers. It was obvious that the driveway several feet behind the dog was "her" driveway, and that was where she perceived that her territory started, and where she became emotionally and behaviorally different.

Although both males and females will defend a territory, intact (unneutered) male dogs roam (use a home range) farther than neutered males or intact or spayed females. About 90 percent of owners whose male dogs roam report that their dog doesn't go as far once he has been neutered. It may be that the airborne odors of a female in heat, or urine marks of other dogs, or other stimuli are just not as attractive to the pooch and therefore not "worth" following after neutering. The same is true for cats.

If you don't neuter your male dog, the effect of the testosterone will be more roaming, more marking, and more assertive forms of aggression. When I was studying behavior genetics and aggression, I helped a colleague with some research involving house mice. The male winners of a number of mouse fights had

Should Your Dog or Cat Be Neutered?

Castration—or removal of the testicles—seems to decrease instances of future aggression in about 60 percent of dogs who are castrated, especially if it is done early, before the dog has a history of being aggressive. If aggression is already a well-established habit, neutering will reduce the fuel (hormone) but not change the behavior.

Neutering does tend to have an effect on what we behaviorists call emotionally reactive behaviors, such as when a person approaches a dog too closely for that dog's taste. The intact dog seems to react by barking, biting, or growling. That reaction is faster and lasts longer for an intact dog than a neutered dog. It will also usually take the intact dog longer to calm down after he has become aroused. Now there's a good reason to consider neutering before he has a chance to have a lot of experience with this kind of behavior.

In neutered dogs, the baseline level of reactivity seems to be lower, the reaction is slower, and the dogs don't reach as high an intensity. In other words, a neutered dog doesn't get as emotional about it. The reaction remains high for a shorter period of time, and the cool-down phase is also relatively quick in the neutered dog.

Neutering will help calm your adolescent dog, reduce aggression, and, of course, decrease the population of unwanted dogs. But neutering alone is not a cure-all for every complaint a male dog's owner can think of. It will set the stage for less testosterone-driven behavior, but positive results will depend on the right kind of training and handling by those around the pet.

Aggression in cats is not only affected by spaying and neutering but also by how the cats are treated as adolescents. Undesirable behaviors will be minimized if your cat is altered before the objectionable behavior becomes a habit. For both sexes, yowling and escape attempts are less frequent. Males won't roam as far if they do escape from the house because other cats' pheromones are just not that attractive anymore. Male cats are also less likely to fight other cats and come home wounded. Female cats don't have litters to protect, so maternal aggression isn't a problem. And male cats are less likely to spray. With all the high-arousal tendencies reduced, people are less likely to become a target for a spayed or neutered cat's emotional outbreaks. And it's a lot easier to build a friendly relationship with cats who don't have these highly reactive behaviors for you to contend with!

higher amounts of circulating testosterone than the male mice who lost. Although it's risky to generalize from one species to another, I can't help but wonder if marking is more frequent for some dogs, based on their history of aggression. It may also be that higher levels of testosterone cause more marking, which leads to that male becoming "top dog," or that the marking is an effect of that status.

At any rate, the hormone-driven behaviors in males include dominance aggression—where the dog wants to control you—as well as that 7-year-old-boy's behavior, and probably possessive aggression, where he is protecting or "owning" objects like toys and food. One study showed males to be more playful than females, but others have not, so that's not something that can be stated with any certainty.

In fact, it's important to bear in mind that for all these gender-based differences, we're talking about tendencies rather than absolutes. In other words, some female dogs are more aggressive than male dogs—even those from the same litter. So, just as there are major genetically related behavioral differences among dogs within a breed, there also tend to be pretty large genetic differences within each sex.

If you get any male dog and compare him to an "average" female dog—whatever that is—the odds are that the male can reasonably be expected to be more aggressive to his owner and to be more willing to attack other dogs, to mark more often, to roam farther, and to do more mounting and embarrassing (to humans) sniffing at people.

Neutering probably won't make a difference if your dog is starting to bite you when you reach for him beneath your bed or when he's submitting to you and you punish him anyway. That kind of defensive aggression happens because the dog is fearful, not because his hormones are raging out of control.

Neutering can make a big difference if your dog shows unprovoked aggression toward people and other dogs. It's true that male dogs tend to get along less well with other dogs. If you don't like the way your intact male dog is looking menacingly at your child or other dogs, or is beginning to show any kind of assertive, controlling, dominant-type behavior, then neutering should make a difference.

❧ The Third Gender

Today we try to take steps to curb the ever-flowing tide of unwanted canines and felines, so intact dogs and cats are usually spayed or neutered at some point along the way. When they aren't, there's often trouble ahead. In my practice, I usually run into dogs who are fulfilling their biological destinies—and making some kind of trouble for their families because of it. And when socio-sexual maturity rears its head—sometime between 1½ and 3 years in dogs and 6 to 15 months in cats—some people (even those with teens in the house, which should give people some kind of fair warning!) have a hard time coping with the changes. The only really sound reason not to spay or neuter in today's over-populated pet world is to carry on a breed. Owners who are not professional breeders need to look at a couple of factors before they take this route.

Want to Breed Your Dog or Cat?

Before you breed your pet, run down this checklist to see if he or she meets all the qualifications of a good breed candidate.

1. Is your cat or dog a purebred? (If you have a mixed breed, there is no way of predicting how those genes will combine.)
2. Did you get your pet from a reputable breeder, rather than a pet store, puppy mill, or elsewhere? (If you got it anywhere but from a breeder, chances are you can't prove that it is a purebred.)
3. Do the papers establish a pedigree of three to five generations? (If not, you have no way of knowing whether the breed standard is being upheld.)
4. Are there several animals with titles in the pedigree? (Your dog or cat should come from a line of show champions, grand champions, supreme grand champions, and the like. You're not being snobby by insisting on this—you need to know if there are conformation, obedience, or field titles to see if the breed standard is being upheld.)
5. Is your cat or dog health-certified? (Passing on genetic diseases or hip problems is not going to improve the breed.)
6. Does your dog or cat have a stable, loving temperament? (If every now and then he gets aggressive or fearful, don't breed him. I have enough work already!)

If you can't answer "Yes" to all six of these questions, then you need to stop and examine your motives. Remember that somewhere between 10 and 25 percent of all the kittens and puppies born annually are unwanted and destroyed. You would be better off rescuing a couple of pets from a local humane organization—which will probably insist on spaying or neutering them as a condition of adoption.

DR. WRIGHT'S INSIGHTS

Many people, unfortunately, try for years to replicate their beloved companion. I had a colleague who knew a woman who loved a cat so much that she got another when Frisky passed away, but he wasn't affectionate enough. So she got another, but he didn't purr and meow just like Frisky. So she got another. She ended up with 20 cats, trying to find one that was like the one she loved so much. And even though she got the same breed, she just couldn't replicate that cat. Neither will you. Instead, as the song says, love the pet you're with.

"Aha!" you say. "But what about cloning?" True, it's been done quite successfully with mammals now, and there are even a few pioneering sites on the Internet advertising cloning services for pets. If you have the money, wouldn't that be a way to make sure Bowser or Snowball doesn't ever leave you? Well, physically you might have a match. But there is a whole lot of difference in the outcome of personality traits and the outcome of behavioral tendencies based on how you rear your pet. So even if you have a clone, unless you rerun the tape of your life with the new dog or cat at your side, the clone will turn out differently—behaviorally speaking—from the original pet, even though they are genetically identical.

It's always best to appreciate your dog or cat for who he is, at the moment and throughout his life. And when the time comes, go out and look for another dog or cat to love in her own right, rather than trying to replicate the experience you might have had with your "perfect" pet.

My personal opinion is that if you are not a professional breeder, you ought to spay or neuter your pet (all of ours are neutered). Period. It's a lot safer for you and your dog or cat, much less hassle for you, and your pet is going to live a longer and healthier life because of it. Even if you were to breed your pet, there's no guarantee that any of the offspring would be anything like the dog or cat you love!

Social and Sexual Maturity

I mentioned that an age range for the onset of sexual maturity (puberty) in domestic dogs is 6 to 15 months, with most dogs beginning to become sexually mature between 7 and 10 months of age. (Some large or giant breeds of dogs take a little longer.)

Note that domestic dogs who live in a group or come together on a regular basis may come into puberty and be sexually mature between 6 and 9 months. However, they don't reach *social* maturity until they're between 18 and 24 months, the age that marks the end of adolescence. Dogs' confidence increases with age (fearful dogs excepted).

From the dog's perspective, he can now say, in effect, "Now that I am sexually mature, do I have the ability and experience to use my social status to be able to use my sex effectively to create offspring? Am I being taken seriously by other dogs? Do I have a better sense of myself and my status to be able to use my role-appropriate behaviors to get the best mate I can?"

One irritating behavior that occurs during adolescence is mounting. But many dogs, both males and females, start to clasp, mount, and thrust (how embarrassing—get off my arm!) in puppyhood. The male dogs are not sexually mature and can do no "damage" to a female dog in heat. So much of the normal mounting behavior you see in puppies is not sexual at all. And if you watch puppies play, you'll realize that mounting is a normal part of puppy development, where one pup is trying to control or manipulate the other.

At other ages, you can see some dogs who use mounting as a direct challenge to another dog ("I'm going to keep you from moving") or just as a communicatory gesture ("You really arouse me"). As with almost all social behaviors, the more socially mature the dog, the more clearly you can interpret the behavior.

Because age is such an important factor in the development of adultlike, male-appropriate behaviors, it's important to look at a few percentages on mounting and other "bothersome" behaviors here. Dr. Peter Borchelt's research on "sexually dimorphic behaviors" (behaviors that differ for males and females) shows that about 50 percent of male dogs mount by $5\frac{1}{2}$ months of age, 50 percent lift their legs by 8 months, and 50 percent urine-mark by 13 months. So urine-marking is clearly not just a sexual behavior—if it were, it would coincide with the beginning of sexual maturity. Marking is also related to the more socially mature dogs' being more confident to lift their legs and mark, especially if there is an older adult male dog present.

Dr. Borchelt's research also shows that about 90 percent of dogs urine-mark by 24 months of age, 95 percent leg-lift by 24 months, and 98 percent mount by 24 months. So it's pretty clear that between $1\frac{1}{2}$ and 2 years of age, many of the canine behaviors that people think indicate sexual (hormonal) maturity are done more often as the dogs also become more socially mature.

If your adolescent is getting a little more confident and you're wondering whether he is going to have these problems when he's an adult, it's a little too early to tell. You need to continue to train your dog and be with him, guide him in the right behaviors, and reinforce the appropriate activities and rela-

(continued on page 158)

DR. WRIGHT'S
CASEBOOK

McDuff

Sometimes gender-related issues can obscure other problems that a dog is coping with in the family. You must look beyond the obvious when seeking ways to understand why your dog is acting out.

There is growing evidence that in neutered males, interdog aggression (but not defensive aggression) declines, roaming declines, and urine-marking declines, but it doesn't disappear. And the more experience a dog has with these problem behaviors prior to castration, the longer it may take for a reduction to be noticeable—unless a behavior-modification program is initiated, in which case the prognosis for improvement is better.

That's what I was focusing on when I met Erin Houlihan, the proud owner of a lovely little Scottish terrier called McDuff. She was a young woman in her early twenties, and lived at home with her parents.

McDuff had been a model citizen until a few weeks earlier, when he began lifting his leg in the house and marking the things that seemed important to identify as crucial to his sense of self. Urine-marking is influenced by testosterone, but it also is related to the presence of other dogs, their urine, and the social experiences the pubescent male has with strange dogs. Male dogs who have never marked to advertise their territory may also begin marking when "strange" human males visit. I would need to find out if any of these factors had played a role as I donned my detective hat.

"That's really Erin's dog," her father smiled pleasantly after introductions. "We don't have much to do with the dog during the day." I noticed that neither of the parents made a move to pet the dog, although McDuff came up to each of them a couple of times while we talked, rear end wagging expectantly, but they didn't seem to notice.

I made a few notes. "Erin, are there other dogs in the neighborhood?" I noticed that the yards on this street were small and fenced in.

"Yes," she answered immediately. "There are a couple of mutts next door, and a shepherd out back behind our house. What do those dogs have to do with him peeing all over the house?"

"Well, that's one reason he is marking his turf," I replied. "He's warning the other dogs away from any females that might be around."

McDuff showed all the earmarks of an intact male dog feeling his oats upon sexual maturity. There was just one small problem with this explanation: McDuff had been neutered 6 months earlier.

Perhaps the marking had persisted because of purely social reasons. Could McDuff be trying to get someone's attention? "I just have one or two more things to ask you," I said to Erin's parents, feeling like Peter Falk in his Columbo trenchcoat. "Since you are around all day and Erin isn't, I thought you might be able to tell me if it looks as though McDuff might actually be forgetting his housetraining. It happens occasionally."

"No, sir," replied Mr. Houlihan. "That dog asks me or Mrs. Houlihan to let him out about 20 times a day."

"Twenty?!" I parroted dumbly. But now things began to fall into place. "So maybe part of McDuff's problem is that he's missing you during the day, Erin," I said, "and he has to mark all around and then go outside to attract some attention from somebody—anybody!"

It now seemed that the problem started out because there were dogs around the perimeter of his turf, which—from McDuff's perspective—"had" to be marked. Then the marking shifted to the house. This happens with intact males pretty often. And that's why we recommend neutering.

Now, if Erin had been around to talk McDuff out of marking the house, he might have been able to give up this unwanted behavior pattern. But instead, Erin left him alone for long hours after the neutering to attend clogging competitions. And so he tried to get attention from Mr. and Mrs. Houlihan. The fact that he bugged them 20 times a day about going outside was a huge tip-off that he was desperate for the kind of love and affection he got from his owner.

"I guess he fooled everybody," Erin commented, hugging McDuff. "McDuff wasn't trying to be a big macho male, after all—just a marshmallow!"

tionships with you, other people, and other dogs. Then, by the time he does get to be about 2 years old, you'll have a much better idea about the kind of dog he is going to be in adulthood.

Adolescence in dogs is a very difficult time, just as it is in humans.

🐾 Adolescence in Cats

Cats become socially mature when they're 2 to 4 years old. Both males and females may begin to feel their oats at this time, and new behavior problems may surface when there were none before. When one cat starts to feel more confident, he may attempt to restructure his relationship with another household cat. The result can be increased cat-to-cat aggression within the household, increased territoriality, and increased marking (spraying).

Neutering and spaying before the cats are sexually mature (and, therefore, before social maturity) can reduce the likelihood of these problems. Imagine that you have three cats living together in a totally "reproductively intact" household: a 2-year-old male cat, a 1½-year-old female cat experiencing her first heat, and a 3-year-old male—the cat who was there first. Everyone had gotten along wonderfully since the female had arrived about 15 months before, and they were all one happy family.

Now, the male cats fight, sometimes for access to the female cat but at other times, apparently, because of previous fights that resulted in a mutual dislike for one another. The female continues to be the source of sexually attractive pheromones that keep the males excited, but she often rebuffs the inexperienced males' attempts at mating, adding to the hostile environment. Together, these conditions are likely to bring down the house, with spitting, growling, fighting, yowling, and scratching, to say nothing of the pungent odor of male urine sprayed all over the walls and dust ruffles. Additionally, you have bites and scratches from trying to keep the boys from fighting.

Why spay or neuter? It quickly becomes obvious. Spaying keeps the female from coming into heat, and neutering reduces the amount of circulating testosterone in the males. After the cats are spayed or neutered, there will continue to be challenges to the stability of the cats' relationships with one another (since social maturity and confidence go hand in hand); however, the number of behavior problems and their frequency will be greatly reduced.

Adolescence is a difficult time for cats, too.

Chapter Eight

All Grown Up

Congratulations! You've made it through the terrors of kitten- or puppyhood and the changes of adolescence. Now's the time to consider what your dog or cat is going to do for the next decade or so. This is the prime of her life, and the next big set of changes won't come until old age begins to offer its own set of challenges.

So, you might be thinking, I'm glad we finally got here. I now know who this dog (or cat) is; his personality has fully formed; he's stable; I know what most of his good (and not-so-good) traits are. I have a pretty good idea about how sensitive he is, and how much correction I need to administer to keep him on the right track. Housetraining and neutering are behind us. He's finished cutting his teeth and testing out his claws, and I can even leave him alone for several hours at a time without his going nuts. It's not his first choice to be alone, but he's willing to accept the household routine. Now what?

Now is the time to pat yourself on the back and let out that deep breath you took almost 2 years ago. While you're at it, give that great cat or pooch a hug, and (getting a bit philosophical for a moment) think about helping him answer the question, as a popular rock band asks, "What's this life for?" You don't need to enroll your pet in the latest "Zen for Felines" course, although I have no doubt one exists somewhere, now that we've got aromatherapy and day spas for dogs. But you do need to think about what your pet's role in the family is going to be, based on everything you've learned about him in these first two exciting years.

What role, you might ask? He's a dog. He lies around and goes for walks and barks. Or, she's a cat. Her job is to lie in the windowsill! Well, that's a start, but it's not what is going to make for a great relationship between the two of you, nor will idle paws make a happy camper out of your dog or cat. And you're going to need to know what pleases him and motivates him if—heaven forbid—any serious behavior problem should arise in the future.

What's a Poor Dog to Do?

I'm thinking about a nice collie dog I met not too long ago. Champ didn't seem to have any real role in the family, although I did see him being used as a "pillow" by the two little girls in the house as they were watching TV. But my questioning of the parents, two very nice computer-programmer types, revealed that the 5-year-old collie had gotten into a rut. About all he did was follow various members of the family around the house or just watch them come and go. No one really involved him in any activities. He didn't play, wasn't much into exercise or retrieving things, and didn't seem to care much about giving or receiving affection. The owners fed him and took him for walks, but that was about it. So why was I there?

Well, it seems that Champ had nothing else to do, so he developed one bad habit. When a stranger would come to the house, what do you suppose Champ did?

Bark his head off and bare his teeth to frighten the guest away? Nope, Champ was a lovely host. He sat quietly or followed the visitor around, trying to get a whiff of this different—and potentially interesting!—person.

That was fine. But when the visitor would get up to leave, Champ would trot over and take the person's pants leg or skirt in his mouth, obviously trying to detain them. Maybe this person would play with him! I'm not sure that's what Champ was thinking, but it made perfect sense to me! I asked the owners if the dog had shown any sign of aggression, growling, force, or controlling behavior with their daughters or their friends. Absolutely not. He just had this one bad habit!

Here is where I had to put on my detective hat. Recall from the beginning of the book that the way to change a behavior from unwanted to "Hey, do that again!" is to have a number of things around that will motivate your dog or cat to want to do something. In other words, you need to have some stuff that makes him happy. Well, in Champ's family, no one could think of anything to put on that list.

No, he didn't really like food treats; he was finicky. He didn't really want to play. He didn't perk up his ears when he heard the magical (for most dogs) words "Let's go for a walk." When I asked the owners why they had gotten Champ in the first place, they explained that neither of them owned a dog when they were young. Besides, they wanted one to complete the suburban picture of two kids, four-bedroom colonial, minivan, and memories of a little boy on TV calling, "Lassie!"

You can imagine how difficult it is to work with a dog who doesn't seem to have much personality—and who has been shaped in the image of a very nice but very bland family whose members haven't introduced him to many activities. To get him out of a bad habit and into a healthy role in the household, I had to start from scratch with the family and convince them that the dog was a real member of the household who needed to be stimulated to care about doing things. Then he wouldn't be stuck trying to keep strangers from leaving because—invariably, it turns out—the strangers had made a little fuss over the dog and piqued his interest.

Everyone in this nice *Leave It To Beaver* type of family was very sweet, soft-spoken, and quiet. It was an atmosphere, quite frankly, that our friend Champ must have found just a tad too B-O-R-I-N-G. The dog was well fed and taken care of, and he had access to a nice backyard. But nothing was happening.

When I gave the parents and the two kids some options for jobs and roles their dog might enjoy, things improved immediately. The strangers were no longer so interesting when Champ knew he was going out in a minute for some fun. The family thought he didn't like to "play," I found out, because no one would throw a ball more than twice in his direction. So he would just go lie down. I gave Champ's family some suggestions that I will now give to you, in case your pooch is also a couch potato.

Give That Dog a Job

Dogs are good at a number of things besides snoozing by the fireplace or tearing up the house. I've found in my years of meeting pets and their families

DR. WRIGHT'S INSIGHTS

One of the things we behaviorists notice when dogs become behaviorally and emotionally distraught is that they have some need that is not being met, or that their owners won't permit them to meet. So if your dog has shown a strong preference to sleep or rest or play in a certain location, why not let him fulfill that need, unless it is absolutely out of the question for you? Dogs that are "spoiled," by the way, have not been shown to have more serious behavior problems than those who are treated with a rigid disregard for their preferences.

that the most successful and happy crews allow their dogs and cats to be themselves, that is, to follow their own natural interests and abilities to take on roles and even "jobs" in the family.

Let me begin by saying that I don't believe we can dictate individual dogs' or cats'—especially cats'—preferences to them as they settle into the daily routine that will take them through their adult years. One pet will want to sleep in the little bed you have so thoughtfully provided, whereas another has to be between the covers with her head on your pillow. One dog likes to lie where he can look out the window, whereas another would prefer resting in a den-like area under a table. One cat will only sleep on a soft, pliable surface; the next prefers cool, hard tile. One pooch prefers to play Frisbee; another likes to amuse himself with a chew toy.

One of the ways we can determine our cats' and dogs' social and psychological needs is to look at what choices our pet makes if given an option or an opportunity to choose. It doesn't mean, however, that we always must fulfill our pet's preferences if they are not consistent with what we want the dog or cat to be in the family or with our ability to fulfill them. But if we know what those preferences are, by offering the dog or cat choices and seeing how he responds, we can at least approximate fulfilling them to build a trusting relationship.

Dogs can be made to stay in one room or area or to sleep where you say. But if you really want to let the dog be part of the family, why not let him have access to the whole house? It's a freedom issue. You are not violating your pet's "dogness" by giving him the freedom to choose some things. (Obviously, the dog can be taught to stay off the furniture.)

Dogs will do their perceived duty—carry out their job—if we let them. And the job isn't restricted to simply guarding and protecting. My four dogs, for example, all have their work cut out for them, and nothing will dissuade them from carrying out their self-appointed tasks. Here's a sample: Peanut's job is to lie at the landing window all day waiting for "Mommy" to come home. The elderly terrier is the picture of patience. When she's out with us in the car, Peanut will sit at attention on the armrest, watching the door "Mommy" disappeared through until she returns.

Charlie, our chocolate Lab, has a different mission. She enjoys sharing. She will take a rawhide chew and try to place it in my wife's mouth, and Charlie won't eat the chew herself until she is assured that it is okay. Roo-Roo, our standard poodle, has made it her business to go around the house fluffing up all the available pillows and then plopping herself down. Lucy is the lover—she has a quota of laps to sit on and licks to give us daily. It's a division of labor all four dogs take quite seriously, and we try not to interfere too much (beyond trying to keep at least a few pillows out of harm's way!).

🐾 Dogs Were Bred to Work

Most purebred dogs and mutts alike are chosen for companionship these days, but the need to fulfill their biological destiny—that is, the things they were bred to do originally—may still have left them prepared to do certain jobs. Whether it's ridding the ship of rats, helping the master retrieve a duck, pulling a sled, or rounding up cattle, many breeds have always shown a penchant for hard work and plenty of it. But today, the most they're usually asked to do involves sitting while the master opens a can of designer dog food.

The word <u>terrier</u> comes from the French <u>terre</u>, which means earth. These dogs were bred to dig! If your terrier displays a tendency to dig up the backyard—or the carpet!—give him a special spot in the backyard where he can get all his digging out of his system.

DR. WRIGHT'S INSIGHTS

Study the AKC breed books to see what your dog's natural proclivities are. If you have a Jack Russell terrier, who was bred as a ratter, let him have a small patch of ground in the backyard to dig. You might even dig a few holes while you're planting your garden—well away from that area, of course!—and bury a couple of dog biscuits for him to "rat" out! Let your hunting dog develop his innate sense of smell by hiding a t-shirt you've worn somewhere in the yard or house and rewarding him for tracking it down. Being allowed to do whatever his genes have programmed him to do will thrill your pedigreed pal.

Dogs are sometimes forced to take their genetic preparedness (many behaviorists have backed away from using the word *instinct*, it seems too "automatic") and do something with it—some job or other that will interest them and perhaps even please their owners. I am not just talking about "real" working dogs such as guide dogs, drug-sniffing police dogs, or trained disaster workers. I mean household pets who are looking for some socially significant way to spend their days and, indeed, their lives.

See if your pooch can sink his or her teeth into one of the following jobs for grown-up canines.

🐾 Sentry Duty

Many dogs will want to take on the role of watchdog or warning dog. This is mostly a "barking and running from window to window" thing, and it seems to be a natural role for many of them, from the tiniest yipper to the big fellow with the deep and scary bark!

The Watchdog

When my sister, Judi, adopted her dog, Scout, from a shelter, she brought the 6-year-old golden retriever home as a 9th-birthday present for my niece, Erin. Judi observed that, during the first week, Scout greeted everyone at his new address with equal enthusiasm—her husband Tom, the UPS man, a neighbor kid, it didn't matter. "He doesn't seem to have any protective tendencies toward us," she mused.

I believed otherwise. "Just wait another week or so," I told her on the phone. "My hunch is that even though he's a typical friendly golden, Scout will start to distinguish between 'family' and 'stranger' pretty soon."

Sure enough, within 2 weeks, Judi reported that Scout had begun barking enthusiastically at any glimpse or sound of "intruders" invading his turf. This was fine with Judi, whose property is completely secluded from the neighbors. So she endured the momentary racket and praised Scout for being a "good boy" when he barked to announce someone in the yard or at the door. Sometimes, she would even go to the window and point out a "stranger" walking up the driveway to encourage Scout to bark. And true to his friendly golden ways, Scout's bark was truly worse than his bite: The intruders invariably would be greeted with a lick or a wagging tail when they were actually let into the house.

Scout's behavior was typical of a watchdog. This is the dog who announces all visitors or suspicious activity on the property. He is permitted or encouraged to bark as long and as loudly as he wants to, in the interest of alerting the owner and providing a feeling of security in the household. It is a nice idea to reward your watchdog with a pat, an encouraging word, or a treat—although if you don't, she will probably bark the next time the meter reader comes up to the garage anyway.

The Warning Dog

If a full-blown watchdog is more than you need, a modification of that job might suit some families a bit better. You may wish to mold your dog's natural behavior—his preparedness to bark his head off at the sight or sound of a possible intruder—toward a very minimal kind of lookout. Teach him to be more of an announcer than a car alarm! I call this kind of canine the warning dog.

The warning dog has a role that will appeal to people who like or need a very restrained or abbreviated type of announcer in their canine friend. People who live in apartments or other close quarters, have sleeping babies, or simply don't like the sound of a dog barking—even when it's their own—will want to teach their sentry to limit his warning barks to one or two.

Unlike the watchdog, who will usually run to the door, the warning dog can be encouraged to come to you when he is announcing a visitor, not unlike a well-trained butler. I've even seen dogs who have been trained to ring a bell, and not bark at all, when someone approaches. That's a little too undoglike for me! What's next, a tray of hors d'ouvres balanced on his nose?

At any rate, taking a cue from your calm demeanor, the warning dog does not assume the emotional mantle of preparing for battle with an intruder. A treat in the pocket and a firm "Rocky, come!" after the first bark will encourage this pattern to take hold. You can add, "No bark!" and instruct your dog to sit.

(continued on page 168)

DR. WRIGHT'S
C A S E B O O K

Lucky

Remember that, as good as you are at reading your dog's communicative signals, the dog is apt to be just as skilled at reading your own—even when you don't realize you are sending them her way! Lucky was one such dog, and I was glad to have some knowledge of "people" psychology to help me pick up on the clues in this case. She was a mixed-breed female with a pedigree that spelled trouble for anyone who might challenge the welfare of her owner, a young woman named Jodi who lived in an artsy area of Atlanta in a fourth-floor walkup. Part pit bull and part chow, Lucky had taken it upon herself to be watchdog extraordinaire. Jodi was afraid that the dog was about to take it to the next level—she had already bitten Jodi's boyfriend when he tried to push Lucky into her crate.

"I don't really want to crate Lucky when I'm away," the young graphic artist explained as we sat in her rather dilapidated apartment with the dog at her feet, receiving loving strokes as we talked. "But the landlord comes in with workers all the time—I've been having some plumbing problems and they're redoing the bathroom—and I'm afraid of Lucky attacking somebody. I can't afford to be sued, either."

"She is drawing complaints for barking when I'm not here if she hears something," Jodi noted. "But it's when I am here and she's not even in the crate that she really goes nuts." She bent down to pat the dog. "It's a scary place, isn't it, Lucky?" she said to the brown dog, who eyed me suspiciously as she licked her owner's hand in agreement.

That aside to her dog interested me. I had a hunch about something.

"Have you met many people in the building yet?" I asked her.

"Not a soul," she confessed. "I'm not that outgoing," she smiled shyly.

"Do you feel safe here?" I had seen better neighborhoods for timid single girls, and her earlier remark had piqued my interest.

Jodi sighed. "Well, if the truth be told, Dr. Wright, I'm really not used to everybody traipsing up and down the stairs like they do here, and I've seen some pretty scary types going in and out. The rent was affordable though."

Her answer was what I had expected, and I told her my thoughts. Her dog was able to pick up cues from her, and they acted as incentives for Lucky to do her "job"—protecting her fragile owner from all possible perpetrators lurking on the staircase. I asked Jodi if this made any sense to her.

Just as I had suspected, the dog was merely doing her job, responding to the body language of her owner as the young woman tensed and glanced uneasily at the door at the sound of a passerby.

"You probably don't even realize it," I told her, "but you are giving your dog some nonverbal instructions to go ahead and scare away that intruder."

This case called for a little behavior therapy for the owner as well as the dog. I suggested to Jodi that she practice some deep-breathing exercises to use when she heard that proverbial bump in the night. This would take away one of the cues her dog was using—the tense, fast breathing of her owner—and Lucky would see that the footsteps were just another sound, not something to react to.

There was nothing intrinsically provocative about the noises outside the apartment. If they weren't treated as anything out of the ordinary, the dog would learn to ignore them. We talked for a moment about how the neighborhood was really very safe, just a bit more active and flamboyant than Jodi was used to.

Jodi was able to change Lucky's overprotective behavior when she changed her own unconscious signals to the dog. And instead of reprimanding Lucky for overreacting to noises, Jodi spent her time calming Lucky and giving her a little treat for compliance, which probably made the dog feel much better than being yelled at for doing what she felt she had to do. To this day, I'm sure that Lucky is not about to let anyone mess with her owner. And I bet no one is stupid enough to try, either!

So the dog says, "Okay, I told you someone is there. Now I'll stay here, and you take care of it."

A warning dog is a wonderful pet to have and is probably safer than a watchdog. And your role as leader of the family—taking care of business, as it were—is reinforced when you ask the dog to fulfill his sentry role in this way. Anyone who has been in the "visitor" role and opened the door to the resident dog barking wildly or jumping all over them in a friendly frenzy—and that's most of us—will probably agree.

The Four Traits of Highly Effective Warning Dogs

What kind of personality makes for a good warning dog? You might think that a very excitable canine would fit the job description to a T. But hold on! That kind of pooch may be high in motivation, all right, but she is also the type you'll have a hard time shutting up. And she's going to be running back and forth raising hell. Most of the toy breeds tend to be quite excitable; maybe they're showing the bravado necessary to avoid being stepped on by the threatening intruder. In any case, you can admire their sentry abilities, but don't expect them to stop alerting you after one or two yaps!

Setting excitability aside, I believe that there are four predictive personality traits that enable some kinds of dogs to assume the role of good watchdogs or warning dogs readily.

1. Enough intelligence to understand the task. This is not to say that some dogs are several crumbs short of a Milk-Bone, but those who rate higher in obedience will have an easier time controlling themselves.

2. Willingness to bark. Don't laugh; not all dogs like to disturb the peace! My standard poodle hears a noise and gazes at me as if to say, "Oh, gosh, I suppose there might be an intruder out there, but why bother? Let someone else tell him."

3. Willingness to stop barking. This is a much tougher issue. So is the motivation to delegate the visitor to you to handle. Some dogs' attitude is, "Yeah, I see him, but I'll be damned if I'll turn him over to you!"

4. Playfulness. If you can make it a little game, many dogs who like to play will respond with enthusiasm to the idea of barking (once or twice), then coming, sitting, and getting a pat on the head, a nice "Good boy!" and a yummy treat.

Of course, either of these common scenarios is better than walking in on a guard dog.

The Guard Dog

With any job a dog takes on, there is potential for behavior problems. As someone who has been called on to evaluate killer dogs upon occasion, it's my opinion that anyone who intentionally or unintentionally hands out the guard dog hat to a pet is asking for trouble. The guard dog's job description calls not only for announcing the intruder's arrival as viciously as possible—maybe he'll be scared away!—but also for attacking anyone who happens to trespass beyond some point that the dog thinks of as the border he is defending. If it's someone's front door, he decides that the paperboy isn't coming in, or maybe that he isn't even going to be allowed to stand on the front porch.

Guard dogs are often trained to attack "human" figures, and they may also be taught to attack on command. In some areas, they do an effective job of keeping businesses from being burglarized. But I must go along with the hu-

There's never a reason for a family to own a guard dog—a dog who believes his job is to attack "intruders." Usually, a warning dog is enough. In a bad neighborhood or isolated area, a watchdog will give plenty of protection.

Jobs for Cats?

In this discussion of roles and jobs, I don't mean to give short shrift to cats. Although guard cats are rare—despite the ubiquitous "This House Is Guarded by an Attack Cat" poster—many if not most cats consider it their solemn duty to spend hours perched on the windowsill looking for intruders of the feline kind. They will let you know about an intruder either with a variety of hisses and growls or by racing from room to room. Human visitors elicit little more than a yawn, generally speaking, from our feline friends.

We can give our dogs more jobs than we can give our cats mainly because they have been in the household and with us as companion animals longer than cats have. The first cat show in North America was held in Madison Square Garden in 1895—it was won by a Maine coon cat—but few felines outside the show ring in those days would have majored in anything but "mousing."

I believe that most cats started becoming more than just barnyard or outdoor scavengers around the 1960s, when cat litter was first marketed commercially. So, literally, the cats came from the outdoors in. And that's when we started to pay attention to the kinds of traits and personality characteristics that we wanted in our cats, and started noticing the different kinds of things they did in the household, like purring for pure pleasure.

mane organizations that have declared that it is simply wrong to keep dogs solely for protection. It is dangerous to people and unfair to the dog, who is socialized to see humans as threatening objects rather than pals. Let businesses and individuals who want protection hire security officers or install a good alarm system instead.

Unfortunately, many dogs choose "guard dog" for their job in spite of the best intentions of their owners, who just hope Prince won't bite the child selling Girl Scout cookies when she knocks on the front door. These owners don't have verbal or physical control over their dogs, and that's not good. The insurance companies don't like it, the visitors don't like it, and the owners don't like it. Some people don't want to admit to themselves that they have a problem, and those are the people I wind up seeing. (They are quite correct not to try to solve this by themselves.) The first time the dog bites someone, they think it's an accident; the second time, hmmmm, maybe we have a problem here; and if it happens a third time, they give me a call. I'll talk about how I deal with the guard-dog mentality in chapter 13 and what you can do if your dog shows this tendency. Luckily, for every canine drawing a line in the sand, there are two or three who jump over it to lick you all over. And lots of cats are giving loving licks with their little rough tongues, too.

🐾 Loving and Caring

Of course, some dogs are lovers, not fighters. And in my experience, they tend to prefer two types of "jobs," which I call the lover-slug and the caregiver.

The Lover-Slug

If you have a dog who plays the lover-slug role, you'll know who he is. We happen to have one. She's our cockawawa Lucy—a combination of cocker spaniel and Chihuahua. If you want to get a purebred dog that has a really high demand for affection, loves to seek you out for licking, and adores having you stroke and pet him, I believe the companion-dog or lapdog breeds are most suitable. Probably the Lhasa apso, the Boston terrier, the cocker spaniel, and the miniature and toy poodle are at the top in their demand for affection, as is the Chihuahua.

With these purebreds and a myriad of mutts, you have a good chance of getting a lap sitter. You'll need a lot of free time because this pet's job is to come up on the bed or sofa for strokes all the time. But when he smiles at you with a silly kind of grin, you feel like the neatest guy in the whole world. These dogs pretty easily fit into the role of lover-slug of the family.

You may find that dogs who need a lot of affection become a bit too "needy."

Some dogs always need to be taken care of. They must be stroked constantly, or they'll drive you crazy pushing your hand with their nose. "Please take me with you! Please give me something! Please do something with me! Please let me do that with you!"

I've met with more than one client who in desperation hot-wired the bedroom just to gain a bit of respite from the overly attached pooch. Instead of installing "underground" electric fences, these desperate owners have been known to install the wiring around doorframes or under the carpeting in areas where they are seeking respite from their needy dogs' demands. Pretty "shocking" idea, isn't it? Not one I would recommend, when you can teach your dog how to respect your limits instead.

DR. WRIGHT'S INSIGHTS

Cats are often lovers, too. The more social cats even tend to vocalize the way dogs do: "Mmrrrro-o-ow . . . Ple-e-e-a-ase! Pet me!!" You can probably recognize much more care-soliciting behavior in cats. When they come up and do a nose check, or they rub you and rub you again, or come up looking for petting and stroking all the time, they are asking for loving attention.

The Caregiver

On the other side of the coin is the animal who wants to take care of you.

Certainly some of our cats and dogs take better care of us than others. If your dog is the caretaker type, she'll follow you, and, of course, if something falls down from the countertop in the kitchen, she'll eat it and clean it up for you. If you sit down on the couch, she'll get up there with you if she's allowed. She will put her chin on your lap and lick your arm. And if you're feeling bad, and you're starting to tear up a little bit or starting to get angry, she'll come around, look at you, and tilt her head sideways, as if trying to figure out how she can help.

Some may think I am anthropomorphizing a bit too much here (attributing inappropriate human characteristics to animals), but at this point in time, we scientists are inclined to admit that our pooches haven't been called "man's best friend" for centuries for nothing!

And though there are more than 400 breeds of dog, they all share 80 to 90 percent of the human genetic code. "So closely entwined are their lives," said science writer Mark Derr in the *New York Times* recently, "that a number of evolutionary biologists argue that dogs and humans have evolved together." So perhaps dogs *do* feel our pain and our joy. They may whine; they may put their paw on you as if to reassure you, asking for reassurance at the same time. So these caretakers are likely to try to groom you and likely to try to stay with you, and they seem to be sensitive about your emotional state.

It is part of the social nature of dogs and cats to take care of the young of their species, littermates, and one another, and it's a natural social thing for them to do the same with us. Nonetheless, some dogs fit much more easily into the role of caretaker than others. Females are more likely than males to take on caretaking jobs, it seems.

Certainly, the service dog who orients the better part of her day around her partner demonstrates the capacity of some special dogs to be devoted to others. You don't want to take advantage of this type of animal's good nature by making her into a virtual slave—fetching slippers, newspapers, and so forth— although the dog would gladly do that for you if a pat on the head served as a tip. Instead, you might consider bringing her to an eldercare facility a few times a week to bond with some of the residents. Both would benefit from this experience—a "real" job everyone could be proud of.

As far as cats are concerned, we have the same sort of thing. Some cats wouldn't be caught dead sitting on your lap or taking care of you or being close to you or trying to lick you. You can't get away from others. They constantly want to lick you, groom you on different parts of your body, follow you around, knead on you (simulating pressing on the mother cat's nipple to draw out the milk), and so on. I know more than one writer who can't seem to put a word on the computer without her "security blanket"—the ever-present cat purring on her lap.

🐾 The Hunter

The hunter is a role both dogs and cats can play. But they tend to view the job differently. Hunting cats tend to be born stalkers, whereas, for dogs, the point of hunting is the thrill of the chase. This behavior is only natural because dogs originally hunted in packs, chasing down their prey, whereas cats in the wild are generally solitary hunters.

The Stalker

Quite often, cats will take on the genetically primed role of hunter. They will capture things like crickets, cockroaches, or mice; carry them; and put them on your step. This behavior is perhaps a combination hunter/maternal caretaker role. We are not sure exactly what the role is, but we do know its job description: to find the most disgusting, smelly, recently dead thing possible and place it on something that the owner would least want it to touch, like his pillow, or up on the bathroom counter next to her toothbrush. Sometimes, the cat will put the dead bird right on your top step so that you are sure to tread on it when you go outside. At any rate, this is something that cats do, and for the most part, dogs don't do. When dogs catch something, they are more likely to try to eat it right there or to bury it for later.

It may not look pretty, but your cat is giving you a gift. Thank her, and dispose of it later.

The Thrill of the Chase

One of the things that dogs do quite well is chase, whether they are after a squirrel, another dog, a car, or a ball. Dogs typically like to look at something and see if they can run at it fast enough to get it, or—if it is running—if they can chase after it.

Cats are more likely to stalk or ambush than dogs are. And their typical scenario is to get as close as they can and then suddenly leap or run at the prey. So, even though both the cat and the dog may fulfill the role of hunter in the household, it may be that the cat is the head ambusher, and the dog is the chief chaser.

The Indoor Hunter

Dogs have been known to watch squirrels from inside the house in a very active mode. The dog will bark, roar, and/or claw at the windowsill in frustration, driving her owners crazy. Cats, on the other hand, will probably lie in wait and stare right into the eyes of the critter. They will just continue to freeze and watch it. So the style of waiting until it's time either to chase or to ambush is completely different, even though both dog and cat may be completely indoor pets. Dogs are more pursuit-oriented, whereas cats have much more stealth technology, trying not to be detected by the prey until the last minute, when they can pounce. Cats' and dogs' hunting behaviors set the stage for the different ways they both like to play.

🐾 The Player: Pets Just Wanna Have Fun

Cats and dogs generally love play. But there are certainly different items and styles of play for dogs and cats, even though you may recognize the personality trait of "player" in both species in your home.

Dog play—like their hunting—sometimes involves running after or chasing something and trying to chew it after they get it. Their credo seems to be: Chew it up, destroy it, rip it apart, do as many things as you can to it. Then eat it, if it is even marginally edible.

Cats, on the other hand, tend to want to chase and bat and walk past a toy and then attack it. They put the object between all four paws while they're on their back, rotate it around, bite it, and run under the bed from it. So they don't tend to wreck things as much, although over time, they can do some damage to even the best-sewn little mice and catnip toys. But I think the strategy of cats is to go after it, pounce on it, attack it, and play with it. Sometimes people say this is "cruel," but a cat cannot be expected to empathize with her prey.

A dog's play involves going to get something and then putting it in his

DR. WRIGHT'S INSIGHTS

I would say that almost all dogs will play, given the opportunity. But some breeds are more dedicated to the pursuit of pleasure than others. These playful pooches are likely to include the miniature poodle, golden retriever, Shetland sheepdog, Cairn terrier, Airedale terrier, miniature schnauzer, and English springer spaniel.

If you really enjoy playing with your dog and you want a breed that takes on that role, don't get a bulldog, a Chow Chow, or a bloodhound. These breeds will probably have to be persuaded and cajoled to go in the backyard for a game of catch. A basset hound is going to just lie down and look at you with his chin on the floor. Then he'll look up with those tired eyes and say, "What, me play?" Saint Bernards aren't known for playing, and neither are the malamute, Akita, or Rottweiller, which are probably way too busy guarding the door to even think of messing around.

Some dogs, like goldens and Labs, seem to specialize in playing with children. Others want the word "kids" removed from their job description. Dogs who are not crazy about kids include miniature schnauzers, Westies, Chows, and Pomeranians. Even toy poodles are not too thrilled by munchkins. Most child-averse dogs are fairly small breeds (except for the Chow).

How to Play Kitty Tag

If you think cats aren't interested in organized games with their owners, think again. Here's a game that my sister, Judi, and her calico cat, Sassy, played for years.

Sassy would get on the landing to the staircase, about five steps up from the bottom. Judi had to slink, catlike, onto the bottom step (well, part of her, anyway). If she could make eye contact, Sassy was certain to crouch down as well, staring intently at Judi's eyes, motionless. Then she would suddenly run down the steps, "tag" Judi's arm or cheek with her paw, and run back up to the landing. This would continue until Judi had to go get dinner, or Sassy walked away, which signaled the end of the game.

After about 6 years, the family moved to a different house, one with a regular staircase that had no landing part way up. Would the game continue? It did. Sassy simply adapted to her new playing field by "hiding" around the corner at the top of the stairs as Judi crawled slowly up them. Sassy would run down and tag her, and the game was on.

Now you try it with your cat. Assume the position (crouched down, slinking slowly up the stairs, staring into your pet's eyes), and see what happens! But before you do, I should explain that Sassy was a gentle, playful cat who was declawed. The "tag" was a soft bat with her paw. So, needless to say, tag is not a good game for the kitty who responds with a hiss, scratch, growl, or bite. You know your cat best.

mouth and chewing it, or chasing it and bringing it back to you. In the job description of cats, you will very, very rarely find a cat who will actually bring you something as part of a game for you to toss it again.

Imagine a cat playing with a favorite toy, something as simple as a string being pulled along the ground, or one of those little plastic milk-bottle tops that she can bat around and attack and bite. A piece of yarn becomes a "mouse," and a rolled-up ball of aluminum foil can send her literally bouncing off the walls with excitement as she bats it around the room.

Now imagine your dog looking at the string you're pulling on the ground, looking up at you, and thinking, "What, are you nuts?" If he sees one of the plastic milk lids on the ground, he might go up and sniff it, maybe put his nose on it, pick it up if he's a rare individual, chew it to see if it tastes good, and then spit it out and walk away.

Both my dog Charlie and my cat Domino like tennis balls. But Charlie likes to chase the ball, bring it to me in her mouth, and put it on the bed. Sometimes she munches on it for 5 or 10 seconds before she drops it; then she wants me to pick it up and share it with her. When it's in her mouth and I pull it a little, it's not a tug-of-war, it's just sharing. Then she will give me the ball, and if I toss it she'll go chase it.

On the other hand, if I put the tennis ball on Domino's bath mat in front of the garden tub on our deck, she doesn't want that thing there. So she will run over, jump on the ball, bite it, take it in her four paws, and then kick it off the rug. I can put it back on there four or five times, and each time she kicks it or bats it off there. It's a routine she loves. As part of her play, that ball does not "belong" on that rug.

Of course, Domino can't put a tennis ball in her mouth—not that she would be caught with a ball in her mouth!—because anything that would fit would be too small for the dogs to play with safely. So we have to be quite careful and think about playing and fulfilling our dogs' and cats' roles differently, while not placing them at risk in the meantime.

My sister Judi's dog, Scout, loved tennis balls, too. In fact, that was the only thing the rescue person told Judi about the dog when she took him home at the age of 6, tennis ball in mouth. But Scout's favorite thing about tennis balls was the fact that my nephew, John, was a ranked Junior and was in the habit of hitting tennis balls against the garage door, with Scout as his ball boy, er, dog. This was the perfect job for a retriever! Scout liked to roam

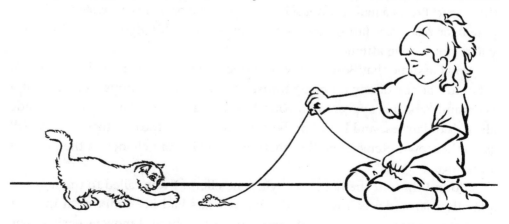

Cats will be eager to play with you if you make the game lifelike. Pull a toy away from the cat, not toward him or you (to avoid redirecting a playful attack on your leg), and pull it in jerky motions, as though the "mouse" were running in short bursts and then pausing. Your cat will love it!

through the pachysandra, tail twirling like a helicopter blade, until he ze-roed in on a yellow ball. He sometimes forgot to return them to little John, but he sure could track them down!

Is Your Dog a Jock?

A subcategory of Players is Jocks. These are the dogs who jog with their owners. If your dog falls into this category, be sure not to tire him—his legs are a lot shorter than yours! Make sure that you provide your dog with fre-quent water breaks because he has no way to cool down easily. Canines lack the sweat glands we depend on to keep our bodies cool during exercise.

If your dog seems to enjoy sports, consider enrolling him in agility training. There are agility trials in which the dog has to go over a series of jumps and perform other physical challenges. Doggie Frisbee is a well-established sport in every big city in America. Competitive types might enjoy a flyball tourna-ment. I have seen dogs pulling skateboards and little kids in wagons. Working dogs (make sure his hips are sound!) make child's play of physical challenges and come back panting for more.

Changing Relationships: Who's on Top?

While you are defining roles and trying out jobs for your adult dogs and cats, don't forget that some relationships between pets are subject to change. If you don't keep on top of things, trouble can erupt seemingly out of nowhere. Things are usually fairly stable in cat/dog households—the cat is the boss and that's that! But in a multicat household, where two or more cats reside and com-pete over resources (and, yes—jobs) things can suddenly get out of hand if you're not paying attention.

Don't assume that Bosco is the boss because he is bigger, or that Snowball "should" be the dominant one because she has ruled the house for the past 5 years and Babe just appeared on your doorstep last week. You need to set aside all rules of fairness and logic and be ready to observe the cats together, as well as on their own. Remember, too, that relationships can change in the blink of a cat's eye.

I found this out the hard way with one client. I was called to the home of a family that had recently moved. The two cats had been mixing it up ever since the family had settled into their new digs about 3 months earlier. The family told me that the black cat had always been the dominant one, and that the tiger cat had become uppity. She had taken over the foot of the bed (horrors!), wanted to eat first, and wouldn't sit still for any less attention than her black counterpart. When I took on this case, I wasted several days trying to reinforce behaviors that had been established at the old house, where the black cat was the lover-slug and the tiger was essentially jobless.

When my clients dutifully followed my instructions, giving ever more attention and greater perks to the black cat, the situation got worse and worse. Soon the cats were having hissy fits day and night. Finally, it dawned on us that there had been a change in their relationship. As soon as things were reversed—now the way was smoothed for the tiger cat to get a social promotion, as it were—the complaining became less frequent, and the cats eventually settled into a comfortable new relationship in which the black one grudgingly relinquished his hold on the top-cat spot.

This kind of role reversal doesn't happen too often, in my experience, but it takes keen observation and often some rewriting of the job descriptions on the part of the family to make it work. Of course, some families aren't going to rewrite the pet's job description no matter what!

It's important to acknowledge the different roles that our dogs and cats play in the family. Allow your pets to try on roles, get something out of them, and get some pleasure out of doing them. We do! It's almost the same as your pleasure in being a grandmom or a dad, a golfer or a gardener, a husband or a small-business owner.

Chapter
Nine

Your Older Pet

You and your pets have shared great times for years. Then one day, you wake up and look at your dog and notice that his golden face has turned snowy white. Or your spunky "kitten" who's older than your teenagers has stopped pouncing at the chipmunks who live under the front step and settles for tracking them with her eyes.

Your pets are growing older. So now let's look at how we can help them make this new and final adventure just as comfortable, loving, and interesting as their cute puppy- or kittenhood, exuberant adolescence, and happy, active adulthood.

Let me begin by sharing the good news that research is going full tilt into improving the health of cats and dogs as we learn more about their genetic and physiological makeup and how to prevent and treat deadly or disabling diseases and slow the aging process. Just as in humans, the search for genetic mutations has speeded up in recent years, leading scientists to identify nearly 400 genetic mutations in dogs (these tend to be breed-specific).

The Dog Genome Project has identified about two-thirds of dogs' 78 chromosomes, and during the last decade more than 20 specific canine disease genes have been identified and characterized. The American Kennel Club's Canine Health Foundation and 60 breed clubs are working toward eliminating all single-gene disorders in dogs through genetic testing and selective breeding. They hope to achieve this goal in the next 20 years. Hereditary blindness in poodles, narcolepsy in Dobermans, and the horrible hip dysplasia in goldens, along with epilepsy, cancer, heart disease, and bleeding disorders, have all been targeted in research labs around the world.

Scientists are working to understand the genetics of cats as well, with an eye toward adding a few more lives to the traditional nine. And, of course, we behaviorists are continuing to observe and study our best friends as we try to figure out what makes them tick. We're also working with owners to improve their understanding of dogs and cats at all stages of their lives. So things can only get better for our four-footed friends as the 21st century progresses.

That's all wonderful, you say, but what about poor Rambo, whose joints are so stiff that he doesn't want to chase a ball, or little Fluffy, whose sparkling eyes have begun to dull with cataracts so that she can't see to leap up on your

One day, you may notice that Gareth's muzzle has turned white—seemingly overnight!—and that Frisky doesn't want to do a lot but sleep. With a few simple behavioral modifications, you can help your pets enjoy their golden years.

Pet Insurance

American pet owners, who will do just about anything for the health of their beloved pets, are spending upwards of $10 billion a year for veterinary fees and related care. And costs keep going up at a steady pace. To take advantage of the latest advances in pet care, you might want to investigate taking out a pet medical insurance policy, as more than a million pet owners have done since 1980. The leading provider, endorsed by the American Humane Association and licensed in 47 states, is Veterinary Pet Insurance. It was started by 750 independent veterinarians and is one of several companies offering this service. You can reach them at (888) USA PETS, or online at www.petinsurance.com. Petshealth, Inc., is another leading provider. You can talk to them at (800) 799-5852. Another useful web site is www.vetmedicine.about.com, which provides links to and information about companies offering health insurance for pets. Coverage options from both companies include wellness visits and treatment for illness, injuries, and poisonings, as well as surgery, hospitalization, chemotherapy, prescriptions, dental care, spaying, neutering, and euthanasia.

lap any longer? What can we do to help these beloved pets hang onto a good quality of life *today*? Let's look at a few issues and see if we can make some positive moves to give our pets a "leg up" as they attain senior status.

Let's start by tackling the most important problems older dogs and cats face as time passes. What can we do to enable mature and geriatric dogs and cats to enjoy the highest quality of life they can in their final years with their human companions? In addition to making their routines regular and predictable, giving them lots of love and attention, and continuing to include them in our lives, we can arrange their living environment to make it easier for them to fulfill their own biological, psychological, and social needs.

Strategies for Failing Senses

Let's look at some specific areas where dogs' and cats' behavior will be affected by geriatric problems. Sure, we're able to keep our dogs and cats alive longer, just as the human lifespan is constantly increasing. And we can offer them a better quality of life through advances in applied animal behavior and veterinary medicine. But in some cases, our pets face a longer period of declining good health than they used to. It's not just a motor loss (a decrease in the ability to move well, or jump or play or even walk) our household pets face in their later years—it's a sensory loss, too.

🐾 Hearing Loss

Sensory loss is apparently something that's often rather gradual in dogs and cats. But when your dog or cat is not responding to magic phrases like "Go outside?" or "Where's your mouse?" or "Get your rawhide!" or "Catnip time!" you'll have a pretty good idea that he's becoming hearing impaired. If he's not running to the kitchen cabinet when you pick up his food bowl, jumping around in circles when you reach into the rawhide canister, or doing flips when you dangle his leash in front of him, he's likely losing his sight.

What to do? Yelling at the dog or cat to get her attention usually isn't the solution. Instead, try to get your cat's attention by walking and talking to her or standing in front of her so she can see you. Some dogs and cats appear to read lips, which is not as unusual as it sounds, since they probably already read your entire facial expression while you're saying the word. Now, they merely rely on the shape your lips make when saying "outside" and your taking a step toward the front door as a pretty good indication that it's time to go pee.

With a little forethought and some common sense, you can make your companion's transition from full sensory capability to hearing loss a thoughtful and fun one for both of you. It's interesting to observe and imagine what your dog is actually paying attention to when you say the important words and phrases to him.

What about your older cat? A cat's hearing is better than a human's or a dog's because he can hear higher tones than either a person or a dog. That doesn't mean his hearing won't get worse with age, though! (Finally, your cat has a retroactive excuse for ignoring you for all those years when you tried to get him to obey your commands!)

Remember that cats normally like to get underfoot—but in a planned sort of way—so they can rub on your legs and be close to you as you go about your daily routine. They still enjoy this if they're hearing impaired, but they may not hear footsteps well enough to get out of the way when you close the door to the kitchen and walk into the dining room. This is especially likely if they happened to be facing the wrong way in the doorway to the dining room. If your cat can no longer hear well, be careful and use common sense. Although tactile receptors in their feet allow cats to feel vibrations from the floor, their ears pick up airborne vibrations.

Six Secrets for Older Pets and Their People

I've learned over the years that there are six things that really make a difference in an older pet's life. Try them and see for yourself!

1. Give them help jumping. Don't expect your cat or dog to be able to leap up on—or jump down to—locations that they could reach easily when they were sprightly adolescents or younger adults. Provide steps for them—even in and out of bed! Don't forget the backyard, if they're used to jumping off a back porch or deck to get to the ground. Cats may need some extra steps getting up to the kitchen counters where they can eat away from pet dogs in the family, or on the way up to their favorite bookshelf to lie on and peruse the area during the day.

2. Respect her limits. Don't force your dog or cat to do things she's not feeling up to just because you want to see if she still can, or you can't stand seeing her not being the same as she was only 2 or 3 months before. Instead, encourage or entice her to do activities she still enjoys that are not beyond her physical or psychological capability. At some point, jogging or walking briskly is not going to be good for her anymore, if she has such problems as joint pain, arthritis, loss of muscle mass, and trouble breathing. Perhaps walking up and down the sidewalk at a time of day she used to jog would be a better idea and something you could still enjoy together.

3. No more soul food. Do not continue to try to make your cat or dog feel good by feeding her some favorite, unhealthy foods. You know that fat causes stress on the heart and joints. Although feeding fatty treats may reduce our guilt that

The Thunderstorm Thing

Imagine a dog who, through most of her life, started panting and pacing and trembling every time it stormed—at least, until she found out that she could hide and mask the sound of thunder by burying her head in the cushion on the couch. That's the way she coped for years. Now, the dog doesn't even know the thunderstorm's coming and seems to watch the rain and have no problem with it at all. You might think that all aspects of the storm become, in the dog's mind, part of "the Thunder," so she'd be afraid of any part of the storm. But that's not necessarily the case. Sometimes dogs only pay attention to the scariest signal or cue, and other aspects of the storm are irrelevant.

Sometimes, old age can be a kind of blessing for your dog. As the dog's we're unable to relieve our pets' geriatric condition completely, it's not in our dog's or cat's best interest to continue to get fat.

4. Spare the rod. Don't punish your dog or cat for geriatric misbehaviors he can't help, even though you're frustrated and want to indicate to him that he should know it's wrong. Instead, enable him to get out of the situation without further stress and strain. For example, remove him from the crate where he wasn't able to "hold it." If he feels more comfortable coming inside and peeing on the wool rug than he did outside on the cement or sidewalk in the city, try finding a grassy area where he can more easily balance on one foot while he lifts his leg.

5. Don't consider touching taboo. Do not be afraid to touch her just because she's not as strong this year, or not as physically attractive, or has a slight loss of hair. Do set aside a time every day to groom, stroke, or massage your pet, so she can look forward to satisfying her social needs and physical comfort, as well as bonding her even closer to you and you to her.

6. Face reality. Don't go into denial about your dog's or cat's infirm condition. Talk to other pet owners online if it helps you to share experiences and feelings about your aging dog or cat. Seek help from your veterinarian, who can keep you up to date on the latest advances that will help maintain your pet's health. Your vet can also refer you to a certified applied animal behaviorist for help in coordinating a standard of care for the geriatric pet in your family.

senses dull with the passage of time—if he has suffered from fear of thunderstorms or other "scary noise" phobias—hearing loss can put an end to his fear. Both my sister, Judi, and I have had dogs who went from being extremely frightened and distraught in the face of thunder and lightning to being completely oblivious of the fact that it was taking place. (I'll discuss these phobias in detail in chapter 14.)

Some have suggested that dogs find storms disturbing because of the electrically charged atmosphere "zinging" them, as it were; however, I have developed a simpler explanation that may account for most of the fear I see in my practice. Like our dogs Peanut and Scout, blissfully unaware of the thunder crashing outside, many dogs give up their storm phobias at about the same time one of their senses fails them. Both dogs have gone from being panting,

DR. WRIGHT'S INSIGHTS

If you think your dog's hearing is going, try this experiment. Change the words you usually say like, "Duke, want to go for a walk?" to something that sounds the same, like,
"Duke, did you know monkeys can talk?" Do everything else the same way as usual.
Chances are your dog will respond as if you actually asked him to go for a walk. Or, recall the many times you opened the closet door to get the leash out, and the dog ran
over in anticipation of his walk from a distant room in the house. Now, you actually have
to show him the leash to get him to enjoy the same old ritual because he no longer responds to the opening or the closing of the door. He still wants to go for a walk; he just
appears less excited because he can't hear the door opening.

whining, and pacing "stormophobics" to approaching storms in a calm, cool,
and collected style as a result, I believe, of now being quite deaf.

Even though dogs also sniff the air for moisture, see the curtains move or the
sky getting dark, and whatever else they do to "predict" rain with their "sixth
sense," my money's on the sense of hearing as the most important factor in "stor-
mophobic" behavior. Guess there's always a silver lining if you look for one.

🐾 Loss of Vision

Dogs and cats who begin to have vision problems—whether it's glaucoma or
cataracts—can pose a problem for themselves and for you. There are certainly
veterinary medical solutions to some visual problems. In the meantime, you
can try to substitute auditory cues, like words, phrases, or claps, for things you
used to rely on vision to convey to your pet—it's just the opposite of the
hearing-loss problem.

If you used to get a kick out of tossing a treat to your dog and having him
catch the treat in the air, or playing Frisbee with your dog (now he can't see
the Frisbee too well, so he's getting hit on the head), try tossing the Frisbee or
treat to his left or his right with a large motion of your hand so that he can see
it a bit more clearly, just from the movement. Oftentimes, he seems less upset
than you are at getting knocked in the noggin, but both of you will probably
be happier if you toss to either side and then enjoy that special treat time you
used to have when everything worked correctly.

In any case, cats have a little better luck with loss of vision than dogs, provided that they don't continue to attempt airborne acrobatics from the armoire

to the armchair and other balancing acts on the living room furniture. Like dogs, cats can have particular problems with sliding glass doors and other inconvenient passage points, but cats have those wonderful whiskers that keep them from bumping into things. (You can help by attaching stained glass or other suncatchers to the clear doors.) So with their good hearing and sense of touch, and their ability to move over, under, and around furniture to keep out of harm's way, cats tend to cope pretty well with visual change.

Age-Related Behavior Changes

As dogs and cats get older, another thing to be concerned about along with sensory loss is the quality of their normally fun or exciting experiences. Now, when they go to the park and play, do they continue to enjoy these endeavors? Or, since they don't experience these events with the senses that they used to

Naming Names

What are we supposed to call our older pets? Just older? Seniors? Golden agers? I'm starting to see visions of shuffleboard tourneys in my head! Well, it's up to you. I know people who have called their dogs "puppy" until the day they died of old age. So it's your choice. But nowadays, when you take your older dog or cat to the vet's, you may encounter a term that has a more specific meaning than you might think. That label is geriatric. Let me explain what that means, because it is not just a synonym for <u>older</u>.

For our purposes, a geriatric pet is one who has problems that are associated with aging. Now, just because your dog or cat has a behavior problem and is old doesn't mean that he's a geriatric pet. And if he's old and he doesn't have a behavior problem, he's not a geriatric pet either. Here's what I mean. If 13-year-old Killer bites you because he has always bitten men in baseball caps, it's not a geriatric problem. But if he bites you because you hurt his sore old gums the last time you played tug-of-war with a knotted rope and he blames your hands, you have a geriatric pet with a problem.

So we behaviorists save the phrase <u>geriatric pet</u> to include those animals who are experiencing age-related dysfunctions related to immobility, sensory loss, incontinence, cognitive dysfunction, and other problems that become more frequent and tend to be more specific in older pets than in the general adult pet population. Now, if your vet says your pet is geriatric, you'll know what it really means!

Sometimes, hearing loss in pets can be a blessing in disguise. Dogs who have been terrified of thunderstorms all their lives may ignore them completely when they can no longer hear the thunder.

have, are they still having fun? Actually, things can become dangerously stressful for your pet while you aren't even looking.

As you leave him at the groomer or take him into the park to see if other dogs are around for him to play with, notice if he seems to enjoy the experience or if he starts trembling. Pay attention to any changes in behavior. Even though they may not make sense to you, they may indicate that his experience is starting to be different. He's not smiling as much. He doesn't spend as much time away from you as he used to in the park. He doesn't want to walk or run ahead of you in the park like he always used to; now he's hanging back and avoiding people who might come up to him and possibly hurt him with a too enthusiastic pat on the head.

🐾 Pain- or Fear-Induced Aggression

Pay attention to changes in your pet's demeanor or personality as things become difficult for him. If you do, you won't be surprised by a full-blown fear of, for example, jumping up into the car to go for a ride. If your pet can't see where he's jumping, or if it hurts him to jump, it can lead to fear-induced aggression. He may strike out against you, seemingly for holding the car door open.

Aging dogs and cats get into biting for similar reasons if they're experiencing discomfort. Pain-induced biting can be a result of forcing them to do things that they're no longer able to do. And this in turn can lead to fear-induced biting if, in their eyes, you're about to force them to do the painful activity. If it's jumping into the car, they become afraid of your reaching for the car door handle and nip the hand you're using to hold them because of the coming pain. Creaky old hips aren't meant to propel a slightly overweight frame onto the seat of an SUV, even with the help of a push from the rear.

Sometimes children or grandchildren forget that the dog is not as young as she used to be and that the cat is no longer the silly kitty who can be picked up in a variety of ways, either. The smaller the children, the more reminding they will need—for their own safety as well as for the comfort of the dog or cat. Many dogs and cats are likely to become aggressive if they are hurt while being picked up the wrong way by an unsuspecting child. Sometimes it's necessary to set new rules in the household for the kids who come over to visit: "Farrell's not feeling well today. Please let him be by himself in the corner," or "Farrell's old, and he's feeling a little grumpy today. Maybe tomorrow he'll feel a little bit better, but let's leave him by himself today."

Similarly, if one of your younger puppies or dogs starts to get into too intense play for your older dog, redirect his activity and play toward you or toward self-play. It's up to you to intervene on behalf of your geriatric pet. If it doesn't look like she can take it any longer, she'll thank you for sure, and your relationship will grow because of it. The same is true of cats. As we saw in "Changing Relationships: Who's on Top?" on page 178, socializing with a younger kitty isn't always fun and games, and the older a cat gets, the worse that relationship might become.

Don't Let Your Older Dog or Cat Get Fat!

Obesity is a very big problem—no pun intended. All dogs—and plenty of cats—are susceptible to gaining weight as they become less active in old age. But some breeds have been reported to be more likely to become obese than others as they watch the birthdays fly by. West Highland white terriers, beagles, cocker spaniels, collies, Cairn terriers, and retrievers are breeds that are noted for obesity in old age.

DR. WRIGHT'S INSIGHTS

You should show children who are old enough to handle a pet how to pick up your older cat or dog correctly. With one hand underneath your pet's chest and front legs (behind the front legs), give the dog or cat support from its hind end and lift the animal by the rear. Support the front of the body and hold it against you, so that your pet feels secure. The last thing you want to do is to lift the dog up by both hands under the chest and front legs and have his hind legs dangling out behind him.

Older dogs are about twice as likely to be overweight as younger dogs, probably because of genetic factors as well as people feeding them high-calorie treats from the table. Also, smaller dogs can run around the house for exercise, but larger dogs need to be taken out. Because most dogs depend on people to take them out, it doesn't always get done.

As adult dogs become elderly, a number of factors limit their exercise—including their physical and physiological condition (and perhaps that of their owners as well, if the owners are also slowing down a bit). So, lack of exercise coupled with the same amount of treats as they had when they were younger can lead to obese pets. Female dogs and cats are more likely to be obese than male dogs and cats, and spayed animals are more likely to be obese than reproductively intact ones.

Here's an interesting statistic: Pet owners who are 40 years old and older are more than twice as likely to have obese dogs as younger owners are. (There's no research yet for cats.) Not only are the owners getting older and probably less fit, but so are their pets! Further, about one-third of owners of obese dogs do not consider their dogs to be overweight; they consider their dogs to be in the normal range of weight.

I suppose the worst-case scenario would be a couple of spayed female Labrador retrievers who are about 14 years old, living with elderly "nurturing" (literally—with food) owners. Both the owners and their dogs are arthritic and overweight, with poor aerobic conditioning. Or maybe the owners smoke and have shortness of breath with even moderate exercise, and they enjoy fried foods and feed their dogs from the table or the couch (of course) or from anywhere else in the house. The owners equate feeding the dogs with giving them love. It's the proverbial "recipe for disaster"!

🐾 Getting Back in Shape

How can you help your older dog stay fit—or lose weight—if he's already obese? One solution is to begin a walking program at least 3 days a week. Both owner and dog can build up to a good aerobic workout. Obviously, before starting such a regimen, both will have to check with their respective doctors. At home, rules limiting the availability of highly palatable and fattening foods and restricting where in the house food is given will need to be instituted.

For example, following a rule that says that food will only be given to the dog or cat in his food bowl in the kitchen will keep treats from being given indiscriminately throughout the house—in the bedroom, on the sun porch, in the living room (with both owner and pet lounging on the sofa and munching treats), and in other risky locations. Habits like tossing your pet a bit of cheese while preparing dinner or scraping all the leftovers into his bowl after dinner can cause a problem. Keeping the food bowl far from the meal-preparation area is the best defense.

Keep in mind that your older dog or cat probably has years of successfully conning you into providing treats, using such tactics as begging and looking sad. "Poor old me, why won't you share that? You don't love me anymore!" He may try all this while whining, crying, moaning, or just drooling. Or he might use the opposite tack, showing some lively pawing for a change with a happy smile in expectation of a treat. All of these behaviors are sure to tear at your heart. How on earth can you disappoint him? Gosh, he doesn't have too much time left, how can I keep him from enjoying his treats? Again, you are equating love with giving treats. Bad owner!

His carrying on is aimed at convincing you otherwise, but resisting the urge to give him fatty treats will only hurt for a few days to a week, I promise. If you consistently don't give him treats or snacks except in his one food bowl, you'll help extend his life and limbs—and your time together—by keeping that weight off!

I recall a dog named Dustie, a Rottweiler who became a geriatric client of mine. Her body looked like a blimp because her owner couldn't help but feed her tidbits—not just any old tidbits, mostly cheese and pigs-in-a-blanket—throughout the day because she felt so sorry for the old girl. Her body was so fat and her joints were so stiff and inflamed, it looked like she was balanced on stilts! After the poor dog went outside to relieve herself, it would take her 30 seconds to figure out how to lie down again. Poor girl.

Obviously, getting up again was quite a task as well, but Dustie fought to get

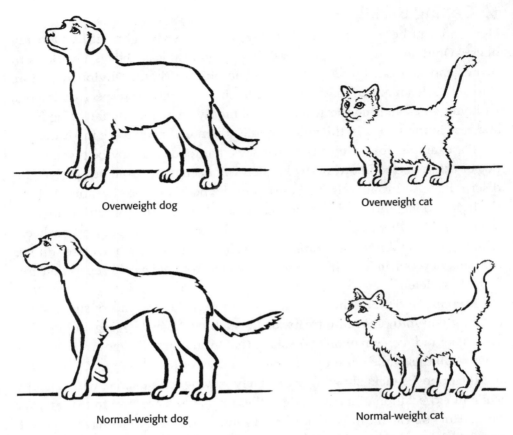

Overweight dog

Overweight cat

Normal-weight dog

Normal-weight cat

If you can't feel your pet's ribs, it's time to cut back on the snacks and switch to low-cal or "lite" pet food.

some sort of treat every time she saw her owner—she knew that she would get food just for trying. This was not the quality of life her owner had in mind for Dustie but rather a consequence of a habit that started as a gift of love. As a result, the owner was reminded daily that a fattening treat was an inappropriate token of her love of the dog she cared so much for. So *don't try this at home!* Instead, learn how to make your pet feel good by providing fun social contact, by giving gentle physical stroking or massage, or, if you must, by handing out nonfat treats, including vegetables and others that are recommended by your vet.

Help for Flagging Appetites

At the other end of the spectrum is the pet who doesn't eat enough anymore, probably because his senses of smell and taste are diminishing. (Not to mention the fact that he doesn't move enough anymore to work up an appetite!) It's often

DR. WRIGHT'S INSIGHTS

If you absolutely <u>must</u> give your dog or cat scraps from the table while you're eating—because you just can't resist his pleas in spite of what I say, or you truly believe that feeding from the table strengthens your relationship and bonding, or whatever the rationale—try this tactic. Make room for a small cup of nonfattening doggy or kitty treats next to your plate at each meal and toss him one when he looks at you with those "ple-e-e-a-a-ase feed me" eyes. This strategy should make you both happy (but will likely horrify any people who continue to dine with you), and he won't put on pounds as a result!

necessary to find something more flavorful for an older pet in this situation to eat.

Another possibility is that he doesn't feel like eating because his teeth hurt. In this case, try feeding him something softer if he has sensitive teeth and he just won't munch the hard stuff. You can switch to canned food or mix warm water into his dry food and let it soften before giving it to him. Jars of baby food lamb, beef, chicken, and veal are usually devoured eagerly by elderly cats.

Another way to up the aroma factor to tempt a dog or cat whose sense of taste isn't what it once was is to microwave his food before serving it. (Make sure it's warm, not hot!) Other techniques include feeding your older pet less food but giving it more often, or making a game of eating. Both will usually get a good response. Dogs love to do something for you ("Sit. Good boy!") and be rewarded with a food tidbit. That's a good way to get the food into him that he needs, if you're willing to take the time and trouble.

You don't want your older dog or cat to lose muscle mass; he needs muscle to be strong as his energy wanes. If necessary, spoon-feed him! If he needs medication, use the old trick of putting the pill in a piece of sticky low-fat peanut butter, a serving of applesauce, or any healthful soft table food.

Elderproof the House

With an older animal, you need to look at your home through newly sensitive eyes. Does the dog need a ramp to get up and down a flight of stairs from the deck to the backyard? Can the cat jump up to the windowsill now that she's 14 years old, or does she need a little step stool placed underneath it?

Older dogs and cats both seem to like to feel soft things, almost as if they're returning to their early lives, when they preferred pliable, warm things, such as their mom's tummy. If your dog is used to a crate or the tile floor in the foyer, or your cat is used to a windowsill, try to provide something soft for him to lie on. As he gets older, he will thank you for it. Even though he can't see the backyard as well, or hear you coming, it will be bliss for him to be able to lie in his favorite place, which is now warm and comfy, rather than on a cold, hard surface. Your old friend needs to protect his joints from hard surfaces and his body from drafts that could cause an uncomfortable chill. So think warm, soft, fluffy stuff!

While everyone wishes that strong, healthy adult pets could stay that way forever, it just can't be. Still, life can be good for your pet! Recognizing and admitting that your pet is growing older and will need for you to make some special accommodations is the first step in solving geriatric problems and improving all the remaining days of our beloved older canine and feline friends.

Cognitive Dysfunction: That Foggy Feeling

Canine cognitive dysfunction syndrome has some of the same symptoms as the mentally incapacitating diseases in older adult humans. Dogs seem to forget things they previously knew well, such as which house to come back to. So

A few simple changes, like providing soft surfaces for your older pet to lie on and easy access to the litterbox, can make all the difference.

DR. WRIGHT'S INSIGHTS

Here's a tip: To find where your pet may have "gone" on your rugs or elsewhere in the house, buy a blacklight at a novelty store. Shine it on the carpeting where you suspect that accidents have occurred. All the spots where your pet has urinated will be quite visible under blacklighting.

just letting them out the back door may be a risky enterprise when they reach a certain age. It's better to have a confined area or take them for a walk. A walk with you is always a better choice in any case than sending them out alone, which is against their social nature.

With cognitive dysfunction, a dog's sleep/wake cycle tends to be unstable. He may become quite vocal and bark or cry or whine inappropriately. Or a dog may wander into a room and just stop, as if he forgot why he went into the room in the first place. And worst of all, he may forget his housetraining.

Fortunately, there is now an FDA-approved drug for the treatment of cognitive dysfunction. So if your pet's going through this disturbing syndrome, you can ask your vet for medication. But regardless of the treatment offered by veterinary medical experts, owners need to adjust their expectations of the pet's mental sharpness and be very patient and, of course, nonpunitive. People wouldn't yell at their grandmothers for forgetting what year it is! They shouldn't yell at their old cats when they forget to find the litterbox, either.

When Doggy Needs a Diaper

Sometimes failure to be able to "hold it" is not related to cognitive dysfunction as much as it is to degenerative joint disease or arthritic conditions. These conditions keep the dog from maintaining a posture that will allow him to urinate or defecate. The poor dog or cat will try to assume the correct posture and attempt to go, but he just can't stay in position long enough to do his business. His attempts often result in trails of poop, or may even involve the dog or cat sitting in feces. These are the times when owners are most tempted to yell, "Sampson!! What are you *doing*?!" But try to restrain yourself. It's not Sampson's fault—it's the geriatric condition.

Some pets leak urine; however, a number of different veterinary medical solutions to the problem, which involve giving the dog a pill, will help tone her sphincter muscles so that they work again. Sometimes an anti-anxiety drug

(continued on page 198)

DR. WRIGHT'S
C A S E B O O K

Crystal

If my clients Nick and Paul, two doctors with distinguished careers and several advanced degrees between them, had taken the time to look at their setup—really stepped back and observed the living arrangements— they might not have been so shocked to find that the way their home was set up no longer met their cat's needs.

"Crystal has been urinating on the oriental carpet for absolutely no reason," Paul said in a panic when he called me. "Please come and tell us what we can do!"

At their lavish home in an exclusive section just north of downtown Atlanta, Paul and his partner showed me the spots in the dining room where the errors had occurred. A quiet, white-haired, 14-year-old cat named Crystal rested on the floor in front of the mirrored bar, watching me intently, her tail swaying slightly. The doctors convinced me that she spent most of the day there, watching her reflection in the mirror. Three younger terrier-type dogs tussled at my feet.

Crystal's story was pretty typical. The owners of a single dog or cat devote a lot of attention to their only pet when no other animals or children are around to distract them. But since the three dogs had taken up residence here, Crystal had been relegated to the "oh, she pretty much takes care of herself" category, complete with cat doors cut into people doors and a couple of litterboxes. Meanwhile, Nick and Paul devoted more attention to the young dogs.

The problem was that no one had really noticed Crystal getting bigger (actually fatter, as happens with older cats) and stiffer. Her physical world didn't quite work for her anymore. Her avoidance of the litterboxes was the logical response of a very sturdy cat who was having trouble jumping through the cat doors. (I use the phrase loosely—the "cat doors" were home-made affairs, nothing more than round holes cut into the doors that led to the dungeon, er, basement, located on the far side of the house, where Crystal had always gone to the litterbox.)

That box was hooded and it was sitting 7 inches from a basement floor support pole. Crystal simply couldn't get into the hooded box without being a contortionist. No wonder she peed on the rug in a wide open space! The hoods, Nick explained, were to discourage the terriers from eating cat feces.

"Let's look at what the other box location has to offer?" I suggested. We retraced our steps, walked through the dining room, and wound up on the far side of the house, down a hallway and into Paul's big walk-in closet, peering past the suits and shoes and bagged sweaters into the darkness.

"Over there, against the back wall," he said. Boy, he wasn't kidding. I found the box. It also had a hood, but the opening was turned away, toward the back of the closet, and again there were only 6 or 7 inches available for the cat to maneuver herself into the box. And in front of the box was a small kitty condo. Surely no terrier would bother with this annoying set of barriers. So that much had been accomplished! The doctors thought that this location would give Crystal a private location to do her thing. Maybe it did when she weighed only 4 pounds.

I persuaded Nick and Paul to give their elderly cat a break and move one litterbox to a discreet location in the living room behind an oriental screen. There would be sandy litter inside and no lid at all. Because the dogs were never permitted in the living room anyway, this worked out perfectly. They put another box in the bathtub in an unused guest bathroom, with a little stepstool leading up to it. The dogs had no interest in being trapped in a bathtub, thank you, so they stayed away from this box as well. Crystal, on the other hand, seemed to enjoy the cool surface under her feet and found this location much more desirable than the back of a dark closet.

Next, Nick got out his carpentry tools and enlarged the openings of the cat pass-through doors so that Crystal could fit more easily. And both doctors vowed to stop feeding their beautiful Persian any more fattening dog treats, substituting some low-cal goodies from a nearby natural foods store instead. Middle-age spread wasn't going to keep this kitty down for long!

If your dog can no longer control herself, doggie diapers will keep your house clean.

is sufficient to enable the dog to relax enough so that he can figure out where he wants to go to relieve himself. This type of problem tends to creep up on you, and behavioral solutions usually involve being as regular as you can about taking your dog out. Sometimes you'll have to add one or two trips outside for him to relieve himself. You can help your cat in this situation, too.

If you just can't seem to make enough extra trips to keep your dog and home clean and dry, consider "doggie Depends." These diapers come in a variety of colors and prints for dogs to wear (cats seem to hate them), they're functional as well as fashionable, and you can select a style for solid waste or solid and fluid waste. Both disposable and reusable models are available in different sizes. They fit both sexes and are available through pet stores, pet-supply catalogs, and veterinarians' offices. Most geriatric dogs need one to three changes a day. Doggie diapers may also be the last resort for people living in communities with scooper laws, if they don't wish to join the shovel brigade.

A Good Life All the Way: Facing a Pet's Death

The question of euthanasia is always a heart-rending one for anyone who has shared their life with a dog or cat. Many of my clients have been in a quandary over whether to consider euthanasia or other options for their dog or cat when he or she is geriatric and failing. Inevitably, I get the question, "Dr. Wright, what would you do if this were your pet?"

My answer involves giving clients options to consider that enable them to make the appropriate choice for their own companion animal. I never make

DR. WRIGHT'S **INSIGHTS**

It has been my experience that dogs and cats don't seem to be as upset with their infirm condition as we owners are. Pets take everything in stride, generally speaking, and that is what's so endearing about them. On the other hand, we owners typically get frustrated and angry, and we feel guilty about the problems that arise in our geriatric pets. We feel guilty because we get angry at our companions of 15 to 20 years, when we know full well that the transgression is not our pets' fault.

People get angry, of course, because they don't like to see feces all over their beds. And they get frustrated because they can't figure out how to pick their dogs up and put them on a counter without making them cry out in pain because their vertebrae have become fused, because their arthritis is so bad, or because of some other physical conditions. This is a rough time.

So it's especially important now for you to continue to have a stable, caring social and physical relationship with your pets. Stroke them, talk to them, entice them out from under the bed or from the closet when they try to isolate themselves. Encourage them to be a part of the family for as long as they can successfully do so. It will be gratifying for everyone, and you'll create some feel-good memories of your old pets to reflect on for a lifetime.

(nor do I ever want to make) a decision for someone else. But I feel a responsibility to encourage them to consider the range of choices facing them regarding their pet's future. My role is to help owners come to terms with their feelings and decisions regarding euthanasia and to provide them with good, accurate information and options about what to do and what others have done in similar circumstances. Good decisions cannot be made in the absence of accurate information and viable options.

Preventing pain and suffering and preventing damage to the individual are reasons to consider euthanasia. That list of reasons includes preventing injury to others as well as risk to the well-being of the pet who can no longer care for herself. Not attending to these issues at the correct time may result in the owners' guilt later, especially if their dog or cat continues to suffer because, behaviorally, he is not the same dog or cat, and his psychosocial needs can no longer be fulfilled. Dogs and cats, as far as we can tell, do not have a concept of their own mortality, but they can and do feel pain. They suffer, and we need to try to minimize their suffering as best we can.

Coping with the Loss of Your Pet

My wife, Angie Wright, LPC, NBCC, is a therapist and pet-bereavement counselor. She offers the following FAQs to help you through the heartbreaking period of mourning the loss of your dog or cat:

Why do I hurt so much? Intense grief over the loss of a pet is normal. Don't listen to anyone who tells you that you're silly or crazy to grieve over your pet. During the life that you shared with your pet, she became a special part of your family. Your pet was a constant companion and a source of unconditional love, comfort, fun, and laughter. Don't be surprised if you feel devastated by your loss.

People who don't understand the bond that can be formed between an owner and pet may not understand your pain. But the only thing that matters is how <u>you</u> feel. Others can't dictate your feelings. Remember, also, that you are not alone. Thousands of pet owners have shared and experienced similar grief.

What should I expect to feel? Different people experience grief in different ways. Almost everyone feels a sad, empty feeling. The following stages are typical when experiencing grief:

- **Numbness.** The numb feelings protect us from feeling the loss. This is a form of denial. Denial makes it difficult to accept that your pet has died. It may be hard to imagine that your pet won't be there to greet you when you come home.
- **Guilt.** Guilt may occur if you feel partly responsible for your pet's death. You may be plagued by thoughts like, "If only I had. . . ." But it's pointless to bombard yourself with guilt for the illness or accident that caused your pet's death, and it only makes it more difficult to resolve your grief.
- **Anger.** Anger may be directed toward the illness that ended your pet's life, the veterinarian who couldn't save your pet, the driver of the car that killed your pet, God, or yourself for not being able to prevent your pet's death. Sometimes your anger may be justified, but when carried to extremes, it can stand in the way of resolving your grief.
- **Depression and anxiety.** Depression and anxiety are natural consequences of grief and can leave you feeling very helpless. Extreme depression affects your energy, motivation, and sleep, causing you to recover more slowly. Anxiety may come in the form of grief pangs, consisting of uncontrollable crying spells, restlessness, and preoccupation with the deceased pet.

How can I cope with my feelings? Experience them. The first step in learning to cope with grief is to allow yourself to experience and express your feelings. Don't deny your pain, anger, and guilt. Only by being honest and learning to accept your feelings can you begin to work through them.

Some people find it helpful to express their feelings in writing by keeping a journal or writing letters, poetry, and stories. Other coping strategies include making a photo album or collage as a memorial; reaching out to others for support; returning to sites shared with your pet, such as a park, pond, or beach where you shared happy times; or donating money or time to animal charities.

Should I get a new pet right away? Generally, no. All family members need to work through their grief first. Getting a pet right away may be a method of distracting yourself from your pain and feelings of loss. But you may resent a new pet for "trying to take the place of" the beloved pet. And children often feel that getting a new pet is being disloyal to the previous pet.

When you do get a new pet, don't try to get one that's too similar to your deceased pet. Don't expect your new pet to be like the one you lost, but allow him to develop his own personality. Never name your new pet after the deceased pet.

Remember that obtaining a new pet is very stressful. It should be done when some of your grief has subsided; otherwise, you may feel overwhelmed.

What should I tell the kids? You are the best judge of how much information your children can handle about the loss of their pet. Often a child's first experience of the grief associated with death comes with the loss of a pet. Their ability to cope with this loss may enable them to handle future losses effectively. Be honest with your children and allow them to grieve. By discussing death and the loss of your pet with the kids, you may be able to address some fears and misperceptions they have regarding death.

Honesty is important. If you tell the kids that your pet was "put to sleep," make sure they know the difference between death and ordinary sleep. Never tell a child that the pet "went away," or your child may wonder what he or she did to make the pet want to leave and may wait in anguish for him to return. Make it clear that the pet is no longer in pain and will not come back.

Never assume that a child is too young or old to experience loss. Never criticize them for expressing their grief and emotions or tell them to "be strong." Don't attempt to hide your own grief, or your kids may think that they should hide theirs, too. Allow your family to work through its grief at its own pace.

❖ The Other Alternatives

An owner may choose to live with or manage a behavior problem, whether it's of a physical or psychosocial nature. She may choose to allow her dog to continue his destructive behavior or house-soiling or to allow the cat to destroy furniture with his claws or expensive rugs with his spraying. Managing a problem certainly doesn't mean curing it or resolving it, it merely makes the problem less likely to occur in certain situations. Certainly, geriatric cats and dogs could be allowed to continue missing the litterbox and snapping at or biting their owners because being touched is painful. Cats could be confined in a small room so that they are less likely to miss the litterbox, and owners could avoid reaching for and picking up their dogs if petting them led to biting. Or the owners could confine a dog in a crate to reduce house-soiling.

Geriatric Dogs' and Cats' Quality of Life Quiz

It might be helpful at this difficult time to take a deep breath and ask yourself and your family some specific questions about whether your dog or cat is still enjoying her life. There are no right or wrong answers, but when you are finished thinking and talking about these topics, I hope your path will become clearer. It's better to discuss these issues in advance than to be faced with panic one day when you realize that you can't lift your 80-pound shepherd off the floor without hurting her, and she just can't get up by herself anymore. You need a plan. My pet-bereavement counselor colleagues suggest the following things to ponder:

1. Is your dog's quality of life a good one? Has she had a happy, meaningful life in this household with you and your family?
2. What is _your_ quality of life like? What is happening in your family?
3. How much improvement (or at least stabilization) are you realistically likely to see if you continue working with or managing your pet and her problem?
4. Knowing the answer to #3, what is your bottom line for deciding that either your own or your pet's quality of life is not high enough for things to continue the way they are if the frequency of the problem behavior continues to increase?
5. Have you definitely done everything that you can do for your pet? Has she continued to decline in behavioral or physical health?

Sometimes psychoactive drugs provide a means of managing a problem, at least on a short-term basis. Other options for dealing with the geriatric problem are to enable the dog or cat to live with her condition, but more comfortably. Both veterinary medical and behavioral means can be used to reduce a pet's pain or discomfort in her old age with the use of drugs such as hormones or environmental enrichment, such as steps up or down, softer places to lie, and other examples we've given in this chapter.

The Adoption Option

But the "I'll live with it" or "I'll manage" approach can't last indefinitely without dealing with quality-of-life questions that involve both the family and the dog or cat. At some point, owners must decide to take steps to resolve a problem, to

6. Is she still having a good time in life, in spite of her infirmities, or is she suffering and are you suffering as well?

7. Are you comfortable with scheduling your pet's euthanasia and deciding when to make that call, or do you prefer to let nature take its course? What if your pet is suffering?

8. How much are you willing or able to spend on heroic measures to save your pet's life?

9. Of the available options, which ones make sense to pursue with your veterinarian because they would add quality to your pet's life, and which would merely buy a little more time?

10. How does your pet react to surgery and separation from you at the veterinary hospital? Would the results be "worth it" in your estimation?

11. What do you need to do to say goodbye? What do you need to do or say that you haven't done yet?

12. When you look back on her death, what memories will you want to have of your last days with her?

13. What kinds of activities can you do with her now and what kinds of special attention can you show her to make those pleasant memories in the near future?

have the animal adopted out, or to consider euthanasia. If the owners can't cope with the necessary changes for their pet, they may choose to place him with someone else. This option works because the dog or cat is not likely to have to endure a particular problem that may be specific to the house he lives in.

For example, if there are very young children in the household, the geriatric dog who has had problems, can't see very well, or can't move around much anymore may be better placed in a home with fewer hands to pull on him, less noise and chaos, and less vigorous exercise on a daily basis. An elderly or geriatric cat who has to put up with a number of adolescent cats may be happier in a single-cat household.

Of course, finding a new home for an old, ailing dog or cat is problematical. And a geriatric animal relies on her owners and on the trusting relationship she has built up over the years for comfort during her final time in the household. The thought of their beloved pet spending its last days in a cage in a shelter repels all but the most hard-hearted (or desperate) pet owners.

🐾 Saying Goodbye

If managing a behavior problem is not a possibility, and curing the behavior problem is not a possibility, and living with the behavior problem is not a possibility, then many owners come to the painful decision that euthanasia is something to consider. And if I were the owner in that situation, I'd be just as confused and in conflict about whether to euthanize as they are.

Of course, if I were talking to my veterinarian or behaviorist about my geriatric pet, I wouldn't want someone to make that decision for me. Instead, I would want them to provide me with the options so that I could make my own informed decision. I would want their support and their confidence that I have considered everything and that the decision I'm making is right for me, my family, and my dog or cat.

It's not unusual for an owner to go through the same kind of emotional turmoil when considering euthanasia that people do when they think about the loss of a human loved one. Guilt, anger, sadness, and other emotions flow when you think of losing your furry best friend. But denying that there is a serious problem is not the right answer. (See "Coping with the Loss of Your Pet" on page 200.)

I've discussed euthanasia with people who had a geriatric golden retriever named Simba, who seemed to have adapted better to his loss of hearing and vision than his owners did. They felt Simba would not be able to make it around a new environment when they moved to a different house in a few weeks. The new smells and strange layout and flights of stairs seemed to terrify Simba on a trip to the new house, and the owners were afraid that without constant supervision, their dog might hurt himself badly. Even with them standing in the next room of the new home, Simba became emotionally distraught, and they were not willing to put him through this in his last days with them. They decided that euthanasia was their only choice for Simba.

But for each owner like Simba's, I see one for whom circumstances make it preferable to let their pet die peacefully at home in the dog's or cat's own time, without human intervention. Still others have moral or religious prohibitions against "playing God" and may wish to let their pet live out his natural life—especially in the absence of major geriatric problems or obvious suffering. Many a fastidious feline has used the litterbox up until the day he slipped away during a catnap. Again, not all elderly pets become geriatric.

For those who choose a nonintervention option, smaller animals will present less of a challenge during this difficult time. A little dog or cat can be carried and attended to more easily than a large animal. Some cats virtually stop eating and look for a place to isolate themselves as the end is near. A thoughtful owner will put a blanket or cat bed in that dark closet and spend time there with the old friend, stroking and talking to her. With forethought, each owner will have the knowledge that he has done everything possible for the animal who shared his life for so many years, and that at the right time, he made the correct choice in considering how to face the final days together.

Coping with Grief

Sometimes taking that huge, awful step of making the appointment for euthanasia is a real problem for owners, but in my experience, most veterinarians have been very open to options people have presented to them for saying goodbye in a private, caring way. If your pet is in pain or for some other reason, and euthanasia is your choice, go ahead and make the appointment. You can always cancel and reschedule if you decide that it's just too much and you have to bail out for now, or if, as may happen, your pet temporarily rallies.

Feel free to ask your vet what choices he or she offers. Some people are more comfortable saying goodbye at home and having someone else in the family take the dog or cat off for euthanasia. Others fare best if a friend who's

(continued on page 208)

Dr. Wright's
E-Mail

Dear Dr. Wright:

Twelve years ago, my family and I adopted two male cats. Just this week, one of them died suddenly. We are all grieving for our lost cat, but we are also extremely worried about the other pet who is suddenly alone. What should we expect from our cat? Do you think that getting another cat is a good idea? Nobody is home during the day, so will our newly single cat be lonely?

Amy

Dear Amy,

I'm sorry for your loss. Like us, cats grieve over the death of a lifelong companion. In such cases, they become both emotionally and behaviorally depressed. They mope around the house, vocalize, and generally look lost. They may sleep a lot or stop eating, self-grooming, or using the litterbox. If this describes your cat, he may have discovered that the absence of his friend resulted in the loss of daily things that "kept him going," such as thermotactile comfort (enjoying the warm touch of his cat pal), social stimulation from playing, and companionship—the same things you miss!

If he is displaying any of these "negative" symptoms, try to reestablish a healthy, active, ritualized lifestyle by providing him with the same kind of stimulation he enjoyed from your late cat. If he ate with the other cat, try staying with him while he eats—have a snack yourself. If they groomed one another after breakfast, try brushing him, or at least stroking him, and allow him to lick or rub you in return. If they played after grooming, try initiating play with him. In general, try to satisfy his need for eating, grooming, playing, and sleeping. Gradually, he should be able to perform these activities on his own.

On the other hand, many other cats seem to look for their absent friend for a few days, and then discover that "life goes on." I know a rather shy, passive cat who lived with a very assertive, take-charge sister for 8 years until the sister's untimely death. The shy cat became more outgoing, sought attention more often from the family than she ever had in the prior 8 years, and even solicited play daily from her humans—an activity she never attempted in the presence of the other cat!

The best reason to adopt a second cat is that *you want a second cat*. Period! Many felines are perfectly happy spending the day alone—gazing out the window, lying in a sunny spot on the living room rug, attacking imaginary critters that wander by—especially if they can count on some terrific time with their caretakers at the end of the day. After all, that's what quality of life is all about at any age!

not too close to the pet drives them and their old friend to the appointment. Owners will make different choices about these issues. Some want to be highly involved in planning a meaningful process. Others will feel comfortable with a goodbye touch on the head as they leave the examination room. And still others want to be with their beloved pet until the very end.

Imagining the inevitable day from start to finish is often helpful in preparing yourself to carry out the final goodbye. You or your children may want to compose a poem or prayer for the occasion and select a spot in your yard to inter your pet's body or ashes. Or you may prefer to have the veterinarian arrange for the details and to visit your pet's grave in a pet cemetery.

One of my clients, Tess, was extremely close to her cat Fudge, who was with her through college and two jobs and three cities. We talked about Fudge's impending demise hinted at by house-soiling and sensory deficits. He was bumping into things because he couldn't see, and she would occasionally find him after work huddled in a corner of the closet, unwilling to move. Tess had decided to take a week off from work to deal with her loss. This would allow her to take Fudge and bury him in her parents' yard in Pennsylvania next to Snaps, the dog she grew up with. These plans were Tess's way of preparing for the death of her fondest friend. Her parents were understanding and supportive, because they knew that Fudge was their daughter's confidante, friend, and family through her "independent" early years.

For heartbroken humans who need help with bereavement and closure, grief counselors are available. You can ask for a referral from your veterinarian or animal behaviorist. You or your vet may wish to make a donation in your pet's name to a humane society, the Morris Foundation, the Cornell Feline Health Center at Cornell University, the Canine Health Foundation of the AKC, or another research organization. Often when people face their fears and feelings about their pet's death and try to decide what they can do to acknowledge and deal with their emotions, they can then help their companion at the end of his life in a humane and caring way.

At this point, pet owners usually swear they will never get another dog or cat because they will never be able to stand going through this pain again. And some of them never do. But most of us animal lovers will keep our pets alive in our hearts and go on to love again. It's not a question of "replacing" the beloved animal—that's impossible—but of filling that huge hole in our lives. The only dilemma is the age-old cry: So many dogs and cats, so little time! For we are mortal too, in this circle of life.

Chapter Ten

Where Good Owners Go Wrong

Even with the best intentions, dog and cat owners can easily head down the wrong path. Thinking that you're doing the best thing for your pet, you can end up totally bewildered—and your pet can end up in big trouble. So think of this chapter as a heads up. Without further ado, here are just a few of the major potholes to avoid on the road to blissful cat and dog ownership.

Myth #1: Say It Again, Sam

Myth #1 says if a few commands are good, a lot must be great! This can be applied to a number of different aspects of life with your dog, including getting him to acknowledge your presence and authority. In training him to respond quickly and happily to a word or command, says this theory, if saying the word once doesn't work—say it 20 or 30 times! By then, you'll probably have worked yourself up into such an angry froth that if your pet has learned anything, it is that he doesn't have to respond until the 27th time he's heard "Come!" or "Sit!" or his name.

Now Sparky doesn't recognize his name unless it sounds like "SPA-A-A-ARKYYYY!!!!" Only then is he likely to say, "Huh?" and look over his shoulder at you, wondering what all the yelling is about. He may start to pay attention to you then. Or he may not. In any event, just stating your commands or his name isn't likely to have any effect at all after this kind of outburst.

Similarly, if he doesn't respond to the first "No!" or "Bad dog," then just like yelling his name over and over again, louder and louder, it makes sense to yell the word "no" and repeat it many, many times, as in the familiar: "NO, NO, NO, NO, NO!!"

When the whole family gets in on the act, your neighbors in all directions can keep track of the great progress you are making in getting that smart dog Sparky to respond to your every whim—day and night. And as far as Sparky is concerned, the name NO-NO-NO isn't so bad; the cat's name is NOPEE-NOPEE-NOPEE.

🐾 Myth-Breaker #1

Remember that your dog and cat will respond to one softly spoken word—their name or a command—if it is followed by something that feels good. They learn to respond to excessive words, phrases, and noises *only* if you don't require a response to a soft word. To show them what's required of them, withhold a pleasant consequence if you don't get a response to the first "Sparky." But give him a second chance soon so that he can learn how to succeed. You—and your neighbors—will be glad you did!

Myth #2: I'll Make Him Stop!

Myth #2 goes something like this: An escalating series of "lessons" will make your pet behave. The second myth deals with what you do rather than what you say. If the first punishment doesn't stop the behavior, broaden and escalate your ammunition—this is war!! Here's how it works:

Day One: Your cat pees on the rug. You give her a dirty look and go clean it up.

Day Two: Tell her she's a "Bad cat!" while you're cleaning up the new spot. That'll fix her!

Day Three: Use "Bad cat!" plus an explanation of why she shouldn't pee on the oriental carpet that's been in the family for three generations, and a detailed accounting of how much it costs to have the carpet guy come in and haul it away for cleaning. The cat usually wanders off about halfway through this lecture.

Day Four: After another spot on the carpet, you are getting desperate and decide to arm yourself. You come upon Satina resting innocently a few hours after the "accident"—ha!! some accident!—and mount a surprise attack, spraying her with a squirt gun while shouting and running through the house until you accidentally knock over a lamp or develop tendonitis in your finger. (Using Myth #1, some people think that if the squirt gun they've read about doesn't stop the misbehavior with one squirt, well then, drenching the cat may work better! Some cats really hate to get wet—and they learn to be not too fond of the shooter, either. So much for working toward a trusting, ever-deepening relationship!)

Days Five and Six: Some random combination of the preceding strategies.

Finally, after a week of finding spots on the rug (last time she squatted right in front of you and did it), shove her nose in it! Works for the dog (also a myth)—might as well try it on Satina!! No? Hmmmmm. Okay, first call the carpet guy and haggle with him for awhile, and then call some behaviorist because your cat is obviously beyond help!

❧ Myth-Breaker #2

Remember that, whether it's a dog or a cat, always just **interrupt the behavior** if it's in progress. Don't try to teach the dog or cat a lesson right here and now. Then teach the correct behavior or location for the correct behavior (in this case, create the optimal conditions for the correct behavior to occur in the litterbox), and add praise and encouragement when it's done to your satisfaction. If you don't anticipate or see the misbehavior, or catch your dog or cat in the act, forget it! Your pet won't know why you are punishing him and will start to avoid you. Think about how you can help your dog or cat prevent the problem the next time, and spend your time and emotional energy on that. (You'll find lots of suggestions for specific problems throughout this book. Just check the index for the particular problem you're concerned about.)

Myth #3: A Dog Is a Dog Is a Dog

According to Myth #3, all dogs are essentially the same, and what worked for training one dog should work for training another. This myth causes a lot of frustration among owners, who don't understand why the second dog is so "stupid" or "stubborn" or "incorrigible" compared to the first dog. The answer may lie in the fact that, in different breeds, there are interesting differences in their behavior and in their responses to different kinds of training. Individual differences even occur within a breed, where one dog may respond better to indulgence while another may respond better to discipline.

One couple, who had successfully reared a wire-haired fox-terrier and noted that it took a lot of commanding and discipline to make the dog obey, were horrified when the same treatment of their new puppy—a sheltie—caused him to lie in the corner and do nothing. The pup simply wasn't willing to take a chance that he would get disciplined again.

Bad to the Bone

Are some dogs just born rotten? Or, are all _____ (fill in the blank with a breed you have read bad things about in the newspaper) vicious? This one's tough because—let's face it—the same handful of breeds heads the list every year when it comes time to tabulate which dogs are killing people or leading the pack when it comes to reported dog bites. But the negative publicity itself feeds the problem. People who want dogs for "protection" read the lists and gravitate toward these breeds when choosing guard dogs and watchdogs. They encourage the aggressive tendencies; they even breed for them.

And so an aptly named vicious circle is created—the more bites and attacks, the more dogs who are selected for those qualities, which leads to more bites and attacks and an ever-growing reputation for the breed as a "rotten" one. Actually, it takes less effort for some dogs to be made rotten than others, but dogs are not predisposed to being vicious. Viciousness in a dog is always a combination of genes and rearing. Early treatment and socialization, experience and exposure to novelty, nutrition, health or freedom from disease, and trauma play their part in shaping the dog's personality and disposition.

Breeds such as the pit bull, German shepherd, Doberman, and Rottweiler have gotten reputations over the years as being "naturally vicious." But often, it's the combination of unscrupulous breeders and vicious or neglectful owners who are to blame, not the dogs themselves.

An early study done by animal behaviorist D. G. Friedman actually looked at the way different breeds of puppies react to indulgent rearing versus disciplined rearing. He and his colleagues reared four different breeds: shelties, fox terriers, beagles, and basenjis. The indulgent rearing consisted of the kinds of things I prefer to do when I first see a dog: They encouraged the puppies to engage in all the activities they initiated, including climbing, licking, pawing, and mauling the owner for affection.

Disciplinary rearing consisted of discouraging free play, requiring the puppies to be compliant, using lots of obedience commands, and working with the leash. When all the puppies were 8 weeks old, they were tested individually by having a person deliver a food bowl to a room. When the puppy approached to eat the food, the person would clap his hands and say, "No!" In other words, he would punish the puppy. The person then left the room and left the dog alone with the food bowl.

As you can imagine, all the sensitive shelties, regardless of whether they were reared with indulgence or discipline, were essentially horrified and did not approach the food bowl for the rest of the experimental trial time, which was about 10 or 15 minutes. A clear breed effect was at work here; it didn't really matter how the puppy was reared. Once disciplined, shelties are likely to say, "Okay, I give up. I'll comply." I should say, however, that too much discipline with a sheltie is likely to cause him never to come out of the corner, as my clients found out. (Of course, with individual differences in mind, some shelties would come out of the corner sooner than more fearful shelties.)

The basenjis, on the other hand, immediately ran to eat the food once the person left the room—regardless of whether they had been indulged or disciplined as puppies. Thus, the kind of rearing didn't seem to matter with basenjis, either, except that the breed characteristics caused just the opposite reaction. They ignored punishment or mild discipline, and once the person left the room, they all rushed to eat the food. In both the shelties and the basenjis, the breed characteristics tended to have more influence on their behavior than how they were reared as puppies.

For the beagles, however, the type of rearing was more important than breed influences. The indulged beagles behaved like the basenjis: As soon as the person left the room, they rushed to eat the food. However, the disciplined beagles, once corrected by the person leaving the room, did not approach the food bowl for the rest of the trial. The same was true for the fox terriers: The type of rearing was important, but because of breed influences, they reacted in the opposite way. Indulged terriers would not approach the food bowl as quickly as disciplined basenjis did; they were very sensitive to correction and more "sheltie-like" due to their upbringing. However, the disciplined fox terriers were more "basenji-like" due to their upbringing!

So when I hear people saying, "Gee, yelling at Bowser kept him from doing that; how come it doesn't work for Frankie? Guess we need to do more of it!" (see Myth #1), the owner is guilty of what we behaviorists call typological thinking. It's almost as though we try to make each dog be exactly like every other dog. If the dog doesn't turn out right, it must be the owner's fault because the owner didn't train him with Method X.

🐾 Myth-Breaker #3

The truth is that there's no "one size fits all" in animal behavior. If you look at two different kinds of sheepdogs, the livestock-guarding dogs are totally disinterested in the sheep they guard from predators, but border collies, which are herding dogs, stare at sheep and have developed the next stage in the predatory hunting pattern that includes eye-stalk and chase. But border collies do not grab, bring down, and kill sheep the way wolves do. It's almost as if some dog breeds have developed no aspects of the predatory pattern of wolves, and other breeds have developed two or more aspects of the predatory pattern, probably due in large part to selective breeding by man. So the herding breeds get some of the predatory behavior patterns but not the final wolflike ones.

One colleague described groups of playing border collies and herding sheepdogs as separating into different "tribes" during play because each breed's playing routine is different. The border collies played eye-stalk chase games, whereas the guardians played dominance-submission games. The two types of dogs remained in breed-specific groups and did not play with one another. And this was within closely related working breeds!

So make sure that you recognize and respect differences while becoming your dog's best friend. Observe what your dog responds to within the general principle of learning—"if it feels good, I'll do it"—and adjust how you treat your dog accordingly.

Myth #4: Once He Tastes Blood, He'll Always Be Vicious

You've probably heard some version of Myth #4. From that first bite onward, Star will be a vicious hunter, she will become wolflike or lionlike, and you'll never get her to stop attacking and feeding on the neighborhood children. You might as well "put her down" the first time she bites, before she lunges at you and rips out your jugular vein. In part, this myth encompasses a second—the myth that you can't teach an old pet new tricks.

🐾 Myth-Breaker #4

There are two fallacies here. First of all, many dogs do not bite again, even if they have bitten and caused someone to bleed. Second, we applied animal behaviorists would not have a practice if we couldn't teach an old dog new tricks. There are a number of instances where people, especially kids, become wounded or are bitten by a mosquito or are scratched during play by a tree branch and bleed, and when they return home the dog licks it! Should we worry? Is the dog going to consider Johnny dinner from then on? Of course not. The dog is probably more interested in caring for you or your wound than in having your arm become her dinner. (This is also true of some cats, who take a lot of time and effort licking their human companions.)

But let's be reasonable. Are dogs more likely to bite once they have bitten if you don't do something to indicate that it's not acceptable behavior? Yes, they probably will bite again if they get something out of it or have a rotten temperament or both. But they are not turned into killers because they lick a wound (taste blood)! If they have been allowed to bite or snap at people for years, it will probably be difficult to get them to stop. Not only will it be necessary to stop the dog from biting and interrupt a well-learned behavior pattern that has become the habitual way he responds to people that approach him, but he also must learn a new way to behave—a new pattern of behavior.

If you're dealing with a younger dog or a puppy and the biting has just begun, the situation is very different. Even if the dog tastes blood, he's not likely to become a vicious killer overnight. Because he's only bitten once, it's not a pattern of behavior, and it should be easier to correct the pup's behavior— all other things being equal—than to change the behavior of an older dog who has been biting for 6 years. You can teach an old dog new tricks, but in some cases it takes more effort than it does with youngsters.

Myth #5: Cats Are Sneaky, Evil Creatures

Scratch the surface of the average cat hater—and people seem to "hate" cats with much more passion than they "hate" dogs, don't they?—and you will find a person who is actually afraid of cats at some kind of visceral, unconscious level. That's my hunch, anyway. Plenty of people fear dogs, but those

(continued on page 218)

DR. WRIGHT'S
C A S E B O O K

Nipper

The myth that all dogs are alike is seen in the belief that if all same-breed dogs are reared in the same environment, they will all turn out pretty much the same. In other words, in the eternal nature versus nurture debate, the nurturers say, "Genes don't count!" And, as many have found to their dismay, that is not true. A recent experience with the hugely popular Jack Russell terrier of TV fame brought that fact home to me.

This case involved a series of phone calls from a rather distraught owner of three Jack Russell terriers. Two of them she had raised from pups. They were now 6 and 7 years old and were great dogs. Mrs. Casper thought that she knew all there was to know about the breed when her daughter presented her with a third puppy, who was now 2 years old. This dog, named Nipper (very appropriately, it turned out!), had behaved very differently from the other two dogs from day one.

Instead of being friendly to the kids in the neighborhood, Nipper would lunge at them and had already bitten two boys. When he sat with his owners on the couch, he would snap at them and bite them. At bedtime it was more of the same: Nipper lunged at and bit whoever was walking by. We agreed that simply being the third dog in the house couldn't account for this kind of aggression.

"This has been going on since Nipper was 4 months old, and I want it to stop," Mrs. Casper told me. I could almost feel her frustration through the phone line. "I thought I raised all these dogs the same way, but I guess I must have done something wrong with Nipper."

This is a fairly common misconception among caring dog owners. If the dog turns out badly, or just different from what he's "supposed" to be, they blame themselves. And they really shouldn't. I told Mrs. Casper that if Nipper's rearing environment and experiences were basically similar to those of her other pets, the dogs' different genetic makeup probably influenced the behavioral differences. I asked her for information about the dogs' backgrounds. The first two had been obtained a year apart from a well-respected local breeder, whose name nearly guaranteed that every attempt had been made to give them a good start.

"And what about the third one?" I asked.

"Nipper came from a place we don't consider to be as good as the first breeder's kennel," Mrs. Casper answered, "but my daughter gave us the puppy as a twenty-fifth anniversary gift."

My idea was that the parents of the dog would hold the key to his misbehavior. So I probed a bit in that direction. Mrs. Casper said that she would have to ask her daughter to find out what the breeding situation had been. She called me back the next day.

"My daughter said that when she went to pick out the puppy, both parents were there," Mrs. Casper told me. "The mother seemed fine, running right up to my daughter, friendly as all get-out."

"But the father was walking back and forth, sort of pacing, and staring at her with a very mean look in his eye. And he was barking the whole time, and even snapped at her."

"The father might help explain things," I remarked.

"Yes, I have to say he sounds just like Nipper," Mrs. Casper agreed. "I guess it wasn't my fault after all." This was scant comfort; the dog was still unsafe.

I consulted with his veterinarian on this case because Nipper's aroused state might respond to veterinary medical intervention, supported by some careful handling from a behavioral standpoint. But I doubted that Nipper would ever be the "same" as the other two dogs of his breed. It is discouraging when dogs are bred in spite of parents' bad traits. That's how "nature" gets a bad rap.

fears are usually based on some kind of bad experience or on the reality of a snarling canine in their face. In other words, their fear is perfectly reasonable. But with cats, there is often no reason for their fear—people will say that they don't know why, but they "just don't trust cats."

People's fears are all wrapped up with their perceptions of cats. We attribute certain characteristics to the cat based on what we have seen in a horror movie, read in a storybook, or heard from our grandparents or scoutmasters in scary tales shared around the campfire. Who hasn't heard Myth #5—that when you bring the baby or grandbaby home, your cat will wait until you're not looking to rest on her chest and suck the breath out of her?

🐾 Myth-Breaker #5

Of course, this old wives' tale is not true, but it's amazing how many people still repeat it. Like many old wives' tales, it probably has a logical basis. Perhaps the cat might be attracted to the formula or milk from the baby's breath. At least, that explanation is a more reasonable possibility for what may appear to be the cat trying to suck the breath from the baby's mouth. But not many people who live with cats and babies are really worried; they know better.

What is the real risk—if there is one—of having a cat in the baby's room? Perhaps the cat will accidentally knock something off the table onto the baby.

Cats may covet Baby's warmth or companionship, just as they do yours. They certainly aren't trying to kill her off!

DR. WRIGHT'S INSIGHTS

When I am visiting clients and I ask whether the cat is obedient and comes when called, they will usually say, "No, I don't think so" or "Gee, I haven't tried it" or "I have no idea." If I say, "Why not try it?" you would be surprised at the number of cats who then run into the room we're sitting in or look around the corner if they're not in the room already. So if you don't try it, it won't happen. Sometimes a cat's sociability becomes a self-fulfilling prophecy. If you treat a cat as though it's dangerous and evil, how social and friendly can it become?

My cat, Domino, is a black cat who sometimes looks like a Halloween cat, but there's one problem: She's terrified of trick-or-treaters! When people come to the door, she hides in the closet. Some evil witch no doubt cast a spell on her because all black cats are evil and suck the breath from babies on Halloween, during the full moon . . . right?!

Perhaps, if startled, the cat will leap off the baby and scratch her, as I have mentioned earlier. But accidents are different from evil intentions. Of course, for safety's sake, it's best not to leave any dog or cat alone with an infant. That's just common sense.

Mini-Myths about Cats and Dogs

In addition to the five "big" myths, ten "mini-myths" crop up regularly. We behaviorists and vets hear them all the time. Here's what they are, and what's real versus what's myth.

🐾 Mini-Myth #1: Let 'Em Breed

Mini-Myth #1 says something like, "I have to let Sweetums have a litter before I have her spayed. It's only fair, and this way she won't end up hyper." The reality is that there is no evidence that spaying has any effect on the temperament of female dogs or cats in terms of excitability. Sorry! As to the "fairness" of letting a pet experience motherhood—I guess that's a question for the philosophers among us to settle. But keep in mind that spayed females are likely to be healthier as they age, and that tons of unwanted puppies and kittens are put to death each day, which seems a lot more unfair to me. What do you think?

🐾 Mini-Myth #2: Spaying Makes Pets Fat and Lazy

Myth #2 may be true. There is at least a grain of truth in that statement because the average spayed female gains about 10 percent in body weight after spaying.

But if you reduce her food intake by feeding her a fixed amount of food each day (or reduce her calorie intake by switching her to a low-fat food) and provide her with regular, daily exercise, the weight gain doesn't occur. I remember a black Lab I saw a few years ago. Her owner complained that the dog had undergone a 25 percent weight gain since her spaying. A clue to why this had happened lay in the fact that every dog from blocks around hung out at the Lab's back door. What was the attraction? The Lab's "mom" was in the habit of tossing her dogs (as well as the eager visitors) handsful of doggie treats all day long. And now, 4 years after losing her female organs, the Lab had gained all that weight. There's a funny thing about spaying: It's often held more accountable for a dog's weight gain than the pet's sedentary lifestyle and too much food!

🐾 Mini-Myth #3: Dogs Can Smell Your Fear

You know this one. Dogs know when you're afraid of them because they can smell fear. We behaviorists have a simpler explanation. It's a behavioral explanation that stems from a pet's ability to see how you act when you're fearful. Stiffening, cringing, running away, even moving your arms close to your body can give you away. Will some dogs take advantage of this fearful-looking state of yours to go on the attack? Certainly, some will. But until we figure out a way to measure the fleeting emotion of fear by means of some hormone, odor, electrical impulse, or what have you, there will continue to be no real way to validate—with scientific accuracy—the hypothesis that dogs (or horses or snakes or any other animal) can "smell" fear.

🐾 Mini-Myth #4: Cats Spray Things to Get Back at You

This mini-myth has several variations: They spray your suitcase because they don't want you to go to that sales conference in Dubuque; they spray the boyfriend because they don't like his looks; they spray the _____ (fill in the blank!) because you brought home that bratty kitten and who needs a sister! Actually, cats like a lot of stability and routine, and all three of these examples are major disrupters of business as usual. Cats also get quite aroused by strange odors. Tell your boyfriend not to take it personally—if he's still around!

🐾 Mini-Myth #5: Dogs Feel Guilty

Here's that old classic: Dogs feel guilty when they've pooped on the rug. Yes, it sure seems as though they are suffering the pangs of guilt when you come home to a cowering, tail-between-the-legs, slinking-away canine instead of the one who usually greets you with exuberance. Yep, sure enough, there's the

When your dog looks like this, he's probably just afraid of what you're about to do; he's not feeling "guilty" about what <u>he's</u> done!

poop on the carpeting, and you go into your ballistics act right on cue. The operative word here is "cue." When he smells that poop (which he has forgotten he deposited there) and sees you looking at it, Lance knows that he is in trouble. He expects to get yelled at or swatted, have his nose rubbed in it, be banished to the crate, or undergo whatever punishment you usually mete out. It's more likely the "chaining"—a term we behaviorists use when one thing leads to another in the inevitable chain of events—that prods him to hang his head in anticipation of the punishment at the end of the line. And it all starts when you walk in the door and spot that smelly pile on the floor. (Remember our experiment with the rubber poop in chapter 5?) Yes, fear and submission are probably at play here, but guilt likely does not enter the picture!

🐾 Mini-Myth #6: Dogs Won't Bite the Hand That Feeds Them

Actually dogs are more likely to bite family members than strangers because there is more than just one kind of aggression. The kind that gets you bitten even as you dish out the chow might be possessive aggression, which may find Snuggles overly enthusiastic about lashing out at you precisely because you *are* the one who feeds him. Get your hands away from his food, he's trying to tell you!! So instead of elevating you to untouchable status because you bring home that pricey designer dog food, your pal is much less likely to consider you dessert if you are the one who spends quality time with him. It's a good bet that the person who fulfills his psychosocial needs—spends time with him, develops trust, and makes him feel good—is a little safer from gratuitous attack than the guy or gal who just opens the can of Alpo.

🐾 Mini-Myth #7: Don't Touch That Kitten or Puppy!

This one's another old chestnut, and it goes like this: If you handle kittens and puppies too much, the mother will abandon them. Ha! No one who has ever given birth will swallow this one! It may contain a grain of truth if you applied it to baby birds in the wild, although even this has been pretty thoroughly discredited. Mammals certainly do not let a human odor or presence deter them from their maternal duties, as long as (1) Mom has sufficient time to establish that the offspring is hers, usually by its smell and taste (from grooming it), and (2) you don't accidentally coat it with some strange animal-carcass odor you just dug out of the brambles in the backyard. You get the idea.

I recently got an e-mail from a young woman asking my opinion about the behavior of her 6-year-old female cat, who had been "found" by a friend at the age of 4 weeks and kept kenneled by the vet until the writer adopted the cat about a month later.

"She climbs on my lap or my chest and puts her little paw on my face to direct it toward her. Then she just 'nurses' on the tip of my nose for a short while. If I try to turn away, she gently puts her little paw back to my cheek and starts again. When she is finished, she usually falls asleep."

I told the young woman that it is not uncommon for cats to continue to knead, suckle, or lick on parts of our anatomy. And I suggested that if she wanted to, she could probably displace the cat's sucking onto a small pacifier—it wouldn't be the first time!

I believe most cats and dogs who have had access to their mothers for at least 6 weeks—unlike this "needy" kitten—will have had their basic bio-psycho-social needs fulfilled by their mother and will be ready to graduate to more "mature" activities like playing, pouncing, nipping, and scratching! So handle your kitten or puppy as much as you please, but don't be in too much of a hurry to take her away from her mom.

Animal behaviorists have studied litters raised with minimal human contact and compared them to those litters exposed to people on a regular basis. It has been shown that as far as people-friendly kittens go, the earlier and more often kittens are handled, the friendlier they're likely to be as adults of either sex. (And the same is true for puppies.) As little as 15 minutes' handling a day has been shown to make a difference in affectionate behavior such as rubbing people and head-butting. But kittens handled or held up to 2 hours a day are even more people-friendly. Although genetics play a part, kitties who are handled a lot are consistently more apt to seek out a friendly lap for sitting and sleeping and have been known to drive their owners nuts with persistent demands for petting.

❧ Mini-Myth #8: Pick 'Em Up by the Scruff

Mini-Myth #8 says that you should pick up the kitten or pup by the scruff of her neck because that's how the mother carries them. Nope! Ask yourself this question: How else is the mother going to carry her babies? It works for her when the kits or pups are newborns, not much weight is hanging from the baby's neck, and the baby's inward-postural reflex still "works." But by the time you get the new kid home and she's no longer an infant, it's a different story. You know that paralyzed look the cat or dog gets when you pick her up that way? Well, there's a reason for that! And it probably has more to do with cutting off oxygen than eliciting a postural reflex. Seriously, always use two hands and support the kitten or puppy under her front legs and behind her back legs. She'll be glad that you did.

❧ Mini-Myth #9: Cats Are Cold and Aloof by Nature

This one has to be the most widespread cat myth. Just this morning I heard a radio talk-show host say that cats are too "snooty" for her; give her a dog that will run up to her and lick her and jump on her—now *that's* a pet! But those fans of our whiskered friends who appreciate the more subtle purring, rubbing, and lap-sitting rituals of felines know that this myth is nonsense. Now perhaps there is some truth to the myth that cats can tell who doesn't appreciate them and prefer to spend their time under the sofa when those folks are around, thereby giving rise to this vicious rumor of a species with a permanently cold shoulder! So this is one myth I am especially pleased to be able to debunk.

Fortunately, anyone who has lived with a kitten and reared it as a companion animal knows how friendly and social and, yes, even obedient a cat can be. Of course, I'm not going to be guilty of typological thinking, so I'm not going to try to convince you that all cats are equally friendly, social, and responsive to people—they're certainly not. But ask yourself these questions: Does your cat greet you at the door when you come home? Does he watch you eat at the table or try to join you there (keep in mind that almost all cats in the wild are solitary eaters)? Does he visit you on your pillow in bed at night? Does he sit next you on the couch? Do you find yourself catching your cat looking at you from across the room, and when he meets your eyes, does he approach

Mother cats and dogs hold newborns by the scruffs of their necks to move them, but that's because they don't have hands, not because it's the best way to hold them! Always cradle a kitten or puppy securely with one hand under her rear end and the other supporting her front paws. Hold her against your body for more stability and security.

you? If you're like most cat owners I talk to, the answer to these questions is a resounding yes.

So, many cats who spend most of their time in the household as companion animals are not cold, unfriendly, asocial creatures.

🐾 Mini-Myth #10: Dogs Don't Bite without Provocation

The final mini-myth is that dogs don't usually bite unless they are teased. Whenever a dog bites a child, the parent usually asks the obvious question, "Were you teasing the dog?"

"No!" cries the child. "I didn't do anything to him. He just bit me!"

Do the parents believe him? They should at least give their child the benefit of the doubt. Teasing is not a major cause of dog bites—at least, not an immediate cause. Sometimes children are bitten by dogs that they get too close to or that they have teased in the past. But less than half of all reported bite victims have had *any* type of interaction with the dog prior to the bite, and in relatively few of these cases could the victim's interactions with the dog be remotely regarded as teasing. Then why *do* dogs bite? I will get into the various reasons in chapter 15.

Chapter Eleven

Dr. Wright's School for Manners

Now that you've read chapter 10, you know that cats don't sit around plotting to suck the life from the baby, and that the angelic-looking puppy you just purchased almost certainly wasn't born rotten.

That's the good news. The rest of the news is that dogs and cats don't behave like paragons of virtue when left to their own devices. So if you want to be able to enjoy your years together, you need to get them started with some basic good manners.

What kinds of things fall into the category of manners? That's not so easy to say because every household and pet is a little different—one dog's rare accident can be another's entrenched horrible habit. But, in general, manners are the things we want to teach our household companions to do or not to do so that we don't end up having to call a certified animal behaviorist to work on a serious behavior problem that might have been prevented. Good cats and dogs will definitely do "bad" things if they aren't taught attractive alternatives. For example:

They jump up on counters or tables—invariably during dinner parties! They nip, bite, and scratch—often "just for fun." Some fun, eh?

They drink from the toilet, unroll the toilet paper, and then run away when you yell, "Come!"

What these pets need is some manners! So let's get started with some basic house rules. Are you—or, more importantly, is your cat or dog—ready? Review the "Manners Readiness Checklist" below to find out!

We covered the motivation part of learning in chapter 3. Recall that you can influence your dog or cat to do or not do certain behaviors by using both aspects of motivation—need and incentive. The internal part is the need ("I'm hungry! It's been 5 hours since I've had anything in my mouth," or "I'm feeling playful! It's been 5 hours since I've satisfied my need for play.")

Manners Readiness Checklist

These guidelines will help you determine when your pet is primed to learn the skills that will make him a valued member of the family.

1. He can settle down well enough to pay attention to what you want to show him how to do.
2. She knows how to pay attention to you. For example, she looks at you for instruction once she hears you call her name. You can teach her this skill if she's not too good at it.
3. He knows how to do the behavior or is bright enough to learn the new skill. You'll know if he's intellectually challenged if it takes him 2 weeks rather than 2 minutes to learn how to sit. So, he's slow, but he's not dumb.
4. She understands that a consequence (pleasant or unpleasant) is connected to the behavior or skill she's just performed. This will be obvious if she increases how often and how fast she sits, for example, after the pleasant consequences for sitting are repeated, or if she decreases how often or fast she steals napkins after you ignore or correct her for stealing napkins.
5. He is motivated to do the behavior you want him to do (or not to do). This is different from being bright. If he knows how to come, it's more like, "Will he?" How willing or stubborn is he? Will you have to use very little effort to move him to action, or all the effort you can possibly muster at one time to get him to merely budge?!

Then there's the external part of motivation—incentive. The incentive provides the resource for satisfying the need: hunger (need) and food (incentive for satisfying hunger). A dog's or cat's motivation is highest when he is in need of satisfying a bio-psycho-social need. If your pet recognizes that you are the source of the incentive (or resource), he will look to you for what to do (how to behave) to acquire or receive the incentive that will satisfy his need at that moment. It places you in a position of shaping your pet's daily routine.

Timing Is Everything!

Motivation is a tool you can use when you're thinking about the most appropriate time to teach your pet a new behavior (manners). And trust me, timing matters. When do you want to begin to teach your dog how to come when she's called, or sit when she's told, or not to jump on people when she's greeted? Should you begin when she's just finished playing or eating her main meal? Probably not, because neither a toy nor a playful person (incentive for play) nor a treat (incentive for hunger) is likely to motivate her to behave, to learn the new skill that she'll need for acquiring the household manners.

If she's already full, food is not as powerful an incentive (unless, of course, she's a golden retriever). As a result, she'll be less motivated to learn or to perform a new skill. A better time to teach a new skill is before her meals or as a part of social play (that's play with someone, as opposed to solitary play with a toy) or prior to social play. Take advantage of her bio-psycho-social needs in deciding when to schedule a training session, to maximize your and her training success.

The Basics

When you start training your pet in my school of manners, you and your pet should know some basic commands. Not only are these commands essential to a well-mannered pet, but they'll make your life much easier and could actually save your pet's life. They include Come, Sit, Off, and, of course, your pet's name. Let's start with the name.

❧ Say Goodbye to "Hey You!"

The first part of teaching manners is to get your dog or cat to realize that they pertain to him. So we have to teach him his family name. Now, an 8- or 10-week-old puppy is probably not wedded to any name yet. But if he's come with a name, why not use it, so that he doesn't have to "unlearn" Max at the same time you want him to learn Rowdy. But what if you just don't like Max? Okay. Let's change it then. The first rule is to use his name before giving a command. So let's see how you might teach Max that his name is Rowdy.

Teaching His Name

I've found two techniques helpful in teaching a pet his name. The first works like this: Call his (new) name abruptly and with great anticipation and expression. Do this when he's a foot or three from you, and you're sitting on the floor with him. When he looks in your direction (we behaviorists call that an orientation response), point your finger at your face, so he'll get used to looking at you when he hears his name. He'll pay attention to your hands because, for him, they are the source of pleasure (he gets strokes and treats from them), and they move! Recall from chapter 2 that critters are prepared to pay attention to important "stimuli" that move. He'll also see your "smile face," which he'll learn precedes things that feel good to him. Now, at the same time that he looks at your eyes, give him some whopping big social praise: "Go-o-o-od Ro-o-o-owdy-y-y!"

Or try this second technique: Call his new name softly, in a whisper ("Ro-o-o-owdy-y-y"). At first, do this when he's lying calmly, then work up to more normal household ambient noise and situations, keeping your volume the same. Repeat it in the same manner—sitting close to him on the floor—and with the same hand gesture you used in technique #1. You may be surprised at how well he pays attention when you make it fun—there's something in it for him—but it also depends on your acting ability. He has to be convinced you're just thrilled that he looked at you when you whispered his name. How exciting!

❧ The Command That Can Save a Dog's Life: Come

Once your pet knows her name, the next step is to teach her to come on command. You can teach a puppy to come using the word "come" or any other word or short phrase you like (such as "Get-over-here," Letz-go," or "Come-on"). Dogs learn words best that are novel and distinctive. This word can save her life because a dog who is under your control will not run into traffic, get lost in the woods, or lunge at an animal-control officer. And she won't make a nuisance of herself when you visit friends. So pick any word you want, but use the same word or phrase each time. Try the following three steps:

DR. WRIGHT'S INSIGHTS

Here is one note of caution: Never punish a dog for obeying you! Be careful not to reprimand or punish your dog if you call him and he comes to you, even if he fails to stop and runs right into your sore knee, or more likely, if you've discovered the shoe he tore up and just want to "get" him! As in the movie <u>Field of Dreams</u>, if you call him, he will come. But it's also true that if you punish him for obeying your command and coming to you, he will not come the next time. So grit your teeth, suck it up, and praise him for coming when you called him. The pain or frustration will eventually go away, but because of your good training, he will not!

1. Call her name. Use her name in a happy, upbeat tone of voice, complete with smiling, to elicit the orientation response we described on page 228.

2. As she begins to approach you on her own, say, "Come," adding salutations as she walks toward you, followed by strokes and gentle handling. "Go-o-o-od Come! Go-o-o-od Come!" The word "come" and the intonation of your happy voice are now becoming connected to her happy feelings, and she will experience that happy feeling in future opportunities to come when she hears the word "come."

The key to success here is to act as though it was Roxy's idea. It's a technique that allows Roxy to connect the action of walking toward you with your giving praise, and that she's responsible for "making" you give her the praise and attention when she hears "Roxy, come."

3. Try whispering the instruction and praise as she approaches. Both of you, be happy!

Make Training a Family Affair

For introducing "Roxy, come" to several people in the family, use the same technique with the following changes:

1. Have everyone sit in a circle around the puppy, within leash length (a 6-foot leash should do). After the first person has finished stroking the pup, toss her leash to someone else in the circle, who repeats the come command.

2. You may extend the distance as training the Come proceeds. You should also do the Come in different areas of the house, backyard, front yard, in "strange" territory like a park or playground, and so on, as you introduce her to those different locations in the next several weeks.

Dr. Wright's
✉ E-Mail

Dear Dr. Wright:

I have a 1-year-old keeshond puppy who will not
obey the Come command. I live on a farm, and she would
rather escape from inside the house to be out in the
field all by herself, catching mice, than be with
me. Most of the time, I go out to exercise her because
I want her to have some freedom, but now that is all
she wants! I have tried many ways to get her to come
when I call, but she won't listen, and I give up. My
fear is that the farmer next door will shoot her
because she will go into his field and disturb his
cattle. She will not respond to anything. I have tried
to run in another direction, away from the field to my
farm, or make excited noises, but to no avail. Usually,
I just wait until she comes back for water or food.
Please help!

Michelle

Dear Michelle:

It is very frustrating when a puppy decides that
there is more to life than spending quality time with
her caretaker, and instead just goes about doing her
own thing (much like a teenager!). Think of how lucky
she is to have you! You keep her warm at night, feed
her that tasty puppy chow and fresh water twice a day,
and give her all those strokes?! You would think that
she'd be more grateful, or at least acknowledge your

presence by raising her head when you go nuts trying to get her to come to you.

Well, so far, you've tried some pretty terrific ways to get her attention so that she can realize that you're just as exciting as that mole in the ground—and from the dog's perspective, that's a really tough choice! Running in the opposite direction often elicits a chase response. That's good. Waving your arms and making "jolly" is also a good attention-getter. I'll bet most readers would run over to you to see what the commotion is all about when you do that (after calling 911, of course!). But relative to that mole and those good earthy smells . . . well, what she's really thinking is, "Don't bother me now!"

On a more serious note, if you're really worried about your neighbor's behavior toward your dog, why not make him a cordial call? Let him know that you're doing the best you can to get control of your keeshond, but that if he sees the dog where she shouldn't be, please give you a call—you'll be right over to get her. (In other words, he doesn't have to shoot her!)

In the meantime, why not take the time to enroll her (and yourself) in an obedience training class? The advantage of taking a class is that she'll be in the presence of other dogs and people—a good socialization thing—and she can learn the basic commands, including the life-saving Come. Since you've done almost everything else to get your dog's attention, why not enlist the aid of a professional to give you some more options and the help you both need? He or she may be the key to helping you save your dog's life. Good training!

Helping the Fearful Puppy Learn the Come

For puppies who are not willing to approach you with a soft prompt (pull) of the leash toward you, don't force them. These seemingly stubborn pups are often quite fearful, and teaching manners becomes, first and foremost, a matter of enabling them to overcome their defensive urges and the overpowering fear that's created by pulling them toward that scary loud person during the training session.

The harder you pull the leash toward you, the harder she pulls against you, and a confrontation results. And that's not the basis for a good, trusting relationship. When calm is achieved, it then becomes just another task of teaching her to come, just like other puppies. We'll address how to reduce some other aspects of fear-induced behaviors in chapter 14.

With a leash on her collar or head halter, repeat steps 1 and 2 of the Come training on page 228, but add a prompt, a little pull toward you to enable her to overcome the inertia of not moving. Then, as soon as she heads toward you, continue to reel her in, as if it were her idea. Reward her compliance with bunches of praise and love and handling!

Teaching the Come without Spooking the Cat

The Come is less of a sure thing for kittens than it is for puppies. This will probably not surprise you "independent" cat owners! For kittens, name recognition depends on the same kind of training, but because cats are not dogs, the kinds of payoffs for kittens—the consequences of approaching you when their name is called—are different in quantity, not in kind. So when kitty approaches, the less arm movement and loud noises, the better. Think of a dog's likely response to your jumping up and down and waving your arms—"Oh boy, play!!!" He'll probably want to join in the fun and jump up and down, too! (Just ask any bunch of kids who play like this in the backyard.)

Now, think of a cat's likely response to the same behavior. Most cats will run off in the opposite direction—it's just not their cup of tea. Some cats will remain in the same room with you, rubbing a chair leg or other piece of furniture a safe distance away from you, and a few cats will run right up to see what the commotion is all about. But remember that there's a reason it's "scaredy cat," not "scaredy dog." Kittens are more likely to respond to and be motivated by treats, strokes, talk, and play (both social play and object play), as long as they're done calmly and with fewer sudden movements.

🐾 Teaching Your Dog to Sit

After your dog learns his name and how to come when called, he needs to know how to inhibit his activity a bit. When he comes to you, you don't want him to learn to jump up on your clothes or your legs, nor do you want him to cower in response to your "No, no, no!" if he gets overexuberant. So let's teach the Sit. If the incentive is food, and he's hungry, try these steps:

1. Position yourself so that you're in front of him, and then show him the treat you're holding between your thumb and forefinger.

2. When you have his attention, lift up your hand slowly and move it just over his head, so that he has to look up and his rear end has to move toward the ground.

3. When his rear end meets the ground, say, "Sit," immediately release the treat so that he can eat it, and praise him as though it were his idea: "Go-o-o-od Sit."

(continued on page 236)

Your dog will sit so that he can follow your hand as you move it over his head. Be sure to say his name and "Sit!" ("Sampson, sit!") as he lowers his rump to the ground.

The Head Halter—Friendly Persuasion

Some puppies may not respond well to learning new manners. This could be because they're a bit slow and need more guidance. Or they may be bright enough, but not sure why they should do that silly thing you want them to do (or why they should stop doing the wonderful thing you want them not to do). Then again, they may have already learned a bad habit. These pups need more understanding, and they need for you to shape an activity gradually so that what is now an undesirable behavior becomes a new, desirable behavior pattern.

A tool (that's <u>tool</u>, as in a device that facilitates doing something, not to be confused with <u>solution</u>, the desired end product of the use of a tool) that can help your puppy or misguided adolescent learn his new manners is the head halter. The head halter is a collar that is used as a gentle communication device—no jerks allowed! The advantage of a head halter over other kinds of collars is that very little pressure is needed to guide your pup away from the "wrong" behavior and into the right one.

Attached to a leash, a head halter can be used to control your pup's head from under the chin, so that a light pull places pressure on the nose and neck loops (much like the halter of a well-trained riding horse), and he learns to stop moving. The result is no running, no jumping, no errors. Any further movement forward, or jumping, or lunging is not possible, and he quickly learns that pulling is futile. You, of course, praise your quick learner.

Another advantage of this kind of collar is that it does not encourage you to struggle against your dog. How many times have you seen a well-muscled dog pull his owner around at the end of a taut leash? The dog's choke chain or prong collar is digging into his neck, and he is straining to move forward, choking, and making gasping sounds. You've seen it too many times, right?! Fortunately, it's neither desirable nor necessary for teaching manners.

So try going for short, fun walks with your pup after he gets used to wearing his halter and learns that pressure on the nose loop means stop, but that he can go forward again once pressure on the nose loop ceases. Then, when he enjoys the sight of the halter (because it signals walks and other fun things), begin teaching him some basic manners with it on.

Walk him through the exercises, offering gentle guidance on the leash here and there as needed, to teach him to Come, Sit, and Say Hello effectively. Be sure to use his name, followed by the command word or phrase, and lots of praise, just as you would if he didn't have on the halter. Make him believe that it's his idea to do those things. When you think that he's "got it," try walking beside him without using the halter, and finally, try giving the commands as if you never needed the halter in the first place.

Some pups learn the commands in a couple of days. But for others, even with daily practice of up to 20 minutes at a time (training time can increase as the pup's age increases—you'll know what he's able to tolerate and enjoy if you pay attention to his emotions!), it takes a week or two to learn the most basic manners. So there is hope for pups and adolescents who need a little more guidance, and the halter is a nice communication device to consider.

A head halter teaches your pup to obey commands by applying gentle pressure rather than by throttling him like a choke chain.

If he decides not to sit the next time around, or he backs up a bit instead of sitting his fanny on the floor, move your hand toward you, so that he voluntarily comes forward toward the treat again. Then repeat steps 2 and 3.

If he begins to jump up or lunge, calmly move the treat much farther above his head, and when he "comes down" from his leap, bring the treat closer (thus reinforcing his movement down), and repeat steps 2 and 3. Be sure to praise his compliance.

✤ The Invaluable, All-Purpose Off!

The earlier you can teach your dog the Off command, the more you'll be able to use it—and to be glad that your dog understands its meaning. If taught correctly, the word "off" tells your dog to remove herself from the present situation. She must either get down from the bed or couch, get away from you or out of your face if she's giving you unwanted licks or walking on you while you're lying on the floor watching TV, or get off your lap if you've had enough cuddling for the present time.

In some sensitive (and well-trained) dogs, it is sufficient just to stare at the dog and say her name followed by the command, as in "Charlie, off!" Wait 3 to 5 seconds for the dog to get the hang of it and try to figure out the intention of your stare and grim face. Because this often causes dogs occupying a subordinate role to become even more submissive, the off-stare is sometimes met with facial licking, which is just the opposite of what you want.

However, if you become still more firm in saying the command, she should try the remaining strategy, which is to leave. At that point, praise her: "Go-o-o-od gi-i-i-irl!" The more frequently you try this, the more quickly she will learn it. But it's important not to overdo the command because your dog may start to think that closeness is not a good thing with you. In that case, she'll develop a strategy of staying away from you as her modus operandi, which is certainly not what you had in mind!

So use the command sparingly, and only when you really want the dog to move away from you or down from someplace she's not supposed to be. Make sure that when you praise her for doing so, you add a nice social reinforcer (petting) to let her know that you still love her, but that it was the behavior you did not want her to do.

Because it's the behavior you're influencing in your dog when you say, "Off!" you may reinforce it in dogs who do not respond to a stare and the grim command by gently but firmly placing them off and praising them as if it was their idea. Dogs learn quickly that if they don't comply in 3 to 5 seconds, they'll become "Off!" anyway!

I've used this technique with all my dogs, and some respond much better than others. (Of course Peanut, our deaf dog, no longer pays any attention to what I say but does respond to a finger-pointing arm gesture downward, which I used along with the vocal "down" when she could hear.) But for the other three dogs—especially Charlie, the chocolate Lab—the Off command is met with their getting up and moving away or stopping an "in your face" activity. Of course, my wife and I add, "Good gi-i-i-irl" at the same time, so she hears the praise coming from both of us. She *is* a good girl, we just wanted her off the bed.

Thus, whether the word "Off!" is used to get your dog down from the couch or out of your face, or to stop his humping before he gets into it, it should mean to the dog to move away and cease whatever he happens to be doing. That's a lot of power for one little three-letter word, eh?

How to Persuade Your Cat to Greet You

Cats don't sniff in private places or knock over visitors when greeting people, which probably adds immeasurably to their popularity among folks who can't stomach such behavior. Unfortunately, many felines don't make it their business to greet their owners or anybody else, for that matter, which is what makes dogs so popular! People, after all, crave attention, too. At any rate, it is not very common for cat owners to want their felines to do tricks or to do obedience commands. So, even though teaching the Sit is not regarded as teaching "manners" to the cat, having her approach to greet you once you speak to her is.

How many times have you felt as though you were talking to the wall when you entered a room and said some kind words to Muffin, only to be met with two blue eyes staring at you (albeit, pretty blue eyes). "Why won't she come to greet me like she does when I get home from work, or like she does when I feed her?" Well, remember that the key to influencing behavior is to make motivation your friend. Select a time close to your kitten's dinner time to practice teaching her to greet you when you so desire. Try these steps:

1. Call her name to get her attention: "Muffin."

2. Place your finger parallel to the floor, 1 to 2 inches above the floor, and point it toward her nose.

3. Say, "Hi, kitty, kitty, kitty" or "key-key-key-key-key"(or your own personal phrase of discrete sounding words) in an upper register (high-pitched voice).

4. As she approaches, hold your finger still, and let her sniff it as you say "Go-o-o-od Mu-u-ufi-i-in!" She'll probably move on to rubbing your finger or leg, the normal "next behavior" for cats when greeting. (If she won't take that final step to touch your finger with her nose, add some tasty wet food to the tip of your finger, and then praise her compliance.)

5. Spend a moment talking to her, sharing your thoughts, and then feed her.

She should begin to connect or "chain" (see "Dr. Wright's Casebook" on page 56 for more on chaining) the greeting sequence with the prediction of food. At some point, you may wish to substitute more psychosocial rewards for food treats after she has learned how to say hello, especially at nondining times of day.

🐾 When Training's Over

When you've finished practicing a command, want to end a "trial," or think that your dog has done the good behavior long enough, give him a release word or short phrase. I use "okay," but others have used "you're done" or "let's go" with equal success. The words should be said in an exuberant way and in an upbeat tone of voice, and they should be fun. They should truly be used to indicate that your expectations of him (what he's doing, his behavior) have changed. The release provides an opportunity for free time, for play, or for him to be on his own best behavior, until you say his name again, which precedes your next instruction ("Duke, come!").

If you would like your dog to learn more commands and to behave quietly

Cats greet each other by touching noses. Encourage your cat to touch your surrogate nose by extending your finger or closed fist knuckles outward and holding the position while calling your cat gently to you.

DR. WRIGHT'S INSIGHTS

Isolation will increase the chances that a dog will become more of a problem. Chaining a dog for long periods of time will almost guarantee a frustrated, unhappy dog. (Think about how you'd feel!) It may also actually result in a greater likelihood of severe or fatal dog bites from your frustrated friend. Remember that socialization is just as important to pets as to people.

on a leash (and who wouldn't?), you may begin formal obedience training when he is between 6 and 9 months old. This will strengthen the leader-fol-. lower bond and complement what you're doing at home to mold or shape your pet's behavior.

After Rowdy and Muffin have mastered the basic commands, you're not out of the woods yet, unless your dog or cat is quite unusual. That's because all pets occasionally use bad manners or develop annoying habits. Others go directly to major behavior problems!

Corrections for any undesirable behavior should be aimed at stopping the activity while it's in progress. Better yet, an instant before she is about to misbehave, draw your pet's attention to you and try to elicit an alternative, desirable behavior. Praising the dog for doing something else will teach her the house rules in a humane way.

Remember that most successful person-pet relationships result from mutual respect among all the parities, be they human, canine, or feline. Getting to know and understand your dog by paying attention to her needs so that she can "give" back to you will help build a healthy social-emotional bond. Dogs reared in this way are less likely to be turned in to animal-control facilities or left out to roam throughout the day.

If physical or behavior problems develop, owners should respond to their pet as a family member and not treat her like an object. Deciding to tie up the dog in the backyard is one way to deal with an unpleasant problem, whatever it may be, but this practice also marks your decision to abandon a friend and to begin to possess "living property." Don't do it!!!

Dealing with Major Turn-Offs

Let's move on now and see what can go wrong after your pet has learned his name, rank, and serial number—and some important commands. We'll start with the big stuff, and then explore some of those little irritants that drive us nuts and send the dog—sometimes literally—to the doghouse.

❧ Teaching Dogs the "Sniffless" Hello

I suspect that very few of us expect our dog to awaken from a sound sleep each morning and walk over to us or a loved one and place his nose in each person's crotch as a form of saying good morning. Thankfully, few dogs do. Instead, they learn to lick sleepy faces or toes that may have found their way out from beneath the warmth of the bedcovers by morning's light. At worst, it tickles. "Whoa! Good morning!"

But more than a few dogs do greet human friends and front door visitors by goosing the crotch. "Come in, Eleanor, this is our dog Row—Oops, oh gee, sorry, Eleanor! Rowdy!! Bad Rowdy!!"

Faced with the prospect that our puppy may develop such greeting tendencies—after all, nosing a person's crotch as a form of shaking hands is a perfectly normal dog-to-dog greeting—if left to his own devices, we need to teach Rowdy the proper way to say hello. He needs to start out with the appropriate dog-person greeting.

The nose-in-your-crotch greeting is a threat both as a frontal attack and as an attack from the rear, so it's best that we address this problem head on. We have to override his natural preference for sniffing the anogenital area as a form of greeting. (See chapter 2 for more on this.) We can do it by using the Sit behavior he has just mastered. Thus, after your pup knows his name and reliably comes when called, try the following technique for teaching the Hello. Two people work best in this training exercise.

1. Enter a room opposite from where your puppy and the second person are sitting, and get your pup's attention by saying his name: "Rowdy."

2. When he sees you, give the Come command.

3. As he approaches to greet you, take a small step toward him and raise your hand in front of you, palm toward your puppy, as though you're about to say stop, but instead, give the Sit command: "Rowdy, sit."

4. As he sits in front of you, praise him with "Go-o-o-od si-i-it!"

5. Squat in front of him (so that you're no longer standing above him) and offer your personal greeting: "Hello!" or "Ro-o-owdy" or "How's that go-o-o-od bo-o-o-oy" or whatever your favorite greeting happens to be. Simultaneously, stroke his chest, the sides of his cheeks, or under his chin, but not on top of his head (not just now).

You can control any attempts on your dog's part to break forward from his Sit or to lunge with your hands, as they are positioned in front of him at his shoulders. If your dog attempts to rise and back up from a Sit, you should also rise and move back a step, stopping all praise. Then repeat the Sit command and resume your greeting. Repeat as necessary. Pups learn to hold the Sit during the Hello because they can't jump up on you or lunge at you. They quickly give up and hold the Sit until you're finished with the greeting.

6. Next, stand up and step back from the puppy; then break eye contact. This ends the Hello sequence.

7. Have the second person repeat the exercise from the other side of the room. Of course, praise your pup's compliance.

When It Happens to You

What should you do if you're visiting a friend who lives with a "nosy" dog, and the dog's nose invariably finds its way into your crotch whenever you visit? You certainly shouldn't turn around and greet him. That's what he wants! Instead, ignore him, and work on maintaining your own self-control.

Angie and I have learned to prepare ourselves for such surprises and our responses to them. We visited some friends whose young dog decided that it was safer to greet us from behind (*our* behinds) at about the same time we were greeting the dog's owners. "HI-I-I-i-i!" we said—fastest nose in the neighborhood. Our friend's response: "Oh, did my dog do that again?" So much for behavioral intervention. I'm afraid that there's not too much you can do in these instances but ignore the dog, and work on anticipating that nose the next time around.

❧ Taming the Overly Enthusiastic Hello

For most visitors, an indiscretion like nosing where a nose is not wanted at first greeting can be quickly forgotten and forgiven, if the rest of your dog's manners are intact. You usher the visitor, your friend, into the living room to have a seat, your dog behaves himself, and everyone is fond of everyone else.

But what if your nosy puppy proceeds to jump up on your friend and create havoc with her pantyhose or, worse yet, her varicose veins? There goes *that* friendship! And your pooch looks at you as if to say, "But Mom, I just wanted to lick her face!" Whatever the dog's excuse, there is nothing hospitable about a host or hostess who allows the family pet to molest guests like this. In fact, it is really rude behavior on your part. But don't worry, help is at hand!

Remember that your puppy wants to know who the visitor is, and she needs to go through some sort of greeting (sniffing or licking) to identify the individual's scent (again, I discuss this in detail in chapter 2). Try the following steps for discouraging your pup from jumping up on visitors.

1. Teach the Hello. Jumping up on visitors is less of a problem when pet owners have already taught their dog to respond by sitting when greeting everyone who lives in the household.

2. Before using your best friend as a guinea pig, have a family member who has already mastered the Hello calmly enter the house through the front door, stand still, and begin greeting the dog: "Luna, come!" followed by "Luna, sit."

3. Instead of squatting, the family member should end the sequence by offering his hand—not with food, but as a source of tactile comfort—held just above her nose. She should sniff the hand first, then the person may stroke her cheek, or cup his hand and give her head a stroke—but no patting allowed. "Go-o-o-od Luna."

4. The family member should then enter the living room and sit down. Most dogs are pleased to have the greeting end with social praise, which in turn makes approaching someone at the door less exciting. This decreased agitation is a good thing, too, because less excited dogs are less likely to jump up.

Should your pup decide to jump up on the hapless family member anyway, he should step back, so that there is nothing for her to place her feet on and simultaneously turn his front away from her. As she places all four paws on the floor, he should give her a Sit command and praise her compliance. Always offer her a way to succeed after she learns that a "bad" behavior won't get her a good result.

5. Repeat the Hello with other family members until your dog is used to walking up to each person entering through the front door, sitting in front of him or her, receiving praise, and proceeding to accompany the person to your living room where other family members are waiting to offer a "good girl."

6. Now use your best friend as a guinea pig.

The key is to do these greetings often enough with family members so that when a "real" visitor comes through the front door, Luna is already prepared to approach and sit, *not* to jump up. If you stand next to a visitor at the front door, you can control not only the "stimulus" your dog is supposed to greet (your friend), but also your puppy, who already has plenty of experience sitting in front of people who enter through the front door. Your friends will love you!

❧ Convincing Cats to Stop Spraying

Although it's only natural for your male cat to let other cats know where he's living by marking the territory—spraying it with urine—most pet owners consider this behavior not only to be rude but also intolerable.

Neutering is one obvious solution for spraying (which is more likely in male than female cats). Other solutions include reducing olfactory, visual, and auditory access to cats coming around outside (or discouraging the presence of cats outside), avoiding contact with outdoor cats or friends' cats so that you won't bring their scents in on your clothing or shoes, and certainly keeping your cat from going outside and bringing these odors into his own house.

If your cat is spraying, peeing, or pooping in just one or two locations, try using a masking scent like a citrus spray in that location with something that sends a message to the cat like "Gee, I don't want to spray (or pee) here." Cats typically smell where they're about to mark or eliminate, so this sometimes helps. Mild chlorine cleaners, such as Soft Scrub with Bleach, can be rubbed with a sponge on the baseboards daily to inhibit the cat from continuing to spray. Or you may wish to toss a few treats or even feed your cat in the location where he's spraying (or peeing or pooping). The goal is to alter the meaning of that location from "here's where I spray" to "here's where I play."

If your cat is spraying, you may wish to rub Feliway, a feline facial pheromone, on the surfaces the cat has sprayed (commonly vertical locations), and rub it on surrounding locations, which your cat may also select. Ask your vet or certified applied animal behaviorist for more information about the appropriate use of Feliway. It's fairly expensive and has been only marginally effective in my experience. But it is another tool to consider in your efforts to change this disgusting behavior before it becomes a habit.

Finally, you may want to consider asking your veterinarian or certified animal behaviorist about the pros and cons of drug therapy. Drugs may be appropriate tools to use in stopping your cat from exhibiting such bad manners in the household. Most successful drug therapies are used as part of a treatment program that also involves behavior modification.

❧ Get Your Dog Off My Leg!

What about that other humiliating canine behavior, clasping and humping? Unfortunately for us humans, mounting is a normal canine behavior. Puppies initially learn how to control another puppy—usually during play—by

mounting and holding on. Mounting directed toward people, especially children, may also be about control or the solicitation of social behavior and attention. Or it may simply be sexual behavior, as mounting and humping are
normal social behaviors in a number of animals, beginning before adolescence,
and may occur in both neutered and intact male and even female dogs.

You'll have an easier time if it appears that this is an attention-seeking device—that is, just another way to get you to interact with your dog, along with
pulling at your clothing, pawing at you, whining, nudging, leaning on you
constantly, or jumping up on you—rather than a sexually motivated behavior.
If your dog is just trying to get your attention, be sure to provide him with opportunities for regular social interactions with you so that he can have his psychosocial needs met, but do it on your terms.

Sometimes dogs are willing to endure being pushed off, shouted "NO!" to,
or even struck by some people as a means of getting some sort of attention—
some social interaction with the people they most want to play with and to
have a relationship with. Ya just have to love them for that, even if they're
grossing you out!

Help for Humping

The solution to the humping problem depends on how far along you and your
dog are in the "I can learn to control you by humping you" sequence, and
whether you're dealing with attention-seeking behavior or sexual behavior. If
the behavior is caused by attention seeking, providing your dog with more of
your attention at other times can stop it. If it's sexually motivated, neutering
can help. However, if this behavior is part of a larger problem of dominance
aggression, or this is yet another aspect of your dog's frequent attempts to control you, you may need to seek professional help.

Generally speaking, the solution for humping is similar to the solution for
any attention-seeking behavior: If you see it coming, try to divert the dog's attention to some other activity. You could call him to go get his bear or other
toy, or you could get him some treats. In short, try to initiate any other behavior that's incompatible with mounting and clasping your son's alarmed 8-
year-old little friend.

Praise compliance. If the dog has already initiated mounting on your leg or
arm, this is where the Off command comes in handy. Say his name followed by
the word "off": "Rowdy, off!" If you don't get a response in 3 to 5 seconds, repeat the command, and then firmly remove your leg or arm so that there's

DR. WRIGHT'S INSIGHTS

Sometimes, trying to push a dog off or stop a dog from humping you can lead to the dog's attacking you. The treatment of this behavior, then, should be part of the protocol for the treatment of dominance aggression, which we'll get to in a later chapter. Owners should not try to solve serious aggression problems on their own. For the most part, however, correcting this behavior sequence can be successfully achieved by following the three steps below.

nothing to control. Sometimes the mounting and humping is accompanied by a lot of arousal, especially if it's sexual or controlling in nature. It may be that you feel uncomfortable pushing the dog off for fear of retribution; dogs sometimes become aggressive when you compete against them for control of your arm.

That's why enticing the dog into an incompatible activity is the best solution. At the first signs of humping behavior, it is appropriate to try to remove your arm or your leg and then give the dog a command he can follow, like "Rowdy, sit."

To summarize, here's the sequence I recommend for teaching your dog that it's not good manners to hump people:

1. Head him off at the pass by diverting his attention to a motivationally incompatible behavior—something like throwing a ball or passing out treats—that is quite enjoyable for him but that places you in control of his actions.

2. At the same time, ask the "victim" to remove her arm, or move your own leg, depending on what the target of the humping is. If there is no stimulus to do something else and the behavior is sexually motivated, the dog may continue to masturbate on his own. Sometimes this is quite frightening to puppies and even adolescent dogs, since they've lost all self-control and don't know what's happening to them. And it can occur even in spayed females, as well as other adult dogs.

However, if your dog is clasping and humping as a means of attention seeking or control, he'll stop in the absence of the stimulus. What this means is that you'll find out pretty quickly after you remove your leg or arm from the puppy or dog whether you're dealing with a sexually aroused dog or an attention-seeking dog who wants to control his interactions with you.

3. Before the dog gets into it, give the Off command. As he begins to release your arm or leg, remove the stimulus, and, of course, praise compliance.

Annoying Little Habits

Now that the "biggies" are out of the way, let's address some of the less earth-shattering examples of bad manners, which can nevertheless lead to short fuses within the household—and to pets being shunned or punished and wondering what they did to make you angry.

✤ Drinking from the Toilet or Faucet

If you're a dog or cat, there are times when a water bowl just won't do, and the dripping faucet is much better (mostly true of cats—I have a good photo of one in action), or a toilet bowl is just fine (primarily true of dogs). So, what's wrong with the expensive stainless-steel bowl you just bought over the Net?! After all, you add to the water every morning! What's the problem?

What constitutes "quality water" is different for dogs and cats than it is for you and me. We wouldn't think of drinking out of the toilet bowl, and we consider it rude to drink from the faucet. But to your cat and dog, whose immune systems are much heartier than ours, a drink from the faucet or toilet is no biggie! The advantages are twofold: The water is fresh and is frequently refreshed throughout the day, and the water is cool. At least, it's cooler and fresher than the water in the metal water bowl that's been sitting in the hot sun for the last 2 weeks. (Adding water to the bowl just isn't the same as changing it!)

The toilet holds an ample supply of water for a thirsty dog. (Beware of the

If your dog's "toilet habits" are grossing you out, make sure you keep the lid down!

blue chemicals you treat your toilet with to keep it smelling fresh—they could be dangerous to your dog.) And surely that dripping faucet in the bathtub that has worn a reddish circle into the bottom of the porcelain tub is an excellent source of fresh, cool water for a cat: lap, lap, lap.

Unless you treat your toilet water with chemicals, there's nothing really harmful about your pet's alternative drinking habits. But if you consider that your pet's drinking from these receptacles is "bad manners," try the following solutions:

1. Close the lid to the toilet bowl! That should make the women in the family happy, too, and it's P.C., anyway. If your dog has already learned to use his nose to lift up the lid (the shorter breeds can't do this), you'll have to baby-proof the lid, at least until he gives up trying (which usually takes a couple of weeks).

2. Provide your cat and dog with porcelain water bowls. They are less likely to conduct heat, and as far as your pet's concerned, it feels similar to placing his head in the toilet.

3. Provide a clean bowl of water several times daily. Dump the water, wipe out the porcelain bowl, and refill it at least once in the morning and again in the evening. If you hand-wash the bowl, be sure to remove all the soap suds: Nobody likes the taste of soap!

4. Place the water bowl near the preferred water source initially, and move it a few feet every couple of days or so until it's located where you want it.

5. Fix your dripping faucet.

🐾 The Toilet-Paper Fetish

Some dogs and cats are finicky: They have to have things just so, they're compulsive, and they can't stand anything that's different from the way it was the day before. And other dogs and cats are more than just carefree: They're downright sloppy with their food and toys; they're frivolous, gregarious, and generally overly expressive. But it's probably too Freudian of me to suggest that the one kind of pet is anal retentive, whereas the other is anal expulsive, even though it's the latter critter who seems to have a thing for toilet paper.

These pets like to spread the toilet paper all over your bathroom floor. Cats love to sit on the commode and use their front paws to unroll the whole thing (that's the whir, whir, whir sound coming from the bathroom), and they seem to take pleasure in watching your expression as you discover their deed. Let's watch as Mommy tries to save the double roll by rolling it back up!

Dogs love to play the unrolling game, too, although they seem to prefer a position on the floor where they can raise their paw and get some good momentum going (whi-i-i-ir, whi-i-i-ir, whi-i-i-ir). Sometimes they love to take the piles of properly stacked paper in their mouths and run with it helter-skelter throughout the house. They make the living room look like the trees some teenagers paper with TP—maybe that's where the kids got it from.

Trying to figure out the need that's being fulfilled by this activity is a tough one. There is no toilet paper in nature, and no one has observed its use in wolves or big cats. Perhaps it's a domestication thing, associated with manipulating and controlling the pet's environment ("Wow! I can make that thing roll!"), not to mention causing Mom and Dad to chase him. What could be more fun?!

Okay, so you could shut the bathroom door each time it's used so only humans can enter. Or, for cats, you could leave the lid up so there's not a real safe place to sit. (So far, no women have liked that one.) But fortunately for us, there is an even easier fix for this problem: Position the TP on the roll "upside down" so that it doesn't roll down but rather rolls up. Not even the most clever dog

Some cats and dogs love unrolling toilet paper. Foil them by reversing the roll so that it rolls up rather than down.

or cat has figured out how to "reverse an operation" (as Jean Piaget, the famous Swiss cognitive psychologist, might have put it). They don't seem to be able to paw up—only down (whir, whir, hrumph!).

🐾 Cats on the Table and Counters

Cats specialize in climbing and jumping up on things, even when they're not sure what's on top of whatever they're jumping on. Most cats are willing to "try, try again," if they have jumped up on a table or countertop successfully most of their lives and it has led to the top of the fridge, a leftover from lunch, or some other incentive worthy of taking the risk. But cats are less willing to try leaping if, when they do, they're met with some kind of unpleasantry.

Although a cat may have successfully leapt blindly onto a countertop for months, you can decrease her desire to do so by making the countertop, from where she sees it from the floor below, look somehow different. Some people have placed a towel on the countertop so that it hangs over the side where she can see it. Others have used waxed paper (a particularly slippery substrate that falls off the counter with the cat when she jumps). And still others have used tape—as in double-sided sticky tape. Cats don't like to jump on sticky tape, and once they stop trying, you can remove all the tape except for a piece that she can see "sticking" over the edge of the counter. ("Hmmm, must be tape up there—just can't risk taking a jump.")

Cats don't like to jump up and receive a scare, either. If you rest a broom against the counter and she jumps, it will crash to the floor. Or place a Snappy Trainer on the new towel on the counter. When she jumps up, it will spring straight up and frighten her off the counter ("Better not try THAT again!"). To date, no one has reported a problem with the Snappy Trainer device harming a cat, only scaring her. Finally, some people have placed on a towel on the countertop a sentry laser or motion detector device, which emits a loud noise when the cat jumps up and activates it.

The idea behind these last two devices is to startle the cat in the presence of something new about the countertop (the towel), so as long as she sees the "new thing" from below, she's not willing to chance a leap. They are all referred to as remote punishers or merely remote devices, because they work to keep her off of things when you aren't around. The cat experiences an unpleasant consequence for jumping up, and it is not connected with you in any way. She still regards you fondly because you didn't have to startle her off the counter.

Ever heard of the old trick of putting a mousetrap upside down under a newspaper to keep cats from jumping on counters? The Snappy Trainer is safer and just as effective.

This technique works for tables and other high surfaces, too. It also works well for keeping cats off your car hood (tired of those dirty stray-cat footprints on your car?). But these devices should be used cautiously if your cat is a bit wimpy or fearful. There's no need to traumatize a cat, ever, just to change a behavior. This scare technique can be quick and effective, but other solutions are possible. One effective solution is blocking access to her "jumping-up place" to a counter, either by placing a 2 × 4 along the edge of the counter or by moving the waste basket (if that's her "ladder") away from her jumping-up point when you leave the kitchen. All these possible solutions will discourage jumping up if you consider it to be bad manners.

Remember that it's important to provide your cat access to a location nearby that she *can* jump up onto to satisfy any need she may have for being on high places. A "kitty condo" may be just what the doctor ordered.

❧ The Nose That Won't Take No

Remember when your puppy was so small that you wanted to pet her all the time? You wanted to take the palm of your hand, place it on her nose and pass it over the top of her head, stroking all the way down the back of her neck and body to her tail, and then do it over and over and over again. And remember the first time you stopped petting her like that and she put her nose under the palm of your hand and gave you a little prod to continue petting her that way? Maybe you realized that she really appreciated you, that she needed you, and that you were really giving her pleasure. And you were more than happy to continue to pass your hand over her head, or to hold your hand in front of her and have her prod you again and again and again. That was great fun—until she reached adolescence or adulthood.

Now remember the first time you were holding a drink in your hand and sitting on the couch with your hand resting on the arm of the sofa. And your dog came up and sat down next to you. You smiled, and holding the drink in your right hand, you looked across the room for a moment to talk to your guests. And then she did that cute endearing thing that started in puppyhood: She nosed the palm of your hand, and the drink poured all over your lap and the couch. Blame it on the bossy-nosa.

That's the last time you thought this behavior was cute! Of course, she was still sitting there smiling, looking at you and expecting you to start petting her on the nose, over her head, and down her back. Much to her surprise, that's not the response she got. It probably should not surprise you that she continued to try to get you and other people to pet her when they were holding drinks or plates of food. And the result was pretty much more of the same: food and drinks were in everyone's laps, and the dog was still smiling and thinking, "Maybe *now* is the time I'll get petted!"

Before this wild display becomes a habit, complete with skyrocketing dry-cleaning bills, make a note to yourself not to reinforce her bossy nose to get more petting. Try the following techniques to keep her from making your lap a repository for loosely held food and drink:

1. After patting her a few times on the nose and head, hold your hand in front of her. When she prods you, make a fist and don't pet her.

2. When she *stops* trying to get your hand to pet her, then pet her. Be sure to wait long enough to pet her (30 seconds or more) so that she doesn't get the idea that she just needs to wait a few seconds after prodding to get petted. If the behavior doesn't lessen in about a week, you're not waiting long enough before she gets the nose-stroke. The idea is to reinforce a quiet nose, to ignore a bossy nose, and to make it impossible for her to move your hand.

In that way, you can still stroke her nose, but when *you* want to, on your own terms.

3. When she's sitting down next to your couch or wherever you're seated, place your hand on the arm of the sofa and let it dangle off the side a bit. The dangling hand often provides an irresistible invitation for a dog's nose to prod. When she prods it, hold onto the arm of the sofa with your hand so it will not move. You'll find that when she continues to prod your hand with her nose and gets no petting, and in fact, can't get the palm of your hand to open or move at all, she'll give up. At that point, it's okay to stroke her (but not her nose!), or just tell her that she's a good girl.

4. If this still won't get her to reduce her persistence, when it appears to you that she's about to prod your hand for petting, just remove your hand. (Keep track of the number of times she tries from week to week, and if it's reduced from Monday to Monday, then your technique is working.) If her nose continues to seek your hand, continue to move it or even sit on it, so that she can't gain access to your hand and get it to move onto the top of her head. You may be surprised at how persistent some dogs are if they've established a habit of doing this over the months or years.

5. Finally, after you have nipped your dog's nosing in the bud or decreased the number of times she tries (from 12 times a day to 5 or 6 times a day, for example), try correcting her with an "Uh, uh, uh!" or "No nose!" When she complies, reward her with the usual "Good girl!" and give her a treat, a few strokes, and other acknowledgments of your appreciation.

It's yet another example of how teaching your puppy good manners means keeping her from learning the behavior in the first place, or stopping the behavior as soon as it becomes apparent that it was not a good idea after all, and praising her for not doing it. This one may be especially difficult for you because you feel sorry for your dog. After all, she's not doing anything "wrong"; she's just being affectionate, albeit annoying. I had one client who became so desperate that she ended up hot-wiring her living room floor and bedroom doorframe to keep her beloved golden away from her. This is just a word to the wise!

🐾 Eating Nasty Things

Your dog or cat decides that he's going to eat nasty things in the household and not display the kind of manners you intended for him to display as he was growing up. But you say, "No way!" The three most common manifestations of inappropriate eating behaviors are eating plants, coprophagia (eating feces), and an eating disorder called pica (eating objects).

Those Tasty Houseplants

From the start, your pet may have shown a preference for eating grass or your favorite houseplants. This is not too unusual. Free-roaming dogs and cats commonly eat grass and other plants as part of their regular diets, even though canines and felines are technically classified as carnivores.

It may be that eating such vegetable matter enables cats and dogs to digest other food they have eaten or supplements their nutrition. Pets may eat plants in an attempt to initiate vomiting when they're experiencing nausea. It may be that some of the leaves of your favorite plants are in your cat's way and become a target of play to him zooming through his escape routes. Or it may just be that this is one sure way to get your attention and cause you to chase the critter about the house!

Eating plants can be dangerous, of course, if the plants are poisonous. So it's a good idea to take the following preventive steps to keep your dog or cat from eating, or in some cases, devouring your plants:

1. Provide your cat or dog with some vegetables in his diet. Several different kinds of kitty grass are available in pet stores for your cat, so you can grow his own garden right at home. Place the "snacks" in a sunny location where they can continue to grow (don't forget to water them!) and make them readily accessible to your cat, perhaps in the same place as a favorite plant he loved to destroy.

2. Keep your valuable and dangerous plants out of your pets' reach if possible. I know, it's easier said than done. But ask yourself if you really need to have plants growing in your house in that particular location, and, if it's a poisonous plant, if it's more important for that plant to be there than to have your cat or dog. Your answers will help put things into perspective. (You can always close off a room as a plant room.)

3. Try blocking access to your plant so that he can't get to it. Especially helpful for dogs or cats is a new citronella spray device developed by Animal Behavior Systems (sold by Premier, Inc) that shoots a harmless yet effective mist of citronella when a moving cat or dog gets too close to a plant or the litterbox.

4. Especially for cats, try making the area below the plant unpleasant to walk on. Placing a plastic 3 × 3 carpet runner "feet-side up" beneath the potted

plants discourages many cats from getting close enough to jump up on the pot to eat the plant. Along the same lines, place some chicken wire or other kind of wire mesh on top of the potting soil inside the pot so that it will be uncomfortable to walk on. These solutions also work, by the way, for cats who choose to defecate in potted plants' soil.

5. Try booby-trapping some of the plants by sprinkling a pet-deterrent product or another safe but unpalatable substance on the leaves. You can find many different types at the pet store. In my experience, sometimes they work, and sometimes they don't. But it's worth a try just in case one works for you.

Dogs Who Eat from the Litterbox

For dogs who eat poop, sprinkling the waste with something they dislike is also a good trick to try. Coprophagia is a very rude behavior that dogs exhibit many times from puppyhood. It's particularly annoying when they come in from pooping, and having "cleaned it up," proceed to let you know how helpful they've been by licking you on the mouth. Yuck!

One way to keep dogs from eating their poop is for *you* to clean it up as soon as they do it. Fun, huh?! People who live in big cities like New York are used to carrying scoops when they walk their pups—they clean up and dispose of unsightly defecations as a part of daily life. This requires, of course, that you walk with your dog while he's doing his bathroom duties and perform cleanup immediately. If the feces aren't available, the puppy can't eat them.

It's not unusual for people to discover that dogs lose interest in eating their own or their littermates' defecations as they get older. However, if this is something your pup's doing, and you choose not to deal with the behavior problem now, it may well become worse before it gets better. As with the reduction of other behavior problems, it's best to nip this one in the (taste) bud as well. The sooner you do, the less you'll have to worry about your dog slurping the face of your visiting niece, nephew, grandson, or baby.

Let me say it again. The best preventive treatment for coprophagia in dogs is to clean up their defecations on a regular basis. Do it early and often. It may also help to put the food treatment substances that you can find in the pet store on their dog food. These substances, including MSG, make the defecations taste foul, and in many cases help reduce the problem over time. (Of course, we may think that it should taste foul enough without seasoning, but what do we know!)

Pica: It's Not Even Food!

Pica—the ingestion of nonfood items—is rather common in puppies. It some-times is a form of exploratory behavior (human babies put everything in their mouths to explore, too). But if it continues and there's a pattern, it could be a more serious problem, one that we behaviorists call obsessive-compulsive dis-order (OCD). The OCD problem typically doesn't show up until the dog is older. It's characterized by repetitive chewing and swallowing of nonfood items that the dog seeks out. It's a likely diagnosis when the cause of the problem is not insufficient diet or inaccessibility to toys and other chewable treats, such as rawhide chews. I recorded one case in *The Dog Who Would Be King* in which a dog moved to a new house and proceeded to eat the house itself. I kid you not!

Fabric is another favorite, especially with cats. Wool-sucking and chewing in cats may be a similar phenomenon. In my book *Is Your Cat Crazy?*, I told about a cat who ate two roommates' entire closet of expensive sweaters—no laughing matter to the young women. One point of view concerning this phe-nomenon and cats is that it is hereditary, with oriental breeds such as Siamese and Burmese cats carrying a gene that makes it easier for this kind of behavior to occur. Another point of view is that the behavior is related to the cat's early experience. They received a certain amount of pleasure from suckling before weaning, and the wool sucking is essentially left over from the suckling period.

Yet another possibility is that cats who have been weaned too early (from 2 to 4 weeks of age) are more likely to exhibit this problem. None of these po-tential causes has been reliably verified, so at best we consider them hy-potheses.

Finally, in some cats, wool sucking generalizes to or begins with sucking and chewing other material, including fabrics such as cotton, polyester, and even plastic. As in dogs, it may be that some of this behavior is due to anxiety, as part of an obsessive-compulsive disorder. If this is the case, providing cats with substances they can chew, like romaine lettuce or extremely well-boiled chicken necks, may provide the release they need. Some cats may also require the help of anti-anxiety medications. Consult a certified applied animal be-haviorist or veterinary behaviorist for more information.

🐾 The Canine Kleptomaniac

The owner of a 10-month-old female black Lab wrote to me recently. His dog was very well-trained. In fact, he reported proudly, she sometimes made Rin Tin Tin and Lassie look "mentally challenged," but (there's always that *but*!)

the dog was a habitual thief! She was stealing everything she could get her jaws on, from the TV remote control to shoes to tissues to car keys, and she did all of it right in front of her owners!

The writer, Tom, said that the dog seemed to encourage her owners to chase her to retrieve the item, and if they did nothing, she would go to them to make sure that they saw the stolen item. She didn't destroy the items and would smile through any verbal scoldings that Tom and his wife dished out. What could they do?

Here is what I told Tom: I feel your pain. I, too, have raised a female Lab who once stole things. Luckily, your 10-month-old doesn't chew and destroy the things she steals. I wish I could echo that on Charlie's behalf! Fortunately, you have a few effective options for reducing this maddening game before it becomes a really bad habit. And I think that you've put your finger on the correct diagnosis—that's exactly what it is: a game. You said she:

1. Never steals things in your absence
2. Takes things in front of you
3. Encourages you to chase her

The fact that you don't physically reprimand her is a good thing. Punishment is less desirable than communicating to her that there is something that is inappropriate about her play, and that there are other, better ways to get social attention.

What's inappropriate is that she's essentially demanding that you join her in a game of keep-away with your valuable stuff, and when she decides that it's time! She probably thinks it's all fun and games because that's the message you're giving her. When you verbally scold her, but provide no real consequence for her noncompliance, she may interpret your loud voice (your bark?) as "kicking it up a notch." More exciting play!

I told Tom to try the following program. And if you have a canine kleptomaniac in your household, this advice is also for you:

1. Ignore *all* demands for play. If she can't get you to chase her, she gets nothing for her efforts.
2. Precede her self-designated playtime with play that you initiate—preferably outside, so that she learns a rule: No play in the living room. After she learns that she can have terrific social play with you, she'll look forward to your inviting playtime each day, and you can select the appro-

priate play items (thus saving the all-too-valuable TV remote for your own playtime!).

3. Offer her a chew toy or rawhide chip as a substitute for her favorite stolen item, while making the latter unavailable.

Your dog should be able to learn, as my Lab did, that she can get her daily social play *on your terms*; that running around the house is no fun if no one chases her; and that becoming involved in self-play with a Kong (a toy with holes in it to hold treats) or other great toys is another cool way to get rid of energy and still be in the family. Dogs are not clueless; they just need some guidance.

Accident, Nature, or Willful Misbehavior?

For both dogs and cats, "accidents" are a part of rearing. When you think of accidents, you'll most likely think of house-soiling: Peeing or pooping in a place other than where you want the dog or cat to relieve himself: in a lit-terbox, or on a newspaper in the kitchen, or outside. There are also the kind of accidents where the dog or cat breaks something. But what causes accidents?

You cannot always determine whether the accident is truly an accident. If the puppy gets too excited and pees, he probably didn't intentionally let loose, so it's an accident. If the litterbox in the closet wasn't available because the door was shut and the cat pees on the rug, it's an accident. If your housecleaner yelled at your puppy or kitten for a past transgression (your kitten playfully batted the salt shaker on the floor, or your Lab's tail knocked the vase of flowers off the coffee table), was it the pet's intention to do so? Did he deserve to be yelled at? How will you decide?

Suppose your dog urinates on the oriental carpet. Is that an "accident"? Is it on purpose? Just to give you an idea of how difficult it is to answer this question without some detective work, here are some of the most common causes of house-soiling in puppies and adolescent dogs:

1. Problems with teaching a puppy where to urinate and defecate and where *not* to go.

2. Marking (the dog's lifting his leg on the furniture), which begins to show up at sexual maturity. (See chapter 7 for more on this.)

3. Submissive urination (part of the greeting repertoire of puppies and subordinate dogs when they say hello).

4. Excitement urination. Submissive and excitement peeing are likely to decrease with increasing age, unless people elicit the peeing by "frightening"

a dog with their bold behaviors, such as speaking in a big, booming voice; engaging in prolonged eye contact or staring; walking toward the dog quickly; bending over to pet the dog; and reaching down with their hands to pet the dog's head.

5. Veterinary medical problems, including incontinence, diabetes, and bladder or kidney problems, among others.

6. Separation anxiety. (See chapter 14 for more on this phenomenon.)

7. Punishment-related urination. This one involves loss of control due to threat or a completely different issue, sneaking off to eliminate when you're not looking because, when you last punished the dog, he learned "Whew! I'm never going to pee when *he's* around again." The sneaking-off problem is frequently connected with failures to pee or poop outdoors while the owner stands there, forever and ever, waiting for the pooch to let loose, but he won't. So, as soon as he comes indoors, after having smelled all the great odors outdoors, the dog wants to pee. As soon as you go to hang up his leash, he scurries to the back bedroom and makes a deposit on the clothes you took off and left lying in a pile on the floor. (It smelled like the right place to him!)

8. Soiled carpets (from prior mistakes—urine is attractive to dogs, and the smell indicates "Here's where I'm supposed to squat").

So, pick a cause, any cause. Are you going to blame and punish the dog for his "accident?" And do you think that will help?

How to Handle Accidents

Actually, whether she deserves it or not, the only good that can come from yelling at your pet or punishing her for a past transgression is that it makes *you* feel better—at least momentarily, before the guilt sets in!. (See chapter 3 for more on punishment and guilt.) Regardless of whether we attribute some purpose to a dog's or cat's misbehavior, whether we call it an accident or a willful act, our response when dealing with "accidents" should be the same.

We need to follow a few rules if our goal is to teach the dog or cat the appropriate location in which to eliminate because it's good manners to do so; if we want to teach them not to steal food (from the table or from the other pet's food bowl or elsewhere) but to eat only from their own food bowl, because it's good manners to do so; if we want to teach them not to play with and chew our possessions but to play with and chew only their own toys, because it's good manners to do so.

We will get into establishing good routines in the next chapter, including the proper way to housetrain dogs and cats and a thorough examination of what can go wrong (which, as you can see from the eight causes listed on pages 257 and 258, is usually quite a few things!). I will go over the most useful steps to take to deal with each kind of litterbox and housetraining situation you are likely to encounter. For now, however, here are a few things to remember.

Whether you believe your pet is accident-prone or sociopathic in his misbehaviors, interrupt him and the progression of the misbehavior—but only if the "accident" is inevitable or is in progress. This can be done by clapping your hands, or saying "Uh, uh, uh" or "Wrong!" or using another auditory signal. Or you can pick him up (this can be risky for you, if he's in "midstream") and take him to a location that leads to the right location for his behavior to occur.

Do *not* place him in the exact location you have in mind. That's because we often unintentionally frighten our pets when we are upset and then pick them up. We don't want them to associate our tension and displeasure with placing them in the litterbox, or on the paper, and so on, since they may learn to avoid the place as a result. Instead, place your pet where he can see and smell the location you have in mind for his "act" and allow him to find it voluntarily for purposes of completing peeing, pooping, clawing, or whatever. As always, praise his compliance. And never attempt to "make him" do the right thing because it's likely to backfire.

If the discovery of any type of "accident" occurs after the fact—even more than a few seconds after the misdeed—don't try to correct your pet. Frustrating as it is, he won't know why you're punishing him. You won't teach your pet anything about his wrongful behavior by running after him or throwing something at him or cussing at him or whatever. Clean it up or repair it, and spend your time and emotional energy on arranging for the problem *not* to occur the next time around. Praise yourself for compliance! You'll be glad you did.

There are many more misbehaviors than I've listed in this chapter. The list is as long as the names of the pets who can dream up new and creative ways to drive their owners to distraction. But with the good foundation you've gained from my school of manners and consistent teaching, you'll set your pet on the best possible path to avoid major behavior problems that require professional help. Now let's move ahead and see how you and your pets can establish some terrific daily routines that don't leave room for rotten manners, sneaky subterfuges, or other misguided misdemeanors!

Dr. Wright's Rewarding Routines for You and Your New Pet

My cat, Domino, is already sitting behind her almost empty food bowl on the bathroom floor, staring at me, as the clock radio turns on and NPR blares forth from the bedside table. Thus our morning begins, as it does each morning at 6:15 A.M. I glance out the window to see if the finches and sparrows have begun their feeding frenzy at the bird feeders. This event marks the official beginning of my morning chores. Charlie, my Lab, follows me to the kitchen where we make coffee; find a spoon to scoop a dollop of cat food from a day-old refrigerated can of "wet food"; prepare a tasty concoction complete with hidden pills for our elderly terrier, Peanut, to snarf; and begin our trek back to the bedroom with two coffees, the cat food, and the Pea pills balanced carefully in two hands. Charlie's toenails click lightly on the hardwood floor behind me as she anticipates the "giving of the goods."

"Here's your kitty chow, Domino," I whisper as I place the small spoonful of wet food in Dom's bowl, trying not to frighten her with my big booming voice, and ridding myself of one chore in the process.

"Raie-raie," she squeaks in return, thanking me for serving breakfast. (Domino has never learned how to meow, but I know what she means.)

"Here's your 'pizza,' Peanut," I say, and she inhales the treat, pills and all, and smiles at me in gratitude.

"Here's your coffee, honey."

"Oo-o-o-a-a-ah," my wife yawns as she reaches blindly for the full coffee cup.

Jollying up, I exclaim, "Okay, girls, let's go outside," adding a clap-clap directed toward our old, deaf girl, Peanut, and walk quickly out the bedroom door toward the stairs leading to the fenced backyard. All four dogs clamber down from the bed, spilling Angie's coffee on the way, and race to be the first down the steps to the backyard. Charlie usually wins, but Roo-Roo will prevail if she gets a head start off the bed.

The dogs do their duties, and after Peanut "finishes" (she's always the last to get around to it), I wipe everyone's feet, and we race upstairs to visit with Mommy. After a few well-placed licks, Roo-Roo makes her way to Domino's litterbox and sniffs to see if she can assist me with my next chore of the day, cleaning the box.

This little ritual is part of our morning routine, as predictable to all of us as the sun rising and setting. That's what you want your dog's or cat's life to consist of: events that can be predicted and anticipated as a regular part of his life. All of us appreciate routines—they allow us to save our brains, emotions, and efforts for the special parts of our lives. And, even more than for people, routines make pets feel safe and comfortable. Let's explore some of the major routines you will want to establish with a new pet, heading off potential trouble spots as we go along.

Introducing Your New Kitten or Cat

The way you introduce a new cat or kitten to the household can mean the difference between a "fraidy cat" who hides under the sofa for the first week at home and a confident, happy kitten who settles in with a satisfied purr. So it's worthwhile to pay attention to some do's and don'ts when you decide to add a new feline face to the family portrait.

Bringing a new kitten or a cat into a household doesn't have to be a traumatic or even a trying experience for either the kitten or the adoptive guardian. Felines are most likely to adapt to new living arrangements and homes if the owner gives some thought to setting up a newbie-friendly environment and reviewing some commonsense scenarios in advance.

No single plan will work equally well for all cats, due to the wonderful variety of felines waiting to be brought home (each kitten or cat will differ in breed, personality, early experience, and so on). And, of course, the same kitten may react differently to two different living situations (some homes are smaller, noisier, and more crowded than other homes).

But regardless of individual differences in nature and nurturing, the new kit on the block is most likely to settle in successfully—with less stress—if you can make the following arrangements:

1. Select a room that the kitten can use as her home base for the next several days. The room should have a door or some other way to "privatize" it and shield the kitten from the hustle and bustle of daily living (including curious toddlers and other pets).

2. About half a foot away from the wall opposite the entryway, place the kitten's litterbox, preferably filled with sandy, clumping-type litter. (If the kitten has already established a preference for a different type of litter, use it.)

3. Along a separate wall, or as far away from the litterbox as possible, place fresh water and food.

4. Select another corner (or area away from the litterbox) for Princess to sleep in. You'll probably have to change this location to Princess's self-selected sleeping area later (how many kittens actually sleep where you want them to?), but that's okay.

5. Place the kitten in the room with you with the door closed. Sit down quietly and allow the kitten to explore the room at her own pace, discovering all its contents and figuring out where everything is. Do not take the kitten to the various locations. Do *not* place the kitten in the litterbox! If you *must* place something in the litterbox, make it a little soiled litter (from the cat or kitten's previous home) so that she can tell where she's supposed to go. (After that, keep the litterbox clean.)

6. Be passive. Allow the kitten to discover the room's delights—like window ledges to perch on and look out of, a comfortable couch or bed, tables to jump up on, and anything else in the room that might be used as an "escape route," should one be needed (from the kitten's point of view). Keep toys manageable. Don't give her more than three toys of different shapes and textures at this point.

DR. WRIGHT'S INSIGHTS

The litterbox should always be placed well away from other motivationally important areas, such as sleeping, feeding, and playing locations. The box should be placed in an area that affords your kitten both <u>privacy</u> and <u>escape potential</u>.

    <u>Never</u> punish your cat for missing the box by rubbing her nose in it, shaking her, or placing her in the box while either you or the kitty is upset. Instead, try to look at the situation from Princess's point of view and think about what might have made the box an unattractive place to use and what made the "wrong" place a more attractive location. (Is the litter strongly scented? Is it sticky clay? Does the box need changing? Is it in the wrong place?) Make the necessary changes. Call your local certified animal behaviorist if you can't cope with an elimination problem.

7. All this new stuff may be tiring and/or stressful for your new pet. You may wish to leave Princess by herself when she appears ready to take a nap, or when she appears relaxed and confident in her new surroundings. If Princess chooses to nap in your lap, consider yourself lucky. Of course, you can always try to place her in front of the sleeping area you've selected, and then leave the room as she takes a few steps and plops down on her new bed. If she cries, wait for a pause in her vocalizations, and then reenter her room.

8. Your new companion animal will let you know when she's ready to take on the rest of the house. It's important for you to allow the kitten to do this at her own speed. A door cracked just enough for Princess to enter or exit from her room is a good starting place.

9. Keep other pets and family members away from Princess's room. Introductions to these eager individuals should be well-controlled. Again, allow the kitten to approach and withdraw (back to her room if necessary) from people or other pets that remain passive. To accomplish this step, you'll need to control (by leash or holding) "The Dog." Princess will eventually establish the house rules with "The Dog," but young felines may need help at first.

10. After Princess has become familiar with her new digs and playmates, you'll become aware of her preferences for playing, eating, sleeping, and other activities. At this point (anywhere from several minutes to a few days), you may begin to change the location of the litterbox and food bowl gradually to areas preferred by both you and Princess. If the areas are one and the same, you win!

Bringing Home Bowser: Introducing the Adopted Dog

"I have just met my new owner and she has brought me to a place I've never been before. But I am glad she got me away from that noisy, unpredictable, and stressful kennel where I'd been for a few weeks. There were a lot of strange odors and too much barking. People walked around looking at me, and sometimes hoses sprayed me when I least expected it. I didn't have anywhere very comfortable to sleep or anything fun to play with.

"I've pretty much given up on seeing my old owner. I've felt a bit abandoned, scared, and uncertain about what I'm supposed to do and whom I'm supposed to please. I kept wondering if those people who looked at me were trying to make me feel good, or were trying to scare me with their eyes and hands.

"Now that my new owner has taken me to her home I have a choice. I can trust her, or I can try to get out and run away—or at least hide—so that no one can get me. She's been real nice so far, and she talks in a happy way, and she smells pretty good, and she likes to touch and stroke my fur—which feels good—so you know what? I think I'll give her a chance and trust her."

Although we don't really know what goes through the dog's mind—though I have spent years trying my darndest to read them!—we do recognize that there are some basic patterns or reactions to being introduced to a new household. I have found that the dog—showing dependence—usually stays near his new owner, looking to the human for guidance. He starts out pretty trusting.

In other cases, the new puppy or dog needs to be encouraged to trust the owner at the same time that he's being encouraged to be comfortable in his new environment. Dogs who are confident and have had good experiences in their past are likely to adapt quite well to a new owner or owners in a new living situation. Dogs who are a bit fearful, however, or who may have suffered earlier trauma, may need a little bit more work on your part to enable them to be comfortable in their new surroundings. And that's the idea.

Do not bring the dog into the house and immediately subject him to three excited children, your husband, the kids next door, and the two resident dogs as he crosses the threshold and then complain that the new pet is "hyper"! It is best to work on making the dog comfortable with you first by stroking or talking to him. Watch to see what his comfort level is in terms of physical contact and try to see if he will respond favorably (that is, in a happy fashion versus an assertive fashion) to being offered a treat. In general, try to find the tools or incentives that are likely to enable your new pet to look to you for guidance.

Here are some suggested first steps.

🐾 Relax in His Bedroom

Walk the dog calmly to the area of the house where you plan to have him sleep. Introduce him to his bed or pillow or crate by sitting next to it and stroking your new dog there. Perhaps you will want to place a treat or two on the pillow or in the crate. See if you can encourage your dog to be with you in a calm social situation, and focus on both of you relaxing and being comfortable in one another's presence where he will be expected to sleep or rest. This spot will become a "safe zone" for the dog, where no one may disturb him.

🐾 Take a Pit Stop

Rather than taking the dog to the rest of the house of this point, I prefer to take him to the location where he can relieve himself. (I will get to housetraining shortly, for those dogs who don't show up on your doorstep already housetrained.) Put on his leash and walk to the location where he can smell the pine needles or other soiled places, and stand around until he attempts to pee or poop. When he has done well with satisfying his biological need and you praise him, you have taken a giant step toward establishing that trusting bond that is so necessary to forming a lasting, fulfilling relationship between you.

🐾 Time for "The Tour"

Now go back to the location where he was expected to lie down, and repeat sitting down with him in a calm fashion for several minutes. When you're ready, say, "Okay, let me show you the rest of your home," and try to catch his eye and coax him to walk with you to the bedroom and into other rooms that he'll have access to in your home. Then go back to the resting area and sit down for another few minutes. Talk about what a good boy he's been and how everything is going to be fine. He should become used to the sound of your voice, and how it feels (good!) to be with you in his new household.

🐾 Add a New Face

The next step is to have the second person in the family (if there is one) come into the room and squat several feet away from your new dog, not looking directly at the dog but encouraging him to approach. Be sure to respect your

(continued on page 268)

DR. WRIGHT'S
C A S E B O O K

Champp

I once had a client whose adopted mixed Chow Chow, Champ, started off their relationship by regularly biting the heck out of her. To the owner's credit, she didn't hit the dog when he was "bad." On the other hand, she never praised him when he was "good," either. So his motivation to learn proper behavior was not helped by any verbal cues from his owner. I literally had to teach Miss Segal how to talk to her dog.

This skill does not come any more naturally for some people than cooing "baby talk" to an infant to make him smile if you're unused to conversing with the diapered set, or if you're of a "serious" temperament. Some people just do it sort of automatically, whereas others wouldn't begin to know how. In this case—as is the case with any new pet where you want to start off on the right verbal foot—it was important to turn up the emotional temperature to "warm and fuzzy."

Champ needed an owner with strong leadership qualities to keep him in line. But he also needed for Miss Segal to be able to get him in a good mood with some nice, interesting-sounding words that he could understand. The owner was by nature a little too timid.

"Miss Segal," I said to the fresh-faced teacher's aide, while the pooch was a safe distance away in another room, "you need to be able to praise Champ when he complies, and use a lot of enthusiasm."

The problem was that Miss Segal's personality was almost entirely emotionally flat. I had to show her how to infuse her voice with some spirit. "You say this: GO-O-O-O-OD DO-O-O-OG!" I enthused. "You want to get that dog's attention and get his tail wagging with your tone of voice. Now it's your turn."

"Uh, good dog," she followed woodenly.

"Well, that's fine, but Champ will feel better about you if you exaggerate what you're saying a bit more," I advised. "Like this:

GO-O-O-O-O-OD DO-O-O-O-O-OG!! Now, will you try again, please?"

"Sure," she said willingly. "Good! Dog!" she barked, frowning with con-centration.

"That's better," I said brightly. At least the volume was up. But this tone of voice would never dissuade the dog from lunging at his owner! I decided to try a different tack. I stood up, trying to show her how to look as assertive as possible while offering a warm, friendly tone to the dog. "Wanna gofer ride?!" I clasped my hands and looked eagerly into her eyes.

Miss Segal looked at me over her half-rim glasses.

"No thanks," she declined politely.

That was a tough one. But eventually this quiet client learned to talk to her dog and reinforce good behavior, and with the help of a head halter, things improved dramatically.

dog's comfort level at this point, and don't force him to meet other people if he's not quite ready. (Signs of discomfort or fear include a tucked tail, crouched posture close to the ground, and attempts to retreat behind or beneath something.) When the dog is introduced to the second person, have that person stay a distance away from his bed area and continue to talk to the dog. In that way, your dog can become familiar with the person's mannerisms, his voice, and the fact that he's not going to come up and hurt him in any way.

🐾 Respect the "Safe Zone"

If there is a third person in your family, have the second person leave while you continue to sit with your new dog at his bed, and then have the third person come in and introduce herself in the same way. The idea is to respect the dog's bed area as a "safe zone" for him. He needs to know that it is a location that he can use to feel secure and where no one will bother him, even when he's done something wrong and ticked you off, or even when you're dying to run up to him and give him a big hug. Be sure that the kids in your household know the rules, too. It's important for your dog to know that he can trust each and every person in the family, and that when he feels discomforted or fearful in any way, he can be safe in the location you chose for his bed.

🐾 Carry On

After the initial introductions have been made, you can proceed to walk around the house and do things in the normal fashion. You'll see how your new dog follows and watches you and where he prefers to lie in the house while he watches you go about your daily chores or do your homework or watch TV or pay the bills or whatever your routine is. If he chooses places to rest while he watches you that are fine with you, glance over at him and say, "Good boy!" If you use an enthusiastic tone of voice, he will know that he has done well and pleased you. The rest will come naturally to you and him as you learn about one another in the next hours and days.

Working Out a Routine

After you've taken your dog from a different lifestyle and living environment, it's important to establish a new daily routine, one that you and he can more or less count on as being "what goes on here from day to day." You've already introduced him to the new household, and he has begun to learn to form new, trusting, feel-good relationships with each person in the family. At this point, he can begin to understand where he belongs in the family social structure (re-

DR. WRIGHT'S INSIGHTS

Feedback from adoptive parents is pretty clear on the best time to adopt a dog into your household. It's Friday afternoon or evening, which allows you just long enough to introduce your dog to the household and to the members of your family. You can then wind down and take advantage of his natural circadian rhythms for sleep on the first night in his wonderful new home!

call chapters 6 and 8 on the dog's jobs and roles in the family). You can begin the process of teaching him what to expect will happen to him throughout the day and to be comfortable in establishing a daily routine.

If you bring your new dog home on Friday and settle him in overnight, he'll then have a full two days to learn the daily routine. The first weekend should go something like this.

After you wake up Saturday morning, you and your new dog can go outside to let him relieve himself and then come back in so that the dog can become part of the family right off the bat. Most people will have until Monday to show the dog small examples of the routines he can expect to face during the next week. The idea for Saturday and Sunday is to allow Bowser to experience approximations of the real routine that he can expect to find during the weekdays.

The first part of the day will probably not be too different from a weekday—at least not this Saturday morning—because you'll probably give up sleeping in so you can get off to a good start. The fond "good morning" licks and smiles and play-bows will be followed by going outside and then coming back inside to watch you get ready for your day, fixing coffee and a small breakfast while he waits in anticipation for more good stuff to happen.

Now on Monday, if you have a job outside the home, this is when you'll think about saying goodbye and leaving him for up to 8 or 10 hours all by himself. It may be that he already has experience with this solo time if you adopted him as an adult. Or it may be that he's already experienced some separation anxiety problems in a different household, which may be why he was up for adoption in the first place. (We'll talk about separation problems in chapter 14.)

🐾 The Role of the Crate

If you choose to use a crate—a wire or plastic indoor cage to confine your dog when you can't supervise her—and your dog is comfortable in a crate, say, "Go to your crate." (*Never* force your dog to experience discomfort in a crate.) Stand and let her go in the crate. After she's crossed the threshold, add, "Good girl!" and perhaps a safe word or phrase like "Stay put" or "Bye-bye." Don't make a big deal out of leaving, just go out the door. Come back in. Do not establish eye contact as you calmly walk over to her and let her out of the crate, and then add your praise. This will keep her from going bananas when she sees you at the front door, coming toward her. Of course, during your real absences from home, you and she will go outside immediately so that she can relieve herself, followed by bunches of praise and acknowledgment from you because she did so well.

If you practice the short lessons with the crate on Saturday 5 to 10 times, and additionally take her out with you a number of times, she'll soon learn the significance of each event. She'll learn to discriminate between going outside with you and staying put by herself until you return with a big payoff—lots of social attention and psychosocial satisfaction in being recognized for being a good girl. (Be sure not to just ignore her for the rest of the day!) What you do throughout the day is to reaffirm her relationship with you through talking, strokes and massage, play, treats, or whatever other incentives you have to offer to help fulfill her bio-psycho-social needs.

To Crate or Not to Crate?

Let's discuss this idea of the crate a bit more. During the day, in your absence or whenever you can't keep an eye on your dog, set up a large crate for him to

When a dog is raised with a crate, she'll view it as her safe haven and look forward to spending time there.

rest in (large enough for him to stand up and stretch out). You can make it comfortable for him by adding a blanket or faux-sheepskin pad, some toys, and (of course) access to his water dish, but don't expect him to remain there for longer than 4 hours at a time initially, working up to 8 hours.

I do not recommend keeping a pooch in a crate as a daily routine for the rest of his life, but only as a tool for teaching him it's not good manners to soil the house. After all, who wants a dog who doesn't learn anything during the day

Tips for Stress-Free Coming and Going

To be on the safe side, it's a good idea to leave the home alone for short periods of time at first, easing the dog into learning he's not being abandoned. (If he has some kind of abandonment issues already, there's no need to amplify them!)

One way to deal with this potential fear is to head out with him 8 to 10 times the first day. You needn't go anywhere in particular; just say, "Charlie. Wanna go?" And, of course, have him sit at the door while you attach his lead, then head out for a minute or so. Then you need to go alone. Walk to the door without him, open the door, go outside, shut the door, reopen the door, come inside, and head back to sit down on the couch. You should acknowledge his presence only when he's calm, sitting or lying down, or otherwise acting like this was no big deal. That's the goal. The difference between your going out with him and without him is that you're saying a word or phrase when he goes with you, like "Wanna go for a walk?" or "Outside" or "Let's go to the car!"

When you don't want the dog to go with you, ignore him. This isn't as easy as it sounds! You have to think about it. Do not establish eye contact, and calmly walk to the door to go on your way. If he starts to follow, you can tell him to stay, or settle down, or you can do something that he will learn indicates that he will be best off if he relaxes, and that trying to follow you will not pay off—he will not get to go with you!

It's obvious, if you think about it, that going out the door and coming back 5 seconds later is best achieved when your dog's already calm. In that way, when you return and sit down, you can praise him as though it was his own idea to remain calm and lying down while you went out the door and came back. That you actually did come back, and nothing bad happened to him in your absence, will be reassuring to your new dog and will mark the beginning of his ability to cope with being alone for longer periods of time.

about being a good family member? What kind of job or role can a crated dog possibly have? What kind or quality of life does he have? Most people I work with don't want their dogs to learn how to be crated forever—that's why they choose to consult with an animal behaviorist. So don't expect your long-term house-soiling problems to be solved by the crate! Your dog needs to learn to prefer "outside" to anywhere inside (unless you have a toy breed or other small dog who would do better using an "indoor doggie litterbox"), and he will if you follow the steps I outline later in this chapter. For an excellent videotape on crate-training your pooch, see *Crate Training Your Dog* by certified animal behaviorist Dr. Suzanne Hetts.

🐾 Let's Talk about Food

Lots of good, practical training can take place in these first few days, as long as you pay attention to satisfying the pup's needs and understand how he learns. (See chapter 3 for more on canine learning.) Play it by ear throughout the first day, until lunch or dinner. Serve your own meal first, and ignore him or tell him to settle down until you're through. Then it's his turn. In this way, he'll learn that he gets fed in his food bowl following your family's dinner. Praise compliance in your own special way, and return to an evening of relative calm. Give him one more time outside and a brief play bout (if he's too wound up to sleep) before settling him into his safe area—or other bedtime arrangement—for a good night's sleep.

Speaking of feeding, please note that puppies require three meals a day, adolescent dogs need both morning and evening feedings, and adult dogs need one meal a day. (Large breeds who tend to gulp their food sometimes do better with two smaller meals even as adults. Consult your vet.)

🐾 The Rest of the Routine

Throughout the day, you'll discover what jobs and activities your dog chooses for herself, which will either be okay with you or in need of changing. Perhaps she likes to snooze during the day in the corner or in her crate. Maybe she likes to lie on the landing and look over her living room while keeping an eye on the front door. She may be into slobbering on that Nylabone or tossing that Buster cube around, or maybe all she wants to do is be within 4 feet of you, regardless of where she is or what you're doing. This is your call, and hers—together. And that's why enabling her to be comfortable while she's alone is so important for you to establish up front. Take small gains at first, followed by larger ones. Remember that learning is a process that takes time.

Your bedtime ritual will be your last opportunity to affirm your relationship with one another for that day. Make it a pleasant one. Give your dog a stroke or two, say good night, do some simultaneous smiling—do whatever you want to do that results in a mutual admiration of one another and good feelings before "Night-night!" If she tries to become active at night, try a simple word like "Settle!" or " Wrong!" or "Uh, uh!" to help her calm down and understand that nighttime is for sleeping. Don't engage her in play or even get out of bed, if you can help it.

On Sunday you'll do more of the same, while staying away on brief departures a little longer, interspersed with shorter comings and goings. Don't forget to ask her to go with you randomly, too, throughout the day.

On Monday, be sure that she gets a good morning exercise session before you leave for the day. An adult dog can be by herself all day, although younger dogs, especially if crated, may need a midday potty break and exercise. Be sure to plan on it, and be realistic. Don't force a 4-month-old puppy to "hold it" for 8 hours and expect her to be a happy camper at the end of the day if she does (or for you to be pleased with her if she doesn't).

Don't set yourself and your dog up for failure, but be realistic and open to options for changing the daily routine if it seems obvious that the one you've selected isn't going to work. Someone's needs aren't being satisfied. Analyze the problem and go about solving it by changing something. Don't change things after only 2 days though (unless it's clear that you have a disaster on your hands!) because it takes a few to several days for behaviors to become patterns of activity.

🐾 He Knows What's Next—Do You?

Most dogs are pretty savvy about the various events throughout the day that are markers for the next event that's about to occur. Your dog's thought is, if it happened yesterday and felt good, I'm going to be prepared for you to do that with me again today. There is good evidence of this canine ability to anticipate. If you think about it, you can probably come up with a few examples of things your dog "asked" you about today that you did yesterday and promptly forgot about. Part of establishing good routines is the satisfaction that comes when you both understand and follow through on what is "supposed" to be happening next.

So if you typically feed him at 6:00 P.M., take him outside to "be a good boy" at 6:15 P.M., and then bring him inside and give him a rawhide treat, don't be surprised if tomorrow after you bring him in at 6:15 and forget his treat, he starts to bug you about something. Your immediate thought will probably be

Dr. Wright's ✉ E-Mail

Dear Dr. Wright:

I recently bought a new puppy and he doesn't seem interested in eating. He takes all day to eat his food. It's as if he doesn't like it but only eats it when he's very hungry. We tried changing his food, but the same thing happened. We mixed it with canned puppy food, but that didn't help. We don't know what to do! Should we remove the food after a certain amount of time or allow him to nibble all day, causing him to miss a meal? I don't know if he is losing weight because he's losing his puppy fur and looks like he's really slimming down. Help!

Julianne

Dear Julianne:

Poor puppy! All that great food available for munching, and he acts like it's broccoli! Although pups typically cut teeth for quite a while after weaning, and eating hard food can cause some discomfort in the short term, his problem doesn't seem to be related to pain. Assuming that your vet has given him a clean bill of health, here are a few things for you to consider.

Factors that affect eating other than pain include ambient temperature (eating outdoors when it's too hot de-

that he hasn't finished his business, so you'll cue him "Outside?" and walk to the door. Meanwhile, he'll still be standing there, no doubt where you left him, tilting his head from side to side, trying to figure out why you're needing to go outside again, and wondering where his treat is!

creases food intake), meal temperature (pups prefer warm food to cold food), food type (they prefer meat over high-protein nonmeat foods), food palatability (beef and lamb are preferred over chicken, as are sweet foods—those containing glucose or fructose, but not saccharine—over bland foods), and social setting (compared to one pet, pups who dine together eat more, and eat more often).

So let's put together a plan to increase your pup's zest for food. Try feeding him in a quiet, cool area of the house, where there is little distraction. Feed him smaller amounts of food, several times a day, and allow him about 10 minutes to eat before removing the food bowl. Find out what the breeder or caretaker fed him— he'll probably eat that. Or find a beef- or lamb-flavored canned puppy food that he'll happily lick off your finger, and then place a small amount of it in his bowl. When he finishes it, try giving a little bit more. Once he's used to the taste of the canned food, try adding a few pieces of hard food, until he's used to the mixture you desire for him—switching completely to a hard-food diet should take about 10 days.

The icing on the cake would be to find a neighbor or friend who has a puppy—a littermate or playmate about the same age as yours—to be an "eating buddy." Feed the pups together a few times in the same location you've chosen for daily feeding. There's nothing like a little social facilitation ("Hey, that other dog's eating my food!") to help jump-start a sluggish behavior.

It can be very frustrating for everyone if the new dog owner doesn't understand that dogs often exhibit what psychologists call a recency effect in learning new patterns of behavior when they feel good, and sometimes when they had a bad experience as a result of the experience. (If you let him outside yesterday after dinner but forgot to let him back in, and he's afraid of the dark, he may want no part of going out that door after dinner the next day. He may choose instead to sneak off to the safe spare bedroom to do his postfeeding toilet duties.) So the most recent after-dinner activity that Bowser experienced is likely to replace his expectation of a less-important after-dinner activity. (For example, if yesterday's after-dinner activity was a rawhide treat, he'll be looking forward to one again today, but if it was getting a bath, he may try to avoid you and it—though many dogs really love baths rather than hating them.)

If Bowser looks like he is trying to tell you something, think about what you did about the same time yesterday, and you'll have a good clue about his expectations for your behavior today. Then, of course, if you can't go to the pond with him today like you did yesterday, think about the best way to bring him down. You certainly don't want to punish him! You may be able to diverge into another activity like playing chase the ball, get him involved in self-play with a Kong or other toy, or give him some kind of brief attention, followed by an indication that you're done. ("Uh, uh! Good boy," when he settles.)

Cats and Food

Cats are notorious for nibbling throughout the day. But actually, that's not a bad thing. It's a good way for them to get their nourishment, as long as the food and water are kept fresh and unspoiled, especially in warmer weather. Some cats have rather charming table manners, like dipping a paw in the bowl of milk or food and eating from the paw. Cats commonly lick their front paws to sample things they're curious about, like a leaf or other "out of place" item they might find on the carpet or even an edible crumb. It's fairly normal for cats to bat at a dead fly or something they perceive to be a prey or play item, and then sniff the bottom of their paw to identify the smell. It's their way of investigating things, and it's safer than risking a nose touch. (Committing your nose to touch an unidentified wasp may have dire consequences for such a sensitive body part!)

Cats sample things with their paws much as we use our hands to pick and smell flowers or to sniff an unfamiliar exotic food before we decide if it's safe to eat. Cats who eat with their paws are using their "hands" that way because it works—it's functional. It's just that many cats have never thought of it! If she does use her paw to eat, she brings food to her mouth rather than sticking her mouth in the food. And there may be another benefit: Perhaps she uses her paw to eat because it keeps her mouth and whiskers from getting dirty in the food bowl. Chances are that, after eating, she'll lick the paw and clean up her whiskers and head with it anyway! At any rate, a cat or kitten who eats with a front paw is awfully cute to watch.

In my experience, finding a cat who attacks his food bowl and finishes it all in one sitting is rare. That doesn't mean that he won't lick the bowl when he eventually gets there, though! A cat owner recently wrote to me about another eating behavior she found odd. It seems that her cat, Mr. Jones, scratched at his dish and placemat after he finished eating. I've seen many cats do this, and there are a couple of possible reasons why. First, he may be indulging in a precleaning ritual of getting every little last bit of food for later enjoyment. Those little crumbs that he'll later wash from his whiskers are more likely to stick to a paw that has gathered them up by "scratching" at the bowl. Thus, he may be collecting his dessert, to be enjoyed in the comfort of his den. Or it may be reminiscent of his wild ancestors who covered their half-eaten prey for later consumption.

Cats also usually scratch before they eliminate. It may be that Mr. Jones was trying to say, "I'm full." When cats feel full, they experience gastric discomfort, and it's not too long until they seek out the litterbox. Perhaps the scratching was his way of preparing for the next "motivationally important" event: relieving himself in the litterbox.

Bathroom Etiquette

In my experience, trouble with toilet training is the #1 problem for both cat and dog owners (followed by aggression in dogs). So I'm going to give this frustrating and important issue its own special section. Teaching your cat to use the litterbox and your dog to use the great outdoors doesn't have to be a problem. Most of the time, misguided people are the real problem when there is a toilet-training problem. Here's how to make sure that your pet's bathroom etiquette is worthy of Miss Manners.

🐾 Litterbox Roulette

Cats that urinate or defecate outside the litterbox are often regarded as being big problems, and their behavior is certainly incompatible with our human lifestyles. If this #1 cat behavior problem goes on unresolved for very long, or worsens, it can become life threatening for the cat. Many box-avoiding felines are kicked out the back door or taken to the shelter where, sadly, their days may be numbered. So it's really important to start out right with the kitten's housetraining, and know what to do if accidents begin.

Teaching the cat that the litterbox is the *only* appropriate location to pee and poop is not difficult if you set things up properly. Here's how.

First, make sure that the location of the box both is private and allows the cat to escape to the left or to the right, up or under, if necessary. The box must also be clean. The texture of the litter matters to kittens, and studies by behaviorists have shown that cats prefer the sandy clumping litter to other litters. When your cat urinates, the urine-drenched sandy litter clumps into small balls that can be readily scooped from the litter along with any solid waste. You should add a little more litter so that about 2 inches remains in the bottom of the litter pan at all times. The newer pelletized newspaper litter is also a good option.

Very few cats like plastic liners. Some tolerate them, but it's best not to start your kitten out with one, just in case the smell or feel of the plastic liner inhibits her from walking into the litterbox. Although the plastic liner may make your life easier, it may do little to enable your kitten to establish her litterbox manners. (In fact, if she tears it with her claws, urine may collect under the liner, making a smelly nightmare for her sensitive nose and a mess for you.)

Another factor to consider in selecting a litter is that litter fragrances are sometimes repulsive to kittens. Like plastic liners, nice-smelling litter may make your life easier, but sometimes kittens have an aversion to the fragrance. It's best to avoid deodorized or perfumed types of litter. Still another turn-off is the lid or cover on some litterboxes. Litterbox lids may seem too scary for some kittens, who in nature would prefer to urinate or defecate in an open area so that they can look around to see if there is any reason for them to stop eliminating and run for their lives! The lid may play no role whatsoever in your

Remember that your cat's litterbox needs to be attractive to her, not you. Use an open box that she can see out of; don't use scented litter (use an unscented clumping sand-type litter, not a cheap clay litter); and don't use plastic liners. Keep the box clean, and you'll have a happy cat!

kitten's preference for using the box, but why take a chance? Why not begin with a box that has the characteristics most cats prefer or that provides the fewest reasons for avoidance?

Even though you scoop out the litterbox as needed, you should probably change the box completely on a regular basis. A good guideline is to change the litter if you can smell an odor when you put your nose close to the entrance of the box after you've just scooped it. The sensitive kitten, otherwise known as the clean freak, will let you know very quickly when the box is too dirty to use, and then you have a problem.

As a general rule, it's a good idea to give your kitten an opportunity to select where to go in the beginning by placing two boxes in different locations in the house. After she establishes a preference for one box, you may eliminate the other. (Of course, she may prefer to use both boxes!) Especially in two-cat households, an alternative box is a necessity, although it may not be necessary to have one box for each cat in a multicat household.

To sum up, keep your cat's feeding area away from the litterbox, avoid playing with her there, avoid harsh or strong-smelling fragrances in or near the box, and make sure that you keep the litterbox clean and fully accessible to the kitten as she gets larger. You don't want one with sides so high that the young kitten has to leap up and over them to get in. On the other hand, you want a box with sides that are high enough that the litter doesn't go flying all over the carpet when she digs in the litter for some covering action. A compromise of 2- to 4-inch side height is probably the best solution for both you and your cat. Praising the cat for eliminating in the box is an option, but I'm not sure cats care one way or another about how pleased you are with their bathroom habits, unlike dogs.

🐾 Housetraining Your New Puppy

To teach a pup who is 8 weeks old or older to relieve himself in the right place and to avoid house-soiling in the future, you'll need to figure out where he will stay during the learning period. Some confinement is reasonable because training takes a while to click in, and you want to minimize the areas that will be used for toileting. Even so, if you are at home, you should make every effort to start taking him out. Don't let papers be a crutch.

If you crate him, the first thing you should do when you arrive home is to take him outside. Avoid eye contact with him as well as you can while you do this, so he'll know you're task-oriented and he has to complete the elimination

DR. WRIGHT'S INSIGHTS

In nature, cats often urinate in one place and defecate in a different location. Many indoor cats will adjust to using one litterbox, and successfully urinate and defecate in the same box. But to maximize the likelihood of success, I usually recommend starting out with two boxes. It's worth it!

task first before any play takes place. Then, after he's relieved himself, go play, tell him about your day, ask him about his, and so forth. Do not return him to the crate (if he chooses to go there, let him go in, but don't close the door) while you're home—he's already done a great job abiding by your daytime rules and needs some variety in his life! (For more on crating, see page 269.)

If you don't want to crate him (many people can't stand the thought of a dog in a cage), you can use a baby-gate arrangement with him in the kitchen. The rules are the same. Be sure you have his paper or specially scented pee-pads placed on the floor (near what will become the "bathroom" door once he can hold it during the day long enough to wait for your arrival home to use the outside area). Don't place his bathroom area near his food and water or his toys or his dog bed (resting area). People often prefer to leave their puppies in this kind of setting because it allows for more different kinds of activity to take place (sniffing, exploration, trotting around with a toy) than the crate does. Of course, the best of both worlds for people and puppies whose homes include a large kitchen or kitchen/breakfast area is to place a small crate inside the gated area where the pup can rest on his favorite puppy bed—and the crate door is left open to provide access to the rest of the romper room.

Be sure to clean the soiled area of feces on a regular basis. Leaving defecation lying around may encourage coprophagia (poop eating), especially if you have more than one pooch. (See page 254 for more on this disgusting but preventable habit.) Papers all over the floor will get old very quickly for owners of larger breeds, and keeping the dog crated while you're home doesn't seem very nice, so you might as well start the real training as soon as possible after bringing the puppy into your house.

Eight Easy Steps to Housetraining Your Puppy

Housetraining (don't say "housebreaking," even if it's what you grew up with) is really pretty easy. Don't let yourself be intimidated by the process. Remember that your puppy wants to please you, and he wants a safe, simple routine to follow every day. Use these eight simple steps to create a beautifully housetrained pup.

1. Know the most likely bathroom times. During the day, your pup is biologically prepared to urinate and/or defecate before or after specific events. He'll need to go when he awakens from sleep or a nap, within 20 minutes after eating, after prolonged play and exercise, and before he retires at night. Keep track of his preferences for time of day and how regular he is so that you can head off veterinary medical problems later (if he becomes constipated or has loose stool/diarrhea).

2. Pick a spot. Decide where you want him to eliminate: inside on paper or in a litterbox for dogs, or outdoors in a box on your covered deck, or on concrete, grass, or pine straw (my preference to save your grass). Many toy breeds learn to use a litterbox and never need to go outside for toileting. Outdoors, select an area he can use as his bathroom (in the yard or in front of your condo) that is separate from his play or exercise area. If inside, place the paper or box away from the areas he eats in or rests in. Keep in mind that if you have or expect to have children, this location should not be readily accessible to creeping crawlers!

3. Select a magic word and go out with him. When he awakens in the morning, say "Outside?" or another word or phrase that indicates you're about to go to that area for him to do his thing. Walk quickly with him to the door (preferably a different door than the one you use for play and exercise) and walk with him to a nearby area you've preselected, complete with signs that say "here's the bathroom"—prior stools, of course. He'll enjoy and feel safer going with you. Stand in the area you want him to use, and wait 4 or 5 minutes for him to urinate and then defecate (most dogs seem to have an order preference). Expect him to take a minute or two between deeds, to sniff the right places to help him complete his act. When he's finished, give him some strokes for a job well done, then walk back to the house for his first meal of the day.

4. Get a doggie door. For more flexibility, consider using a doggie door if you're not concerned about tiny burglars or other unwelcome critters entering your house. (I know people who have had stray cat and raccoon visitors make themselves at home!) Teaching your pup to go through the door both ways can be done initially by baiting him with treats, with you standing on the other side of the door to provide an additional social incentive for using it.

If you choose to use a doggie door, when it's time for your pup to do his business, go outside ahead of him, stand opposite the doggie door and wait for him to come out, and then wait until he's through eliminating. Do not play with him or give him any other attention until he's done. (Some dogs just won't go if you look at them. It may be that eye contact is so captivating to some pups that it makes them tense with anticipation—"Play?" "You want me?" and so on—and a tense puppy isn't about to be able to relax his sphincter muscles enough to relieve himself.) When he's done, head back in the house, but encourage him to use the doggie door by standing on the inside looking out, and coax him back through. "Go-o-o-od bo-o-o-oy!"

5. Get used to his cues. Be observant and learn to recognize the signs he gives you for needing to relieve himself. Ask him, "Outside?" and he should trot to the "bathroom" door. If he trots to the other door (the one you use to take him for a walk and other pleasurable activities), say, "No," or otherwise let him know that's not a choice, and sit down. Soon, he'll be reliably indicating his need for "relief." Expect a puppy to take about 4 to 6 weeks to get the hang of it. Again, some will learn faster; some, slower.

6. Fine-tune his feeding schedule. If you feed your pup three times a day (you can reduce feedings for 6-month-olds to twice a day), he should begin to get "regular" in no time. Free-access feeding (ad-lib feeding) is okay, and his digestive system will become conditioned to his own eating schedule, which is fine in most cases. But if you find that your pup is eating at strange times, and having too many accidents at inconvenient times (I know, all accidents are done at inconvenient times), you may wish to switch him to three 20-minute

DR. WRIGHT'S INSIGHTS

Some pups will take less than a minute to finish their toileting, whereas others need longer than 5 minutes; so pay attention to your dog's specific needs. Beware, however, of the dog who cons the morning restroom attendant into staying outside longer and longer before he does his thing. Some people take a full 30 minutes before work to make sure Gonzo has "done both" so that he won't leave a gift on the rug in their absence. That's why I recommend "Dr. Wright's 5-Minute Rule to Prevent Stalling." This 5-minute outside guideline keeps puppies from delaying their duty. If the pup doesn't do it in 5 minutes, bring him back inside and try again in 5 minutes. If he begins to sniff the carpet, head him back outside—he's ready. Praise compliance enthusiastically, which shouldn't be difficult for you, considering the alternative!

feedings a day after a 24- to 48-hour fasting period (consult your vet first). Just remove any uneaten food after the 20 minutes, and he'll quickly learn that there are three mealtimes during the day. If you still eat sit-down meals at home, why not schedule them to coincide with his meals? Many people and their dogs enjoy the social facilitation of eating in the other's presence!

7. **Expect occasional accidents.** Accidents will happen. It's important for you not to punish the dog if you don't catch him in the act. It's more productive, I believe, for you to spend your time and energy recognizing the signals he gives in preparation for "going" so you can reflexively say, "Brutus, outside?" and quickly walk to the "bathroom" door. If you *do* catch him in the act, try to interrupt it by clapping your hands, or saying, "Uh! Uh!" or "Wrong!" After he stops and you have his attention, quickly walk to the "bathroom" door as you say, "Outside?" Don't forget to praise his compliance.

If he won't stop, go to him and pick him up. (This is not without risk! Some pups continue peeing. Others don't, although you probably have the former!) Place him facing the bathroom door, and then say, "Outside?" Remember that the *last* thing you want your dog to learn is that he'd better not pee or poop in your presence. So don't punish him by sticking his nose in it or by otherwise humiliating him. You're the professor! He's the student! Enable learning to proceed by allowing him to learn from his mistakes—he will have some mistakes, regardless.

8. **Go off-duty overnight.** For nighttime arrangements, consider crating your pup in the bedroom area or providing a small, enclosed kiddie play area (like a playpen) with the opened crate inside the area, and a newspaper or dog litterbox inside. The play areas allow him to relieve himself if necessary at night—but a walk just before bedtime and no bedtime treats are good hedges against nighttime "bed wetting" or pooping. Sleeping in Mom and Dad's room provides for a nice, safe, psychosocial experience for him as a puppy, and you can gradually move him to a different location at night if you decide later to have him sleep somewhere else. Most of my clients, however, reveal that they allow their pups to sleep on the bed, at least part of the night, and there's absolutely nothing wrong with that!

An Absolutely Essential Routine: Exercise!

Every dog's daily routine should include some form of exercise—preferably with someone in the family—since building relationships with your dog is an ongoing process, and exercise is good for you, too! Some good types of exercise include:

Dr. Wright's Do's and Don'ts for Walking

Walking your dog will be a great experience for both of you—something to look forward to, not dread—if you start your pup out on the right foot. Follow these do's and don'ts and get walking!

• DO have a friend or neighbor approach you when first walking with your pup. She needs to learn that approaching people signal happy things. Try walks off-territory (out of your usual bounds) a couple of times a week, so that she gets used to seeing people and other dogs in different contexts and environments.

• DO walk your puppy or dog daily. Place your dog on your left if possible, if walking down a road facing traffic or on a side street, so that you're the one who gets hit by the car, not him. (Actually, drivers will see you more clearly than your much shorter dog in the same spot. It's also easier to nudge him with your knee to get off the street and to teach him to not cross in front of you during a walk than it is to pull him toward you as he bolts across the street on the right.)

• DO allow your dog to greet people, but say, "One at a time, please," if more than one person comes over to meet your dog. Look to see if she's staring at the person. If so, break the stare by pulling gently on her head halter and having her look at you. Give her a happy word if necessary.

• DON'T stop if he appears frightened (his tail is tucked, his hackles are raised). Keep walking with him on your left, the person approaching to your right, and say, "He's feeling grumpy today, sorry," as you walk by the person. If his emotions change to "angry" or "fearful" at just the sight of an approaching person, take charge of his mood and behavior. Head off in the opposite direction, or increase the distance between you and the approaching person. Move off the street as you walk

• Playing flyball
• Playing Frisbee
• Chasing a stick, on the ground or in a lake or pond (Go Labs and retrievers!)
• Walking your dog
• Jogging with your dog
• Tossing the ball or stuffed bear in the house (for toy breeds) or outdoors (for larger dogs, or any dog if you don't allow running and hectic play indoors)

by the person, "jollying up" your dog but being in control at all times of his access to the person. Break his stare at the person until he "forgets" she's there. This technique works the same way whether it's a person, a dog, or both approaching.

• DO notice to see if she begins to regress the older she gets (about 1½ to 3 years, when she becomes more sexually and socially mature). If she does, work on "retraining" her <u>not</u> to take responsibility for you or the house when people approach. Ask a friend or relative to approach her with a treat. Your friend should toss the treat underhanded in front of her as he approaches (while not looking at the dog initially and walking past you and the dog), and then retrace his steps and walk up to greet the dog with the treat.

• DON'T watch your dog become emotionally aroused—either fearful, with hackles raised and tail tucked retreating behind you, or staring with tail in the air swishing, crossing in front of you—while you wait to see what happens. Waiting to see usually results in, "Yep, he bit the guy!"

• DO always take responsibility for your dog during a walk, and take charge of her emotional state and her behavior as people approach. <u>You</u> make the decision about what she's to do and not do, <u>especially when she's emotionally aroused and you don't have a good feeling about her—or about the person or dog approaching</u>. She may even feel your uneasiness if you're scared of the person, and take on the responsibility for keeping the person (and maybe his 5-year-old son) away by biting. Although we all like the feeling of knowing our dog will protect us if we're in trouble, none of us like the feeling we get from knowing that our dog bit someone and we could have prevented it. And we surely don't like the feeling of being sued!

- Exploring trails (great odors!)
- Self-play with toys

I've even seen harnessed dogs pulling their owners on skateboards! The idea of allowing the dog to get some exercise is precisely that—stretching his muscles, using his senses and muscles together, even getting aerobically fit, all in the pursuit of being a dog. During play, enable your dog to enjoy his dogness. This is not the time to require him to do obedience stuff!

Please don't make your dog stay all day in a fenced-in backyard by herself. She's likely to sleep all day, await your arrival home by sitting alertly at the entrance to the gate, or even attempt to escape her 40-acre enclosure and wind up lying at the front door. That stuff is not good for her! She won't care about exercise if you don't make it happen. You have to allow her to discover the appropriate way to get rid of her energy, to be a dog.

Sufficient exercise allows your dog to get rid of all that energy she stored during the day awaiting your arrival home, awaiting her freedom from a crate, or doing other jobs, such as looking out for your car coming in the driveway, or watching the 1-year-old play in his own "crate," or the cleaning lady clean, or keeping the sofa cushions warm until you walk through the door. Exercise in the morning and again in the evening makes most dogs willing to get through the day with other routines that don't involve such a wonderful being-alive activity.

At the very least, be sure that you make time for a morning and evening walk, or some form of exercise to enjoy with her as part of your daily routine.

Going to the Vet and Other Adventures by Car

Let me start this segment by warning you that it's never a good idea to feed your dog or cat just before you go on a trip, unless you want a Trip from Hell. And that it's a very good idea to let her relieve herself before you travel. I learned the hard way years ago, by having my cat, Turk, jump on my lap as we backed out of the driveway and pee all over me. That was a nice-smelling car! I was pissed—literally!

Okay. So going to the vet is not a daily routine (hopefully!). But it's a regular enough occurrence that you want to make your pet comfortable leaving the house to go for rides in the car, especially rides that include a stop at the vet's.

First, it's a good idea to decide if you want to put your dog or cat in a carrier before placing her in the car to take her somewhere. I hope that you will do so if you don't have a car set up with a grille for pets to travel in, because in an auto accident, a pet becomes a flying missile just as a small child would. If you do decide to use a carrier, you must first make it attractive to your dog or cat. It's important to do this up front, if you can, so that the first experience of going to the vet is a pleasant or at least a nontraumatic one.

🐾 Teaching Your Pet to Love the Carrier

Training cats and dogs to enjoy being in their carriers is not really difficult, but it needs some advance planning to work. Find a carrier that is stable

The secret to success with a carrier is to get your pet to feel comfortable with the carrier before you have to use it to go to the vet.

enough and strong enough to hold your dog or cat but that has holes in it for breathing and ventilation. Most do.

You need to get your pet used to a carrier. If you don't, she'll bolt at the sight of it after the first time you are "forced" to stuff her in it! Fortunately, this is easy. Just leave the carrier in her room or in the den or someplace with treats or her favorite toys in it for a week or more before you plan to use it. Be sure to show great interest in the carrier, but never try to place your dog or cat inside it.

After a while, she'll jump in herself to investigate—especially if your pet is a curious cat! You can close the door and stay with your cat or dog for several minutes and then open the door to allow her out. If you make a game of this a few times a day, you'll find that your pet will actually prefer to go into the carrier. When she enjoys this game, try picking up the carrier and carrying it across the room and then bringing it back again and setting it down. Wait 30 or 45 seconds for the dog or cat to get her bearings; then open the door. Be sure that you don't open the door if the dog or cat is trying to get out or you'll probably unintentionally reinforce that behavior.

If you need to entice play, use a milk-bottle cap, twist-tie, or other small object that you can place in the holes and move back and forth to elicit a play response from your cat. Play will take her mind off trying to escape. Or you could try a treat. Chances are that being carried across the room and back will be no big deal to your pet, or it will be regarded as just another part of what happens in the carrier.

Follow this by taking the carrier (with your pet inside) to the car and shutting the door. Then open the car door and return with the carrier to the house. Use your own judgment as to how fast you think you can go with your particular pet. Other steps in the sequence include turning the car on and off; backing out of the driveway and then into the driveway; and then going for a ride around the block, in different directions in the car, stopping somewhere briefly, and then returning home. Make sure that you proceed through the steps slowly to get your pet used to them. Don't try to do them all at once.

Fido and Fluffy should learn that this is a nice social event that ultimately ends by returning home and getting lots of attention, playing, getting treats, or running around to get rid of some pent-up energy. If your pet begins to show discomfort at any step along the way, end the trial, and return to the start location.

After you have your dog or cat used to going in and out of the carrier and riding to different places in the car, you may wish to consider going back a few steps. Pick up the carrier with her in it, walk across the room, and then place it back where you started and let her out.

🐾 Car Trips without a Carrier

If you decide not to use a carrier, allow your pet frequent visits to the car for short trips to determine if he's likely to get carsick; wedged beneath the brake or accelerator pedal; permanently affixed to the top of your head (cats and tiny dogs only); transformed into a loudly yowling or barking airborne object for the entire trip; or turned into a vicious "protector" of the car, vigorously defending it from joggers, bike riders, and others you pass. Your mild-mannered dog at home may become a dangerous (to you) Cujo, leaping back and forth over seats, and growl-barking at passersby as you try to drive.

Be careful of that. Don't act amused. It could cost you your life as well as your dog's (and potentially, an innocent third party's). Instead, immediately stop the car and wait until either he calms down or you can draw his attention to you. Some people have had success (in dogs only) by using a happy word to elicit a happy emotional state in the dog when he sees a jogger or biker, essentially preempting the dog's aggressive reaction by connecting the sight of the jogger to a happy mood. Praise the happy face as if it's his idea, as you've done at other mood-altering times.

As the novelty of car travel wears off and your attitude of not tolerating misbehavior in the car becomes consistent, your dog may be able to calm him-

DR. WRIGHT'S INSIGHTS

Here's a word of caution! The single worst thing you can do is put your dog or cat in the carrier, and because time is precious, only use it to go to the vet. If you do, your carrier, the car, and a ride in the car become strung together in your pet's mind, producing more and more fear because they predict impending doom!

self enough to take short trips unrestrained. If not, go the carrier route if you can, or consider installing a gate in the back if you have a station wagon or SUV. Within a couple of weeks, you should be able to go anywhere in the car, and trips to the vet will not be obvious before you get there.

🐾 The Practice Visit

Now it's time for a "practice" visit. After you arrive at the vet's, ask if the doctor will come to the waiting room for a moment to greet your pet, who will soon be coming for his first appointment.

Most vets are glad to do this, as it helps to prevent problems down the road (so to speak). Perhaps the vet or receptionist will give your dog a nonfat dog biscuit. Then go back to the car and either continue on a ride or return home. Gradually introducing your pet to the carrier, the car, and the veterinarian really pays off. And I have many clients who didn't do it and now have dogs with vet phobias or biting problems! By spending those extra couple of hours initially, you allow of a lifetime of fewer worries when those times approach when you must see the veterinarian.

This kind of behavior training can be fun for both you and your dog or cat. Many of my clients have indicated that one of their dog's favorite things is to go on trips with them. And yes, my clients have even said that the cat loves to go with them on trips to their parents' house or their home on the lake. You may enjoy having your cat lying on your shoulders or sitting on the back ledge or in your spouse's lap along the way. To each her own! But be safe—a sudden jolt can result in a concussion for your pet and some permanent scars for you.

However you like to travel with your dog or cat, please do consider taking your pet with you on trips—or merely on daily car rides as an end in themselves. Soon car rides and all the other routines we've talked about in this chapter will become the "good stuff"—part of weaving the rich tapestry of your lives together as real companions.

Chapter Thirteen

New Relationships, New People, New Problems

You have worked very hard to build a relationship with your companion animal. You have gotten to know each other's needs and expectations. You've worked out the minor glitches. You trust her and she trusts you.

You're pleased to have her for a friend. You feel a sense of pride when you introduce her to people who come over to visit, who meet you in front of your house for a chat or on a daily walk, or who watch when she decides to show off by leaping and climbing all around the kitty condo. You enjoy people's glances and smiles when they see her walking with you or riding in the car next to you with her favorite tennis ball in her mouth. You look forward to greeting one another first thing in the morning, and you can't wait to play with her when you come home from work or errands. It's great just to be around each other every day, sharing your special bond.

And now, you decide to screw it up by bringing "the new person" into your life—and into your pet's life.

Some pets take an immediate dislike to the interloper, whereas others seem to cope well with the new person's presence, perhaps hoping that he or she will just disappear eventually. One thing is certain: Fitting a new person into your lives can be a traumatic experience for both of you. It often results in serious behavior problems that are as surprising to you as they are upsetting—because "Angel just isn't like that, and I don't understand where this is coming from." (Does that sound familiar?)

I've heard often from my clients that "Except for this one small problem—well, it's actually a BIG problem because it involves [the new person in the household]—my dog/cat is perfect. She's wonderful. But I can't continue to live like this, and I can't expect my new [boyfriend, wife, college roommate, live-in nanny, elderly father-in-law] to keep putting up with this stuff either. What's going on, Dr. Wright? Is there any hope of getting back on track? And how can I get this new person in our lives and my best friend to get along?"

I can't guarantee that they'll never hear the dreaded words: "EITHER THAT *!#* ^#%!# !! DOG/CAT GOES OR I GO!!" But there's almost always hope—after we do the diagnostics to figure out how the cat or dog sees the situation and commit to taking the steps that can make the pet more comfortable.

There are, clinically speaking, a variety of different causes for dog and cat misbehaviors that are related to the presence of the new person in your life. We will deal with the addition of a baby or toddler to the family in chapter 15. The other new people in your house all have enough in common for us to discuss what your dog or cat probably perceives to be the problem, which in turn will enable us to suggest different options to resolve the misbehavior.

The new person might be a close friend, spouse, foreign exchange student, housekeeper, elderly parents, or freeloading Gen Xer. It can be a short-term or long-term arrangement that brings you all together, but that doesn't matter. After all, the dog or cat doesn't know that Uncle Nick is going back to Italy in 3 months. What all these people-pet arrangements have in common—whether they love the dog or are afraid of the cat—is that by the time you see unwanted behaviors unfolding, the dog or cat is probably too stressed out to cope.

Stress: Not Just for People!

Stress and its effects are hot buttons for everyone in our 21st-century lives. But we don't always think about the fact that animals feel the same kinds of pressures and upsets that we do. And they can't go to a spa and get a massage at the end of the day! Instead, they may cry, bite, or pee on the rug right in front of the fussy new person who doesn't seem to appreciate dogs much in the first place. I know that "stress" is a very vague concept, but from observing and studying dogs and cats in their relationships, we behaviorists have zeroed in on three specific aspects of stress that most often seem to send our feline and canine friends into a tailspin (so to speak).

After you and your pet have gotten to know each other and settled into your life together, it's easy for a new face to upset the apple cart. Any or all of the three stress-boosters can throw your pet for a loop when you bring that new person into the circle.

🐾 The Three Signposts of Stress Ahead

The three components of stress that dogs and cats experience when attempting to cope with a new person in their lives include novelty, unpredictability, and uncontrollability. Let's look at each one.

Novelty

New is bad—at least as far as your pet is concerned! Novelty means simply that the dog or cat isn't used to the new person and finds newness upsetting. Some cats and dogs are much more alarmed by something "different" than others are. They are just like people! Think about a group of friends being served a new dish at a restaurant.

One person's eyes will light up: "Ah! Something I've never tried before. Great! I don't know what it is, exactly, but let's see if I like it!" And he pops it in his mouth without another thought, asking the chef later just what that new taste treat was.

Some people—probably the majority—will call the waiter over and ask what the "thing" on their plate is. Then, unless it sounds like a staple on the *Survivor* TV show, they will at least sniff it, poke it suspiciously, and then take a small bite to see if it's anything they want to consume.

And a third type of diner will not even have the curiosity to ask what "it" is or—heaven forbid!—try a little bite. They know everything they need to know. It's new. It looks new, it smells new, and it might hurt them! It might taste bad. The texture might be unappealing. It might be overcooked or too raw. They tried something that color once before, and it made them sick. It might be spoiled. They might be allergic. Whatever "it" is, it's something different, and they are absolutely not going to eat it, no matter what.

This, unfortunately, seems to be how a *lot* of dogs and cats react when a new person is added to the "menu" at home. The more novel or unique the person, the more the pet wants to resist, and the more potential stress the dog or cat has to cope with if the new person is still there the next day.

Unpredictability

For your pet, unpredictability means: "When you're around, I don't know what's gonna happen!" As we learned in chapter 12, dogs and cats need to know what is about to happen next in their daily routine. The more unpredictable life is from day to day, the more potential stress the dog or cat must cope with.

Your pet depends on his daily routine and the rituals we have talked about. He knows that when the sun comes in the dining room window, the school bus will be coming soon, and Jamie will run up the walk and greet him at the door and toss toys he can chase until he's worn out. So he watches for the sun to come in the window. But if Jamie never gets off the bus because the new nanny has picked her up and taken her to a piano lesson, his expectations are foiled. He's confused, and an important social need is not fulfilled. Chaos is always more stressful than routine, until or unless it can be managed into something predictable. And that brings me to the third signpost, lack of control.

Lack of Control

The third stressful element often attached to the new face at the dinner table is the lack of autonomy and independence a dog or cat has in determining what she is able to do next. The new person might "clean up" the floor and put the cat's favorite jingle balls in a box where she can't get to them. Or the new boyfriend might try to make the dog lie down in the corner every time they meet in a room, instead of letting him sit or lie down where he wants. His owner never makes him do that! And the list goes on.

Novelty, unpredictability, and lack of control together with a new person in the household can easily create a situation where your pet's coping mechanisms are stressed to the limit. She is no longer able or willing to try to live with the changes in her life, and misbehavior of some sort occurs as a result. Some dogs and cats immediately dislike the new person probably due to the person's novelty, experienced through one or more of the dog's or cat's senses. The animal seems to be saying, "This new person stares at me a lot, he has a big, booming voice, and he keeps sitting next to Mom, which makes me really ticked off." Following this line of thinking, the pet continues, "I think I'll growl [dog or cat] or hiss [cat] at him, or maybe I'll just lift my leg on his pants leg [dog], or jump up on his lap, and then turn around and spray him on the chest [cat]."

But take heart—it might not be that bad. (Remember that rare diner who liked "everything new"?) Because dogs' and cats' temperaments and their ability to react to changes in their lives differ, so do their abilities to cope with and accept new people in their lives. People who have two or more pets in the family may find that Jimbo loves the new guy in town, whereas little Warren practically freaks out right off the bat. Other dogs and cats are able or willing to put up with the changes for a while, but at some point, they probably think, "This routine is really messed up, and this guy ain't leavin'!" or "Not that woman again! I've had it; enough is enough. I can't stand any more!" Then they proceed to appall you with some dastardly deed.

So some dogs and cats are quite sensitive to any changes in their routines caused by the introduction of a new person in their life, whereas others could not care less. Those dogs and cats who are most sensitive to stressful changes may be able to cope with those changes for some time before a behavioral breakdown is apparent, but others may react as soon as the intrusion occurs.

🐾 Stressful Acts

Let's look at some common examples of events that tend to be disruptive to dogs and cats who must deal with new people. Sometimes spending time with the new person in your family is disruptive enough to change your pet's routine in ways that just don't cross your mind.

Perhaps that new person in your life likes to go for long walks, and you decide to take Frankie with you. He's not used to it. Too much exercise may make him tired and anxious about trying to keep up. Or perhaps the new person likes to walk alone with you, without Frankie (gasp!), cutting down on the amount of exercise your dog normally gets. So too little exercise can also be a stressor and usually results in too much leftover energy, otherwise known as asking for trouble!

Maybe you go through obedience commands on a regular basis with your dog because he's "in training." Some dogs don't do very well with placing a lot of demands on their behavior outside of class. Or taking "orders" from someone they've never even met before—like the new person in your life. Don't forget that the collar and leash are to be used as a communication device, not as a means of forcing your dog into compliance.

Suppose for a minute that you are Frankie. Perhaps you have been lucky enough to go with your owner to her new significant other's home. But she forgot to tell you where you're supposed to relieve yourself. Or you just can't access that designated place to urinate or defecate because her boyfriend trapped you in some mudroom, and you have no idea what you're supposed to do or when you'll get out. What's up with that?

Being alone can also be stressful for a pet if he's left at home, while Mom is out gallivanting around with her beau and not coming home at the "right" time like she usually does. Remember your pet if you go out for the evening with your new friend! Don't forget the routine you have for allowing Frankie to relieve himself. Imagine trying to "hold it" until your guardian decides to come home so that you can relieve yourself in the proper location. It's not a nice way to feel or to be treated, huh? (Of course, bigger problems can result from a pet's being left alone. He could inflict damage to the home or to himself. We'll cover this syndrome, called separation anxiety, and its symptoms in chapter 14.)

Finally, Frankie may feel stressed because he thinks that he must constantly keep track of Mom, and what this guy is doing with her or—heaven forbid—without her. Never being able to relax because of the changes in your daily routine brought about by the new person in your owner's life is quite stressful. One reaction to all this sress is to strike out at the people in the house.

Three Kinds of Aggression

The stress of having a new person join the household can bring on three kinds of aggression. All three can result in the same unpleasant and even dangerous behavior. They are possessive, protective, and dominance aggression.

What exactly do I mean by aggression? No, aggression is not just biting! Aggression is usually considered a range of nasty behaviors that include lunging, snapping, clawing, or biting directed toward an individual. They can be preceded or accompanied by a dog's growling or a cat's hissing and spitting.

Many things influence aggression—from the dog's personality and role to what the interloper does, where the owner sits, and so on.

Fighting for something that is "his"—you, a bone, a toy (in other words, control over a resource)—is sometimes referred to as competitive or possessive aggression. People who play tug-of-war with their puppies may end up with

DR. WRIGHT'S INSIGHTS

All the advice in this chapter is aimed at preventing aggression. If it does occur, don't try to handle it on your own. This is not a do-it-yourself project, even though you know your dog better than anyone else knows him. Contact your vet for a reference to a certified applied animal behaviorist who can help you resolve the problem in a way that will be effective and safe. Aggression is dangerous!

dogs who display possessive aggression toward their other toys (and, of course, toward the people who try to take them), if rules for play aren't established early. (Play never includes snarling, snapping, or biting).

Protective aggression rears its ugly head when the dog blocks a visitor's access to a family member and becomes aggressive to the new person because he perceives his role in the family as that of "the Godfather." He bites when someone reaches for, touches, or—heaven forbid—hugs a family member he feels must be protected. It's not a dominance issue, which requires a closer relationship than he has with the new person. It's not a possessive aggression issue. (You protect your child when a stranger approaches you because you fear for her safety, not because you fear losing her as a valued resource.) He ends up attacking to protect "his person," as a canine mother might protect her young from strangers.

Some dogs perceive themselves as occupying a dominant role in relation to someone in the family, as we talked about in chapter 4. Dogs who have dominance issues attempt to control physically a person they know well, or to limit that person's access to the dog himself or a desired resource, like another family member's affection. If the person tries to interfere with the status of the "top dog" by overstepping the boundaries the dog has set, he usually risks getting nailed. This behavior isn't directed toward visitors or strangers to the household because the dog doesn't have a relationship with those individuals. The rules don't apply to them, only the one family member.

❧ Triggering Aggression

Let's look at the poor guy who comes to visit often enough to begin developing a relatively compliant (subordinate) relationship with Brutus. He pets Brutus on demand. He walks around Brutus in the house and yields to him in general (perhaps to please you, initially). So, as many resident dogs do, Brutus takes on a dominant role with the not-so newcomer.

And now your friend decides to sit next to you several nights in a row, hug you, and stroke your cheek. You reciprocate. It happens to be when you usually pet Brutus, but at this point in the goings-on, you're oblivious to your canine buddy!

This new guy's messing with Brutus's source of affection; and what's this? Your dog hears you emit a strange vocalization in response to the guy touching you! That's all your dog can stand. If Brutus could talk, he'd complain, "Here's this not-so-new guy doing something I didn't give him permission to do (a violation of my dominant status), taking my affection from Mom (a valued resource), and attempting to hurt her. (Didn't she just make a strange noise?) It's my job to protect her!" So dominance, possessive, and protective aggression meet in one ugly moment—all three motivations for biting rolled into one.

"Grrrrrr—snap-bite!!"

"Wow, where'd that come from?" you gasp. Now you know!

The Fine Art of Groveling

Fortunately, most dogs who are introduced to new people in their households are either submissive enough or young enough that dominance aggression toward the newcomer is not a viable strategy—at least, not at first!

To get your attention, then, your dog may do what worked in the past, which dogs learn from a very early age. They become quite submissive, complete with tail down and wagging and with head close to the floor, in what

Young or submissive dogs tend to seek your attention by groveling rather than becoming aggressive. You may think it's adorable at first, but if you don't ignore this attention-getting behavior, it can quickly drive you crazy!

can best be described as a "grovel." This bid for attention is most likely to occur when someone new enters the equation and takes away some of the precious resource—you! Or it can happen when there are a number of other contenders for affection; in my house, for example, there are two humans, four dogs, and a cat.

My response to groveling by my youngest dog, Lucy, was to do what most people would do: I said, "Oh, poor girl," and stroked her on the head and paid attention to those poor sorry eyes, letting her climb onto my chest for some affectionate stroking. Of course, all this attention reinforced the groveling. This situation was not good. My wife also fell for Lucy's "Poor me, whatever will I do without your love, since you seem to love Charlie more than me, even though I'm certain that I'm the sorriest little girl of less than a year old you'll ever see" routine. Now, whenever we gave affection to one of the other dogs, groveling Lucy pushed in and intercepted all the petting.

Danger! Know When to Back Off

If the new person in your life violates the rules set by the resident dog, the dog may attempt to escalate his assertiveness and force compliance. See if you can figure out which kind of aggression is at play in these three risky behaviors: The new guy reaches down to pet Arnold on the head. The new guy rings the doorbell, and when Mom answers the door, he enters the house, even though Arnold is standing in the doorway blocking his access to Mom. Arnold positions his favorite tennis ball between the new guy's front legs, and in an attempt to play fetch, the guy reaches for it. We see dominance, protective, and possessive aggression. There are no surprises here.

If the dog growls at you in these circumstances and bares his front teeth so that you can see the armor with which he's prepared to defend his "resource," what's next? You're faced with signs that your dog is going to bite, and if you don't respond appropriately, you risk attack. What should you do? Keep in mind that should you not respond appropriately to these signals—if you don't defuse the situation or if you unintentionally provoke the dog—you may in fact expect an inhibited bite or nip. (By "inhibited," I mean he's holding back on applying full-scale force.) It's his "level one" escalation of aggression, which will hopefully get your attention. Unfortunately, some dogs skip right to red alert—DEFCON! Should you attempt to do anything other than "the right thing," you may have an all-out battle.

🐾 Strategies Our Poodle Taught Us

We learned one technique for curbing this bad habit from Roo-Roo, our standard poodle, who on more than one occasion has let Lucy know her limits. Roo gives a brief growl, stares at Lucy, and Lucy retreats—albeit slowly and with careful, purposeful steps—in the other direction. To our knowledge, Roo has never followed through or escalated her warning to include anything more than a stare and a grumpy growl.

So I decided to stare at Lucy when she tried to intercede between me and Peanut, and added, "Do-o-o-on't you come over here," my warning-growl equivalent. Sure enough, Lucy lay down and watched to see what that was all about. I helped her decide that this was a neat thing to do by paying attention to her 5 to 10 seconds later, thus praising her for not groveling.

We have discovered that variations, such as "Uh, uh," or "Lucy, off!" work just as well, as long as we tell her she's a good girl immediately after

The safest thing to do as you size up the dog's motivation for threatening you is to make yourself a less attractive meal. Remain motionless, divert your gaze (look up or away) so that you avoid eye contact, and smile happily as you use your most expressive voice to convince him that you're ready to "Goforaride?" or "Getchyerbear" or any other activity that usually puts him in a happy mood. As soon as he's changed from grumpy to confused—or even happy—figure out whether he was experiencing fear or anger from the emotional signals he was presenting you.

Remember that fearful dogs want to be left alone, and that retreating from him while continuing to look happy will probably defuse further aggression. But if he's into control and is angry, do not move. You can breathe easily after he moves away from you or when you see his muscles loosen up and his tail wag (down, of course), and his demeanor change in such a way that he appears to be ready to go for a ride, get his bear, or do whatever joyful activity you suggested to him. Now move happily out of the situation, and try to figure out which of his needs were being threatened by your presence.

We'll address some options later—things to do and not to do to avoid dog bites—but if you'd feel more comfortable placing yourself in the care of an applied animal behaviorist or veterinary behaviorist, by all means, consult one.

Dr. Wright's E-Mail

Dear Dr. Wright:

My 1-year-old pit bull has taken to head-butting everything in his way. Mojo knows that things move if he hits them hard enough. He butts the refrigerator to knock the cereal off the top. He butts his sister. Why is he doing this?

Carolyn

Dear Carolyn:

Mojo sounds more than a little "headstrong." It may be that he is asserting himself to get what he wants. It could signal the beginning of a pattern of controlling behaviors, as he learns to control the activities of those who interact with him. He may like his sister, but he probably loves making her back away from his favorite resting area.

You will probably want to neuter Mojo. Then if things don't improve, you may wish to get a head halter to direct his head to the side when he starts to butt. As he lowers his head and steps forward, let the leash do the correcting. It's attached to the halter under his chin, so if he continues to move forward, the only thing he can butt is the floor. When he stops his approach and backs up, praise him extravagantly! He'll soon see that this kind of "controlling" behavior won't get him what he wants.

complying. One of my fears in adding a mild correction to her groveling was that I would increase her submissiveness because she is in a subordinate role to me. The appropriate response of subordinate dogs to assertiveness is to become even more submissive. Be careful not to be too assertive in correcting your groveler, or you may find that the problem becomes worse! If it does, try "Plan B": She wants to please you, so offer something to do that keeps her from coming toward you, yet provides her with something that she can do that normally feels good as a consequence.

"Lucy sit, good girl," followed by, "Lucy go get your bear, where's your bear?" usually works well, in that she can't sit and continue to push in between Peanut and my hand at the same time. (As we behaviorists say, these are incompatible behaviors.) Now I can continue to be affectionate with Peanut, and Lucy has become involved in searching for her toy bear. Mission accomplished.

Lucy has learned that when she wants something from me, all she has to do is sit and not grovel, and something fun will result from sitting. My responsibility is to make sure that I meet her expectations. If you've been following along in my school of manners, the plan may sound familiar: To extinguish undesirable behavior, give her a desirable (from your point of view) task that she can do in its place and then provide her with a pleasant consequence for doing so.

Granted, groveling can be heartwarming in puppies and smaller dogs, but how cute can an adult, drooling, 100-pound bloodhound be when he crawls between you and your new boyfriend and winds up getting slobber all over both of you? Talk about testing your boyfriend's love!

You should work on disrupting signs of pathetic behavior early on. You don't want your adult dog to exhibit either one of two extremes: whining, wailing, crying, whimpering, head-rubbing, tail-lashing, pawing, groaning solicitations for attention or staring, growling, tail-switching, jumping-up with teeth showing threats, demanding that the new person cease all further contact with you "or else." It's funny how some dogs can be so stingy with sharing your affection, regardless of which set of symptoms they display.

Helping Your New Family Bond with the Dog

With pets, most behavior problems can generally be traced to some disruption in routine that deals with a bio-psycho-social need that is not being met. So most treatment programs for behavioral problems caused by fitting a new person into your pet's life involve redefining the ritual and routine to include

A Pet's Checklist: Seven Nasty Things to Do When You Can't Cope

If I could give some of the animals I've worked with the power of speech, here's what they might tell your pet about letting you know she needs help with the appearance of a new person:

1. Be nervous and jumpy. Become more reactive or skittish to normal sounds like the TV or the abrupt appearance of someone coming into your room. They may think cats always act this way, but you know it isn't true!

2. Overreact. Leap up and bark, or tuck your tail and run under a table or sofa and hide. Make the new person say, "This dog really hates me."

3. If the new person tries to pet you, bite him. If your owner or the new person comes over to console you, interpret that increase in proximity as a threat and lunge and bite them or offer a defensive, fear-induced nip. In other words, if you behave aggressively, your owner will tell Dr. Wright that this is "out of the blue" and uncharacteristic of you.

4. Become depressed. Mope around the house. Decide to stop eating—they're bound to notice that!

5. Begin barking or howling at the slightest noise. Let them know that you feel threatened.

6. Chew up the new person's stuff. Go back to chewing the things you wrecked when you were a puppy, or scratching the furniture like you did when you were a kitten. Chewing "his" leather shoes with his odor or scent on them is a great way to get rid of your frustration and a creative way of coping with his presence, even if Mom doesn't like it.

7. Scratch or lick your skin raw. The situation has driven you to act in a compulsive, self-destructive way that they will have to deal with. Maybe she will kick the freeloader out the door so that things will get back to normal. (Or maybe she will kick <u>you</u> out instead—don't let that happen!)

the new person. The new person should become a resource for the pet rather than his competition.

Let's say that you get your child back after 4 years of college and 2 years of high rents, and now Junior seems to be a permanent fixture in your home. At first, your dog might think this is terrible. But thanks to some good planning on your part, your dog now has twice the affection he had before. And he gets to share himself with two people—you and his new "brother." Or maybe you've invited your elderly father or mother to join the family. Or you may have just gotten married and added a spouse to the household.

The idea here is obvious. Because your dog is already a family member—the only family you have had living at home up until now—fitting the new person into the family means having him and the dog do together things that used to be reserved for just the two of you. The list includes any and all activities that occur from the time you get up in the morning until you retire at night.

❧ Adding "Him" to Your Dog's Routines

It's really pretty easy to work the new person into your dog's existing routines. Some of these activities, like walking the dog, feeding the dog, talking to the dog, playing with the dog, or other "jobs," can be done by both of you. In this way, the whole family is involved in the same activity for motivationally different behaviors, like fulfilling the need for exercise, food and water, socialization, and caregiving.

In our household, Angie is the source of the most secure affection for all our dogs (she's the primary caregiver), and I'm the source of the most interesting activities—play, walks, and romps around the lake or in the park. We both give hugs and play with our dogs daily, but they seem to have decided that Mom is better at giving and receiving affection than I am, but that Dad is better at walks and playful adventures than Angie.

Soon, the new guy in your family will find his own "niche" with Duke, and preferences will unfold naturally. Such is the development of social structure in families. We all do some daily events together at the same time, like eat or play ball, but some combination of us do different daily activities together. Then we all have our own special, separate time. I do my writing at home while my dogs lie about me gazing at me, playing with or grooming one another (yes, mostly them, but me occasionally), or sleeping. It's a nice, social, family thing we do without Angie (because she goes to her office to work during the day).

The exception for us is Peanut. She spends her day lying at the front door in the foyer, looking out the window watching for Angie to return. Angie won't be returning anytime soon, but this is Peanut's job, and she will do it to the best of her ability. Peanut was Angie's dog before we became a couple, and I became "the new guy." I ask Peanut if she needs to "go outside" about the same time daily, and we exchange glances as I traverse the foyer to and from the kitchen, but she chooses to do her thing, which she regards as an important daily activity, instead of joining the rest of the family. However, the three dogs we adopted since then, Roo, Char, and Lucy, are all lying in the study with me—Charlie on the couch, Roo on the tiled bathroom floor (she loves cool surfaces best), and Lucy beneath the computer table.

Strategies for making the new person part of the family should involve activities that result in Duke's feeling good, and that you are already doing with him but want your new house member to join you in or to do more of in the future. For example, the new person can take Duke outside in the morning to relieve himself. First, all three of you can go out together. When Duke becomes familiar with the new guy's presence and he becomes familiar with Duke's peculiar preferences (like sniffing the pine straw beneath the cedar tree for 20 seconds before trotting in the opposite direction to poop behind the hedges— heaven forbid you try to get him to go to the hedges first!), you can let them go out together on a more permanent basis. (How you get the new guy to agree to do that is, of course, beyond the scope of this book!)

🐾 Special New Routines

Another strategy is to add something to your dog's daily routine that he enjoys and to make it something that Junior and Duke do together. If Duke likes to go for rides in the car, get Junior to take him for the rides. But before your son takes him, be sure he says, "Wanna go for ride?" In that way, Duke can start to pay attention to Junior's facial expression and voice, as well as predict an activity that he'll both enjoy and associate with Junior, the new guy in town. If this activity can be done on a daily basis at about the same time of day (translated, right after Junior gets home from work but before dinner, another important event), it will become a regular, predictable routine that will enable Duke to form his own relationship with your son.

Basically, the introduction of a new person to the family should enrich your dog's life as much as it does yours. What about cats? Let's see.

Do's and Don'ts for Friendly Cat Encounters

Cats decide whom they are going to be friendly with on an individual basis, beginning with their first impressions at the first meeting. You can call it a reactive temperament, being fickle, a multiple personality (as Emma does in "Dr. Wright's Casebook" on page 308, to explain Phoenix's very different relationship with her versus Brandon and all visitors), or just being very selective in deciding who sees and gets to enjoy what side of your cat's very complex personality. You should, however, remember this: Cats are like elephants; they never forget.

When you first bring that new guy home, take time to let him know the do's and the don'ts for meeting your cat. Who knows, maybe this will be the one who winds up "in the family," and you'll want to be sure that *your* "Phoenix" will get off on the right paw with your new friend.

🐾 Do's for the Potential Spouse, Significant Other, or New Housemate

Here are the eight best things you can do to get off to the right start with a cat.

1. DO slow down. With cats, it pays to think "slow." Move slowly and take time with everything you do in your first meeting with the cat.

2. DO keep your voice low. Talk quietly to the cat and the cat's owner. Cats tend to be alarmed rather than thrilled by too much enthusiasm.

3. DO keep your distance. Mind your own business with relation to the cat and the owner at your first meeting. Sit in a chair in a different location from the owner, just in case you are about to meet a protective, territorial cat.

4. DO introduce yourself in an open area. The living room, den, and dining room are good places for the cat to see you from a distance and choose to advance or retreat as needed.

5. DO know the proper greeting. Review the finger-point "handshake" on page 118, and offer it when the cat appears interested in approaching you.

6. DO allow the cat to set the pace. Be sure to allow the cat to greet you in an active sense, while you remain quiet, slow to move, and relatively passive while sitting in your chair or on the floor.

7. DO let him do his thing. Allow the cat to jump up on you, smell you, rub against you, and investigate you and your scent. The most you should do is to offer your finger for him to rub against.

8. DO initiate play after greeting. Ask the cat's owner to give you one or two of the cat's favorite toys, and initiate play by gently tossing the toy or moving the dangler away from you and past the cat, who by now is interested in finding out if you know what to do with a toy.

🐾 Don'ts for the Potential Spouse, Significant Other, or New Housemate

With a cat, what you *don't* do can be as important as what you do. Make sure you avoid these eight behavioral faux pas:

1. DON'T move quickly or with large, exaggerated motions.

2. DON'T say hello to the cat with a big, booming voice or with a loud, screechy, excited voice.

3. DON'T walk over and then reach down to pick up the cat.

4. DON'T pick up the cat, period!

5. DON'T follow the cat if she chooses to retreat into a bedroom or other enclosed area.

6. DON'T involve yourself in delusional thinking that, in spite of the best advice, you'll know when the cat is ready for you to go over and pick her up because cats like you.

7. DON'T clap your hands or slap your lap and make faces at the cat, as if you're trying to get a dog to jump on your lap. You will just frighten her and look silly.

8. DON'T press your luck—if you ignore the first seven don'ts and a cat is reckless enough to jump on your lap anyhow, don't pat her on the head or try to hug her, as it will be quite a while before you get the opportunity again!

Stress and the Sensitive Kitty

Although many of the stressors for cats are the same as those for dogs, some stressors are unique or pertain most clearly to cats. They include introduction and handling issues, as we've already discussed, and other aspects of the cat's daily routine that may change directly because of the new person—or because of the time constraints the new person places on you, the owner, because of your new relationship.

Anytime you change your routine, your cat's daily routine will undoubtedly be affected, too. Cats respond differently to changes in routine. Some cats adapt quite nicely, whereas others become rather aroused and consider the unstable routine a challenge to adapt to, and yet others become frightened by the uncertainty of it all and begin hiding from life. Behavioral symptoms that result from an inability to cope with change due to irregularities of routine include irritability leading to biting or scratching; loss of appetite; whining, crying, or incessant vocalization; compulsive licking or other repetitive behaviors; and almost any other misbehavior cats are capable of. One of the more serious and common problems is aggression—biting and/or scratching—directed at the new person. But by far the most common result of stress and changed routines is the failure to use the litterbox. And what havoc that can create!

What to Do after You've Blown It

Assuming that your new housemate has failed to follow any of the do's and done a number of the don'ts, you and he can try the following steps to undo the damage resulting from the bad impression that he's already made on your best friend. Don't give up! They work.

1. Begin again from the get-go. But do the introductions in a new location that does not yet have any negative associations for the cat. Reintroductions are more successful when done in a neutral or relatively positive context.

2. Follow the do's and don'ts on page 306.

3. Use treats or toys to entice her to approach closer to you. Calmly and without exaggerated movements, toss a treat to her left or right. Toss the treats closer to you, as long as she responds favorably. If she responds better to play than to food, toss a toy attached to the end of a string to her left or right, and pull it toward you, allowing her to capture the critter every now and then.

4. The more often you do this, the faster she will learn to change her perception of you from "nasty" to "not bad" to "neat human."

5. Keep it short, say, from 1 to 5 minutes at a time. If you ignore her after your invitation for social contact, she will be less nervous and more likely to consider that you no longer pose a threat to her.

6. If all else fails, take off that baseball cap! Many cats and dogs are afraid of hats or "hate" everybody who's wearing one. State can still win without your free ad, honest!

7. Even if things start going well, remember that you're probably still a bit scary to her, and that progress should be measured in weeks, not days. You should notice that she is spending more time with you today than she did a week ago and that she remains closer to you today than she did last week. She also probably initiates contact with you when you walk into the room or at least follows you into the room. With any luck, she should be rubbing against some part of your body following a finger-nose touch before too long. In no time, your cat will no doubt fall in love with the new person and learn to see him through the same eyes that you do. (If not, well, remember who was there first!)

DR. WRIGHT'S
C A S E B O O K

Phoenix

For cats, first impressions certainly matter. Cats seem to build upon their first exposure to a guest entering their habitat, and each experience adds to the previous one. I was reminded of this when I recently went to consult at the home of Emma Martin. Emma had adopted Phoenix, a 1-year-old, recently declawed and neutered male Cornish rex, shortly before her marriage to Brandon only 2 months earlier. Phoenix was raised in Florida in a cage for most of his first 10 months of life. His sisters were caged together next to him at the breeder's facility, so he could see and hear them but not interact with them. Cornish rex cats are commonly outgoing, fairly confident, and exploratory cats, but because of his earlier kenneling experience, Phoenix tended to be a bit shy and standoffish, according to Emma.

The living room was very tidy, except for a large throw that was crunched up on the left cushion of the leather sofa, next to Emma. Because it was July in Atlanta, the air conditioner ruled the day and the nearly bald (okay, hairless!) cat had curled up in the warm throw. When Phoenix awoke, he noticed me immediately. He came over to the chair where I was sitting for a brief introduction. All I did was stick out my finger, which the cat proceeded to touch with his nose.

While Emma and I talked about him, Phoenix proceeded to show off his toys and his hunting and pouncing skills. In contrast to his endearing behavior, according to Emma, the Cornish rex had two big problems: inappropriate elimination and aggression. He had decided to pee on the Martins' bed the previous week, he had taken to licking himself continuously, he had been pawing on Emma's blouse with excessive kneading, and he had begun to bite at her fingers and nip at Brandon anywhere he could reach. What happened to that shy little cat she first brought home?

Emma noted that Phoenix had never fully accepted anyone in the household other than herself. Brandon was regarded with suspicion. Other people's attempts to meet the cat were always met with a hiss, or with Phoenix invoking his escape procedures: He ran down to the basement to hide in the crate where he had lived for most of his life.

Brandon's first meeting with Phoenix went relatively well. Then the newlywed started treating the cat like a dog, playfully batting the cat on the side of the head and roughhousing with Phoenix in what Emma described as "not gentle" play. Cats don't normally understand rough play. Phoenix didn't. And not surprisingly, Phoenix's reaction to Brandon is quite different from the cat's reaction to Emma. She continues to enjoy forming a new relationship with Phoenix, with the cat coming to visit and solicit attention, play-rubs, and stroking from her. But the Cornish rex bites Brandon's ankles and, when in a playful mood, jumps on Brandon and bites his hands or anything that moves, which elicits rough play from Brandon. So we now have a vicious circle forming.

What seems to have happened to Phoenix's perception of people is that hands may signal something to bite rather than something that always feels good, and that strangers in the household should be regarded cautiously from the get-go. So when Phoenix immediately took to me, Emma was amazed and astounded at first. Then she covered her confusion with, "But of course the cat would respond to you like this, you're the pet psychologist!" The truth lay somewhere in between: Anyone can learn to let a cat approach him so the cat won't be overwhelmed.

I was able to help the Martins with their cat's behavior problems. But if less stress had been put on the new pet in the beginning by Brandon's faulty approach, the difficulties might have been avoided altogether.

🐾 Boycotting the Litterbox: A Quiz for Cat Owners

If your cat has stopped using the litterbox regularly, before you pitch both of them out, review these five causes for avoiding the litterbox and see if one of them fits.

1. Litterbox problem: cleanliness. Perhaps you do not clean out the litterbox as often as you once did. You have more to do at home now, and you just don't have enough time to clean it out twice a day. The cat is stressed out because she has to put up with that ammonia smell coming from her litterbox. Yuck. Maybe she'll decide to pee somewhere else.

2. Litterbox problem: location #1. Now that the new person occupies the spare bedroom and bathroom, your cat's privacy is no longer respected, and she has to worry about time-sharing the bathroom facilities with a stranger. Yuck. Maybe she'll decide to pee somewhere else.

3. Litterbox problem: location #2. Now that you've moved the litterbox to the laundry room, there are still problems. Your cat has to put up with a noisy dryer or washer, potentially frightening the heck out of her, and risk the new person walking in on her with his dirty clothes when she's in a precarious position. The closed-in laundry room limits her access to an escape route. There is too much noise, no escape route, and too much stress. Maybe she'll decide to pee somewhere else.

4. Litterbox and food bowl problem: location. Do not move your cat's feeding location too close to the litterbox in a feeble attempt to keep the "cat things" in one more confined part of the house so that you and the new guy can enjoy the rest of the house. Sure, Kitty can learn to eat in the laundry room, but if the food bowl is too close to the litterbox, she may decide to pee somewhere else.

5. New person's stuff problem: novelty. Because any change is potentially stressful for cats, the introduction of new furniture or clothes, which smell different (clean or soiled), provides additional sources of stress. Maybe she'll decide to pee on those smelly clothes in the laundry basket, next to her ammonia-reeking litterbox.

Quiz Question for Cats
Where will you pee? (Answer: Somewhere else!)

🐾 A Simple Soiling Solution

Despite your cat's discouraging answer to the preceding quiz question, there's still hope. Maybe she's the laid-back type. But if my many years of helping frantic cat owners with the number one cat problem are any indication, most cats' laid-backness stops at the bathroom door. At any rate, a rather quick fix to the problem is possible—as long as other factors are not involved (that is, physiological or veterinary medical troubles), and as long as you will make the changes and have a little patience. This opton beats replacing all the carpeting.

The solution relies on treating the relocation of litterbox and food as if you were introducing your cat to a new home for the first time. (See chapters 6 and 12 for more on this.) Not only is the new guy disrupting the household, but he's also keeping your cat from satisfying a basic psychobiological need: to urinate and defecate in a clean, quiet, secure place. What was once a great place for the litterbox is now a lousy location. In short, the solution involves placing the litterbox in a private place that gives your cat the option to pee or flee if she detects the impending interruption of her "duty."

You should clean the litterbox as often as you did before the new guy showed up. (If you don't have time, assign that duty to him!) And provide your cat with a second litterbox in a different desirable location. If you cannot follow these "ideal" steps, do the best you can. Your cat will let you know in no uncertain terms if you've followed them closely enough! Remember that stress is additive (one stressful thing in a cat's life is "no biggie," but don't ask him to cope with the new guy, a new routine, *and* a scary litterbox location!). So taking care of this one important quality-of-life concern may lessen your cat's anxiety—and his potential to act out in other ways, like scratching and biting.

How to Keep Kitty from Mauling Grandma

Dogs aren't the only pets who suffer from aggression. Stressed-out cats can become aggressive as well. And though an aggressive cat isn't exactly life-threatening, that biting and scratching sure can hurt! Here's how to calm Kitty down.

🐾 Biting the Hand That Pets Her

Biting Grandma's hands when she is only trying to be nice is no way for a friendly cat to welcome her owner's mom into her home. So what if Grandma opens her hand wide as she tries to pat you on the head—big deal. There is no need to latch onto her arm and bite the heck out of her finger! How could she know that you considered a pat on the head a threat?

Grandma didn't know the proper cat greeting, and she certainly didn't know that she was supposed to let you run the show. Was anything but your dignity really hurt? It really makes you mad to be petted like a dog, eh?

Poor Grandma. What can we do to keep this from happening again?

Observing the following suggestions decreases the likelihood of attack toward hands for folks of all ages, not just grandmothers. Try them!

1. Reduce the cat's overall arousal. An aroused cat is likely to interpret his arousal as anger or fear and strike out accordingly. He's either biting defensively out of fear, or aggressively because he's just plain ticked off. He's getting too much stimulation.

2. Ask your mom not to move her hands in front of the cat's face. This movement is tantamount to asking your cat to scratch and bite your mom's forelimbs. Instruct your mom (in a nice way, of course) to tuck her hands under her armpits and not to wave them about when the cat comes around. It's okay to remind her to do this as often as necessary.

3. After your cat appears to have lost interest in your mom's motionless mitts, ask her to toss a toy (underhand) on the floor so that your cat can light into something less human. Just for safety's sake, ask Grandma to keep her hands under her arms after she has tossed the toy until the threat of attack is clearly over.

4. Praise compliance (Grandma's and the cat's!).

🐾 Grandma's Ankles: The Perfect Target

Viewing Grandma's slow-moving ankles and calves while she walks down the hallway into the kitchen to get her morning orange juice and cereal can be the source of some excitement to a cat whose daily routine has been upset by the very person to whom those legs are attached. Your cat has read the script. (Haven't they all?) In case you haven't, here's how it goes:

Lie in wait for legs protruding beneath a frilly pink robe,

crouch in preparation for attack,

quick-run,

pounce,

latch on,

hang on until you can't stand the screaming any longer,

release,

run,

and hide!

The attack is great excitement and great fun! Who needs passive aggression when they can have this?!

Consider the cat's motivation—need plus incentive. The need may be sev-eralfold, including the need to attack those novel-looking legs (the predator in her), the payback incentive, and certainly play (aggressive play, no doubt, but play nonetheless).

Grandma's solution to this problem involves standing still, which keeps her legs from being the stimulus for the attack, much as we asked her to keep her hands from moving. If there are no moving legs to attack, we've removed the cat's incentive. We haven't, however, removed the cat's need to attack in play.

To satisfy your cat's need for play, ask Grandma to provide him with an-other stimulus to attack—one that she provides as she walks down the hallway to the kitchen. Recall from chapter 3 how we displaced a cat's playful aggres-sion onto a moving object that the victim tossed in front of her while stopping just short of the location where the cat initiated its quick-run, pounce, and so on? (If you don't remember, see "Dr. Wright's Casebook" on page 60.)

By stopping, Grandma saves her legs from attack. And by tossing a ball of

Moving legs can be an irresistible target for a playful or aggressive cat. And if those legs aren't too stable, their owner can risk broken bones as well as a few scratches. Make sure that everyone in your house knows how to divert a cat attack.

foil or other moving object away from her, in front of the cat, and toward the kitchen, she provides the cat with a satisfactory object on which to redirect its aggression. (Be sure to equip Granny with two or three such balls in case kitty doesn't see the first one.) Now Grandma can safely proceed to the kitchen to enjoy her breakfast.

Your mom should not try to kick at the cat, scold the cat with pointed finger waving, or race quickly to the kitchen, cherishing the illusion that the cat won't notice her moving quickly by or that she can outpace her fleet-footed predator. These ideas will no doubt result in Grandma's providing the cat with even more exciting moving body parts to latch onto. If only he didn't have to put up with those bothersome screams!

A Final Word to the Wise: Owner, Read This!

In spite of the guidelines in this chapter, and everyone's best intentions, there are bound to be some rough edges when you combine a settled-in pet with a new person in the household. So here are a few final dilemmas you're likely to run into with your dog or cat and the new person. I'm not certain whether they should be classified as pet behavior problems or human behavior problems. Perhaps they're both. I've listed them as recommendations for the new person in your family because your dog or cat is not likely to read them.

I'm certain that the people who have just moved into your household and already have a relationship with you are well-intentioned and don't mean to overstep the bounds of their "newness" with your pet. And I'm certain that it's no mystery that you can do certain things with your dog or cat that other people cannot do, even that new person in your house, because you and your pet have a history together. You love and trust each other.

At times, however, your mother or father or that new person in your life must realize that the relationship the two of you have does not translate into the same kind of relationship they have with your pet. Because it's okay for your dad to treat you like his little girl does not give him the right to treat your pet in the same way—at least not from your pet's point of view! Your dad can still get away with lecturing you, but he may not survive lecturing your Rottweiler on why it's better to give than to receive, just because he's related to you!

Added to this misconception that "your pet equals you" is the new person's responsibility for your pet in the household. You need to make it clear that the new person is responsible for your pet's behavior in your absence. For example, if the FedEx person comes to the door and the dog starts to run out, it is the human at home, not the front door, who is responsible for keeping the dog from bolting and chasing the FedEx person, or from jumping on him. The dog is certainly not going to stop herself.

You will need to go over "the rules" just as if your pet were a child, so that everyone involved—including the pet—knows what is okay and what isn't, no matter who is in charge. It's your antique settee; you can let the dog on it or not. But it's probably best if your new lover or your father-in-law goes along with the same plan.

Nonetheless, taking responsibility for the pet in your absence does not mean that your mom, for example, should attempt to hug your dog just because she hugs you. And it doesn't mean that she should attempt to kiss your pet on her forehead or rub her belly just because you do that with your pet. Thus, for the safety of that new person in your family and for the safety of your dog or cat, I offer the following recommendations.

🐾 New Guy, Read This!

I've found that when a new person is added to a family, the new person often has a few personality problems in coping with the beloved household pet. After all, people have stresses and insecurities, too! If you're having mixed feelings about Duke or Princess, here are some ways to resolve them.

Tips for New People Who Feel That the Dog Is Competing with Them
Try these six tips for getting along better with "your" new dog—and liking him a lot better, too!

1. Do not grab the dog by his collar. It's a convenient way to keep him from doing what you don't want him to do, but it's also a good way to lose your hand. Use your voice to communicate your desire, shut the door, or both.

2. Do not use your foot or leg to reposition the dog. Yes, it's a great, low-effort way to keep the dog away from the opened front door or to keep him away from the tray of food you set on the coffee table, but it's also a good way to lose your foot. Use your voice and communicate your desire to the dog rather than resort to the lazy man's way of controlling the pet's access.

3. Do not hit the dog. Don't yell at the dog in anger or raise your hand with the newspaper in order to teach the dog a lesson. The dog's owner would never think of hitting the dog, and you shouldn't either—unless you want to risk losing your nose or some other body part that you place in the dog's face while scolding him. Physical punishment is not necessary to change behavior.

4. Do not tease or make fun of the dog. That demeans him and the trust he places in people. Your goal should be to engender mutual respect in building your relationship.

5. Learn the rules. Be sure that you understand that house rules are the rules the dog's owner sets for the household, and they are not your own idea of what's "okay." If you're not sure what the rules are, sit down with the dog's owner and clarify this point. It may be okay in this household for the dog to get on the couch, even though it may not be where you used to live.

6. Give Bud a chance. Don't move in and tell the owner that dogs smell and you don't want to be jumped on. Learn to give him a bath and help with his socialization training! Find something you and the dog can enjoy doing together. I know I can always use another "friend for life," and I bet you can, too!

Tips for New People Who Are Suspicious of Cats

These five simple tips will work wonders in helping you overcome your fear or distrust of cats and create a warm, friendly relationship. And isn't that a better thing for both of you?

1. Do not pick up a sleeping cat. Not even to cuddle with her. Let sleeping cats lie—unless you like facial scarring and freaked-out cat noises.

2. Be gentle. Do not shoo the cat off the counter or any other location by pushing her with your arm or other body part. Give her a chance to retreat on her own terms. Or you could walk up to her with your finger pointed toward her nose and, after a nose touch and rub, pick her up. Place one hand firmly yet gently behind her front legs under her chest, and use the other hand to support her rear end; then, lift her gently to the floor. Don't ever grab her by the scruff of the neck. How would you like to be picked up that way?

3. Let her come to you. When she needs to be given medication or to be taken for her bath, do not wait until she goes into the litterbox to pick her up. Call her to you, or better yet, let her owner take on that responsibility.

4. Know the rules. Be sure that you understand that house rules are the rules the cat's owner set for the household; they are not your rules. Find out what the owner's rules are. If you don't want a cat under your covers at night, just shut your door. She won't understand being shoved off the bed when under the comforter is her favorite place to sleep in the owner's bedroom.

5. Get to know the cat. Don't walk into the house and announce, "I'm not a cat person." I'd hate to see you miss out on one of life's most wonderful sensations: the happy purr of a newly found friend napping in your lap!

Chapter
Fourteen

Home Alone: When You Both Work

Your dog or cat has now had plenty of opportunities to learn what to do in relation to family members, visitors, and prized household possessions. He has also learned to get through the day with an acceptable routine that you have mapped out. Most of these routines take place with you there, so they're socially acceptable routines that help to improve your relationship with one another. Your dog or cat has also learned how to cope with new relationships and how to cooperate with the new people in your life.

But what happens to that well-established routine when everyone's gone? You leave, your spouse leaves, the kids are off to school, and for the first time, for most of the day, your pet's home alone. I have had cases of dogs who were confined in the home and continually broke out. Then the next home I would visit would have a dog who was kept outside in the yard and who kept breaking *in*! It has become quite clear to me over the years that it's not just the place you leave your pets that really matters. It's the fact that you have left them alone, and many animals just can't handle it.

Who Can Cope Better, Cats or Dogs?

For your dog or cat, these can be the best of times or the worst of times. How he reacts to your prolonged absences depends in some cases on how well he was taught to experience brief departures, and then longer departures, as a puppy or kitten. Kittens and cats typically have significantly fewer separation problems than dogs. That's not to say that cats cannot or do not get into some fairly bad habits when you're gone. But because cats are less socially dependent on you than dogs are, they seem to be able to handle social changes in their daily routine better than dogs do, at least for the most part. Alone time can be, for cats, the best of times: days spent sunning, resting, hunting imaginary critters, exploring taboo countertops, and the like, all without the insult of interruption from young Johnny or Judy or even from Mom or Dad.

But for dogs, being alone is not always a good thing. Being left without guidance, having to decide right from wrong with no one there to help him make those decisions, or to praise or smile at him in his moment of need, can be scary. When he's home alone, no one is there to let him know whether to take charge when someone comes to the door, how many times to bark at the mailman, whether it's okay to jump on the windowsills to keep the squirrels off the deck, or what to do when he's just got to pee and no one is there to take him outside.

An Ounce of Prevention for Dogs

Not all dogs have problems when you go off for the day. But those dogs who have difficulty with your absence have several things in common. Avoiding these red flags won't guarantee a problem-free pup, but they will reduce a dog's chances of developing separation-related behavior problems in your household. These common factors can reduce your dog's ability to cope with prolonged absences:

1. Very early adoption
2. Very late adoption (if socialization is not ongoing)
3. Too little exposure to a caring, active human caretaker as a puppy
4. Too little exposure to different people and different places as a puppy
5. Quality of experience before and during the owner's leaving, and on return home
6. Genetics

🐾 Maybe the Problem Is in His Genes

Let's discuss the last factor first. Genetics can contribute to separation-related problems, but, thankfully, you cannot do anything at all about them! You can't change the genetic makeup of a puppy (at least not yet). I mentioned the importance of both nature and nurture in chapter 2. For problems of separation anxiety as well as behavior problems resulting from separation-related situations, some dogs fare better than others. And for some dogs, medical conditions (skin conditions, in particular—it doesn't take much stress to send an itchy-skinned dog into a bout of scratching, especially when there is no one home to interrupt it) interact with separation to increase the probability of problematic behaviors. Not enough good research has been done to identify which breeds are more susceptible than others to separation-related problems.

🐾 Bringing Her Home Too Soon Can Be Disastrous

Removal from her mom, littermates, and rearing environment prior to 7 weeks of age can interfere with a puppy's ability to cope with separation later on in life. A better time to separate puppies from the breeding environment is between 8 and 9 weeks of age. At about 3 to 4 weeks of age, puppies begin to follow one another, and social interactions proceed naturally in the form of playing, chasing, pouncing, ambushing, grooming, developing an inhibited bite (that is, holding back), and other behaviors that result in secure relationships among puppies in the litter. In addition, puppies essentially become attached to the location in which they are reared, including the setting's layout (what is located where); what it looks, smells, sounds, and feels like; and how they need to get around it to fulfill their needs.

Disruption of these experiences is likely to result in physical problems (like weight loss) and behavioral problems due to the destabilizing effects of the pup's being separated from his home and companions. The quality (how fulfilling the experiences are) and quantity (how often the pup experiences being fulfilled) of these early-attachment experiences matter. They may predict the pup's future social relationships, reactions to novelty, and ability to be comfortable in a home setting. Placing a puppy with a new owner immediately after the puppy has been weaned from her mother is a no-no. Weaning usually takes place between 5 and 6 weeks of age, and puppies indicate their distress at being separated from familiar circumstances and littermates by crying constantly (we behaviorists call this distress vocalization) at this age.

Although a certain amount of stress is desirable as puppies develop, too much stress interferes with learning, including learning how to be with and without someone and learning how to be calm, cool, and collected when left alone in a home environment different from where the pup was reared. And no matter what you've heard, soft, furry toys don't do much to soothe the pup's negative reaction to separation. Too much stress experienced by puppies adopted very early may result in severe upsets and have profound behavioral effects later on.

❧ On the Other Hand, Don't Wait Too Long!

Very late adoption—removal from Mom, littermates, and the rearing environ-ment after 14 to 16 weeks of age—can also interfere with a puppy's ability to cope with separation later on in life, provided the pup's daily routine in the kennel setting remained unchanged. Puppies can become overly attached to the location in which they were reared, as well as to one another, and they have a tendency to develop a prejudice against anything or anyone unfamiliar. They also may begin to work together functionally as a pack of dogs. Where one dog goes, the others go. When one puppy rushes toward the caretaker, they all rush to the caretaker. If a person runs by the outside of the kennel enclosure, they all chase in pursuit. Litter-reared puppies also become fearful of (and less likely to explore) unfamiliar objects, locations, and people. Thus, it is usually very difficult for puppies who are strictly kennel-reared, for example, to form stable relationships with people or to feel calm and secure in a nonkennel setting.

❧ Make Sure Someone Cares

Too little exposure to a caring, active, human caretaker as a puppy is probably the most important factor leading to separation-related behavior problems in the adoptive household. (It is also one of the most important rearing factors in dogs that bite people.) Without exposure to a kind, active, human caretaker on a regular basis from about 3 to 14 weeks of age, the domesticated puppy is not likely to fulfill his destiny as a companion animal in the human household.

Puppies need to become familiar with all aspects of a person, and what she stands for in the puppy's life from day to day, throughout this period. A person's voice and what it predicts has significance for the puppy, including changes in how a person talks to the dog—soft or loud, high or low pitch, static or modulated tone, continuous or discrete words, and other aspects of voice I covered in chapter 2. These changes enable a puppy to recognize "voice" as part of personhood—how the person says the dog's name and looks at him prior to correction and consoling; how she pets the dog and physically contacts him in general (stroking, massaging, patting); how the person's physical appearance looks to the dog (male-female differences, wears glasses, walks softly); how the person smells to the dog; and other aspects of personhood that the dog uses to recognize her. "Ah ha! Here comes Ellen, smiling, walking with a relaxed, carefree gait, and singing my name. Now that's a good thing—I'm about to get hugged!"

And although the puppy may not be able to respond behaviorally to such niceties until he can stand and approach or avoid the person at about 3 weeks of age, it doesn't mean that you should wait until the pup is 3 weeks old to handle him or talk to him or enable him to experience personhood in a fundamental way prior to this age. That's because it's likely that what the puppy learns about his caretaker precedes the pup's ability to respond to the person behaviorally. (This is similar to children, who need to be exposed to language before they're physically able to speak words.)

If Ellen had remained passive in the presence of her puppy all those weeks; if she didn't react to her puppy's invitations for play, physical contact, protection when frightened, or other activities; or if she failed to initiate similar kinds of activities with her puppy, she would not figure significantly in the pup's life as an individual with important bio-psycho-social resources the pup can use to satisfy his needs. And the likelihood that Ellen and her pup would develop a caring, stable social relationship would be slim.

Timing Is Key

If a pup is not successful in forming a primary social relationship with someone prior to being placed with an adoptive owner after 14 weeks of age, there is only a slight chance that he will be able to learn to be responsive to human family members' handling, talking-to, and other appropriate social behaviors as an adolescent. These dogs exhibit reactive temperaments in the household. They wind up at either end of the "temperament distribution": around people they have too much arousal, resulting in either aggressive and hyperactive tendencies or fearful and defensive tendencies.

In short, in the absence of the development of a stable, person-pup psychosocial bond prior to about 14 weeks, the adopted puppy will find it very difficult to feel part of his adoptive family. He won't be able to feel the bond of affection or the trust necessary for the psychosocial experience, he won't be responsive to learning household manners, he won't meet new people well, and he probably will be a source of disruption in the family's life. For these dogs, separation is not the only problem; the daily routine of living among people is the real problem.

So, now you're going to leave him alone? Hmmm. If there are no people, there will be no problems, right? Forget it! Try living with escape attempts and the damage that results each day when you return home from work. He will be digging—breaking off his nails and ruining the woodwork. He will be chewing—breaking his teeth trying to chew his way out the window. (What price, freedom?!) And he will be finding food in the trash, the bread box, and the pantry. Is it separation anxiety? No, the dog is not anxious. Is it a separation-related problem? Yes, now that no one's home to bug him, from his perspective, it's either "I'm outta here," or "The house is mine!"

Even one caring, predictable relationship with a human being during a puppy's early rearing is sufficient in most cases to equip the pup with the perceptions and behaviors required for establishing a secure relationship with a similar kind of adoptive caretaker later on. But "similar" is an "eye of the beholder" phenomenon for dogs. You can never be certain which sensory aspect of the initial caretaker the puppy is most likely to select as "standing for" Ellen. That's why the next category of factors is also quite important for the proper early socialization of puppies, and why early socialization to different people and places is often a hedge against separation-related behavior problems.

🐾 Expose the Puppy to All Life Has to Offer

The less familiar a puppy is with different kinds of people and places, the less secure his psychosocial attachment is to people other than his initial, primary caretaker. He will also be more dependent on a particular individual's presence or a particular location for making him calm. If he's taken to a new location, like the adoptive owner's home, the dog will experience more distress, especially if he's left alone. Severe anxiety and the likelihood of behavior problems can result. He may begin to act out after he can no longer cope with the stress.

Although a small to moderate amount of stress is a normal result of separation from a beloved person and an indication of the psychosocial "value" of that person to the dog, too much stress is not a good thing. It is interesting that puppies about 8 weeks old seek out people rather than other puppies for relief of separation distress, another indication that 8 weeks may be the most opportune time for placing a pup in a new home with his new owners. By this age, most puppies are quite receptive to wearing collars and head halters in preparation for puppy head start. (See chapter 6 for more about puppy preschool.)

Varying the Routine

Now, if Buffy had been reared for the first 4 or 5 months by her caring caretaker, Bob, in the backyard kennel, along with her littermates, she would have

Do Dogs Prefer Other Dogs or People?

A student wrote to me not long ago asking if animals really want to be people's pets, since all the research he could find pointed only to the benefits to humans. Well, we're also beginning to gather evidence on this subject from the dog's point of view, and it should make us people feel pretty good!

I told the student that because dogs and cats have been domesticated for many generations, they start out tame and socially responsive—so they've got a leg up on nondomesticated animals who start out "neutral" toward people. As a result, puppies and kittens are genetically prepared to form trusting relationships with their human caretakers, and most do so—with the right kind of early socialization. Dogs and cats learn to depend on us to meet their biological, social, and psychological needs in a predictable way. To the extent that we fulfill our end of the bargain, they prefer to be with us rather than with other animals or by themselves. Here's some pretty convincing evidence.

Four comparative psychologists from Wright State and Ohio State Universities compared the behavioral and physiological stress responses of pairs of dogs in their home kennel and in a novel environment. The dogs had also formed a relationship with their human caretaker. When one dog from each pair was removed from the home environment for several hours, the remaining dog from each of the four pairs was fine. However, when placed in a novel location, the lone four dogs showed significant stress responses and increased activity. These responses to novelty did not improve, even if the dogs' kennel mate was present.

But the activity and stress levels were not elevated when the lone dogs were in the presence of their human caretaker! Even when their kennel mates were present in the strange environment, the dogs maintained closer proximity to and solicited more attention from their human caretaker than from their kennel mates. These results highlight the importance of human companionship for our pets, and help answer the question of whether animals like to be pets with a resounding "YES!"

no problem getting used to a daily routine in the backyard in the kennel looking forward to spending time with Bob. In all likelihood, Buffy would prefer relationships of the canine variety, but she would be accepting of Bob, who is accustomed to interacting with his puppies in the kennel setting.

If, during the first 14 weeks, Bob instead brought his kids (an 8-year-old boy and 13-year-old girl), Grandma, and his cigar-smoking brother, Joe, into the kennel, Buffy would get used to interacting with people in different categories of physical characteristics. Which is to say, Buffy would become socialized to people who differ in gender, age, reproductive status (remember the importance of pheromones to dogs), height, weight, and a variety of other factors. So now that Buffy has been well-socialized to different kinds of people, she'll be a successful candidate for adoption, right? After all, she may easily form a secure psychosocial bond with her new owner because she was exposed to different kinds of people as part of her early experience. But the answer is, not exactly.

Bob did not think about exposure to different kinds of settings and objects, those not found in the backyard kennel setting. Imagine a puppy going through the period of socialization for whom exposure to novelty is met not by fear, but by excitement, exploration, and figuring out how to classify such objects in terms of their importance to him, the dog. And to the extent that the puppy has been familiarized with different kinds of objects, such as ceramic food bowls, a water pail, grass, a wooden doghouse, and the small concrete-block enclosure leading from the entrance of the whelping box to a grassy run, all surrounded by chain-link fence, he'll be less freaked out by different kinds of objects as he grows older.

But in the kennel he's not likely to encounter the breadth of items that he will run into as an adolescent and adult dog. What happens when you put him in the car for the first time when he's 16 to 20 weeks old, he discovers it moves (he may also discover that it moves what's in his stomach right out of his mouth!), and some strange person comes to the car and reaches in the window to pet him? What happens when you take him to the veterinarian's for the first time? What happens when you bring him into a house complete with carpet, slippery vinyl floors or tile, and cushy soft comforters to jump up on? Well, you're probably looking at a disaster. Puppies who are not exposed to novelty outside of their rearing environment on a regular basis until well after the socialization period have a hard time remaining calm and nonreactive when placed in a new location.

Home Alone

What happens to a pup's perception of the sights, sounds, smells, and other novel sensory input in your home when you leave him alone? What happens to a pet-store or puppy-mill puppy who knows only life in a cage, when he's placed in a home with objects and rooms that provide him with lots of different sensory experiences—and then he's left alone? Chances are, he'll freak. He'll become anxious. He'll act out. He'll damage or destroy stuff, maybe even himself. So, prevention of separation-related behavior problems in adopted pets must also involve exposure to novel settings and objects, early and often, until the puppy is lucky enough to be adopted.

If socialization is done in this way, with successful exposures to different kinds of people and places, the puppy will take with him into his new family a hedge against separation-related behavior problems. He'll know how to form a stable, secure relationship with his new caretaker, he'll be familiar with and relaxed in his new environment, and he'll be able to more easily learn the rules of the household and some appropriate manners. Now there are fewer ways to mess up, although it can be done with inappropriate treatment of a new dog by a well-intentioned but unwise new owner, as we'll see in the next category of factors.

🐾 Think It Doesn't Matter How You Come and Go?

Your pets care about your comings and goings. Assume that your dog has never been left alone for longer than a couple of minutes, and now you have to go to work. You'll be gone for 8 hours. You've had all weekend to introduce your puppy or dog to your family and different parts of the house, and you've gotten to know him much better than when you picked him up from the breeder a few days ago.

Although most good breeders and some shelters will take a puppy aside, play with him in a new location daily, and allow him to get used to being alone prior to adoption, don't count on your new pup coming equipped with "home alone" experience. It's just as likely that he's never been alone in his life! So the "cold turkey" approach to leaving him for 8 hours by himself may undo much of the early experience the breeder or shelter workers accomplished during his early days. Even preventing access to a family member or members when the dog is in the house, but in a separate room, is sufficient to cause separation anxiety in some dogs.

A dog's response to being separated from a person with whom he has a relationship falls along a continuum—he will surely have some reaction, ranging from mild to acute stress. This form of abnormal behavior may be relatively

Dr. Wright's
✉ E-Mail

Dear Dr. Wright:

My cat, Tequila, has been acting very strangely ever since I went away for a week and had a friend take care of her. My friend would come over twice a day to feed and let Tequila out. Since my return, Tequila has been aggressively attacking her tail—hissing, growling, and biting it. She has never done this before, and it is starting to scare me. She does it constantly throughout the day. Why does she do this, and what can I do to stop it?

Travis R.

Dear Travis:

Your first move should be to make an appointment with the veterinarian to rule out any neurological conditions that might account for Tequila's strange behavior. My guess is that the stress associated with your absence contributed to the onset of the problem. An obsessive-compulsive disorder, or OCD, is an anxiety disorder that occasionally reveals itself in tail-chasing, biting, and eventual mutilation. An obsession is a thought, and a compulsion is a behavior. Together, they

benign in some dogs, resulting in panting, eye squinting, and little else. In other dogs, or even in the same dog, given inappropriate treatment for the problem, the anxiety can result in panic and a dog who is out of control. So the sooner you recognize and treat separation anxiety, the simpler the problem will be to nip in the bud.

create a pattern of activity the cat cannot *not* do. If the cause of the problem (the anxiety) is not treated, the behavioral sequence may show up in another kind of repetitive act, like excessive licking. But don't worry, I am not going to recommend removing Tequila's tail—it wouldn't help anyway!

This behavior problem is best taken on by a specialist—an applied animal behaviorist. He may consult with your vet about using a psychoactive drug to help Tequila with her anxiety. For your part, you can try to reduce the "life stressors" that may contribute to her inability to cope, and reduce any activities that may cause her to become irritable. And when Tequila looks like she is about to begin her destructive pattern of whirling, growling, and biting, nip it in the bud!

Buy a "clicker" or a whistle and use it to get her attention (but not scare her), and then involve her in a pleasurable activity like playing or running down a treat you toss a few feet from her. The more you can divert her attention from the compulsive pattern and toward something she enjoys, the less obsessed she should be about beginning the pattern you describe. If you can keep her on the same daily routine throughout the week so that she doesn't experience additional stress from changes in her schedule, you will have the best shot at helping the professionals help your cat return to being calm, cool, and collected!

Pinpoint the Tip-Off

I ask clients whose dogs have developed separation anxiety to note the specific event that the dog seems to pick up on that tips him off when they're about to leave home. You can and should do this at home! Maybe Buster stands up and begins to follow you when you turn off the bedroom lights. Maybe he begins to pant and look worried, pace back and forth, and begin his anxiety trip when you pick up your purse and rummage around for your keys.

Commonly, but not always, separation-anxious dogs build their anxiety from that first "leaving" event until they are finally alone in the room after you shut the door. In theory, if they experience anxiety greater than their threshold for coping, they act out in some way, with destruction, physical distress, "accidents," or even, occasionally, aggression. Dogs are the worst offenders, but even cats can react to the stress of separation with excessive licking or self-mutilation.

Separation anxiety is diagnostically marked by disruptive behaviors that occur within about 30 minutes after a person leaves home and that do not occur due to other reasons. ("My dog's just rotten! He rips up the sofa right in front of me!" is *not* a sign of separation anxiety!) Some research has shown that how much attention you give him before you leave—ignoring him versus saying a long goodbye—is not necessarily related to the start of separation anxiety. But we do know that turning the problem around is based on reducing the pet's arousal and attention to anything social before you leave.

Recipe for an Anxious Puppy

If you want an anxious, neurotic dog, just be mean to him! If a puppy is physically reprimanded or inappropriately and unpredictably punished in his new family ("I'm going to be sure I'm dominant to this dog. I'll just shake him around every now and then so he'll know who's boss!"), the likelihood of a stable social relationship forming is also slim.

Sometimes puppies who grow up with an abusive human parent become very socially attached to the person, but in an unhealthy way. These puppies experience so much anxiety trying to please their owner and avoid punishment that it's as though they are walking on eggs when the owner is around. When the owner leaves, the dog is left hanging without the secure knowledge that things will be okay when he arrives home. If the owner punishes the dog for chewing something inappropriate while he is gone, and it hurts so much that the dog runs to him for consolation, and he punishes the dog again, the dog will be even more anxious than before. What's the point?

What's Making Your Dog Anxious?

If your dog's all wiggly (a combination of submission, fear, and happiness to see you), it may be a sign of separation anxiety—but only if arrival anxiety has been ruled out. In other words, if Rufus shows extreme arrival elation, and punishment never follows his owner's arrival home no matter what he has done, then the overactive greeting is a sign of separation anxiety.

But if the dog's wiggly behavior is done to elicit owner reassurance, the greeting is related more to arrival anxiety than it is to separation anxiety. The dog has learned that when he approaches the door when his owner comes home, he gets randomly punished. Sometimes she punishes; sometimes she doesn't. And the dog's anxiety about this unpredictability and his relationship with her leads to acting out—doing things like chewing a couch cushion or digging a hole in the carpet—which leads to further punishment when she arrives home.

In a sense, the dog not only experiences separation anxiety on his owner's leaving, but he also experiences arrival anxiety—anxiety related to her arrival home and the inability to know whether punishment will be forthcoming or not. Imagine yourself as the dog, bowels full, hearing the garage door open, unable to hold it anymore: You just let loose. Now your owner comes in the front door and confronts you with a scowl on her face, and you know you're facing impending doom!

She assumes that you did this to get back at her for leaving you alone, and she wants to "teach you a lesson," so she shoves your nose in it. And so it goes. You're both caught in a vicious circle. The more she punishes, the more likely you are to "lose it," which makes the persistent owner continue to punish you. This problem will never be resolved as long as the owner continues the random punitive behavior.

So although Rufus was well-prepared to learn how to be alone in his new home with his new guardian, there is no such thing as a done deal. Training continues after the first 14 weeks, and the quality of the person-pet relationship must be maintained and continuously evaluated as the dog matures and ages.

How to Prevent Accidents That Are Waiting to Happen

Now let's look more closely at those "accidents" that all too frequently await the two-paycheck couple after a hard day at work. Sometimes dogs and cats are so excited to see their people come home that they do something stupid. They may get so excited that they pee all over the floor, bounce off the walls and knock over your favorite family picture, or get underfoot and try to maim you as you try to negotiate the stair steps. What these accidents have in common is what we behaviorists call arrival elation: the dog and cat are thrilled to see you. You make their day. They've been waiting for this moment for hours on end. And sure enough, they blow it, you become upset, they become distraught (or totally oblivious to the fact that they've done anything wrong, like my dog Charlie), and it ruins everyone's day. This series of events is a shame.

Can anything be done? Of course it can, and it's related to the way we got our pets to learn "greeting manners" in chapter 11. Try these six tips.

1. Play the shell game. Your dog or cat stations herself by the kitchen door, gearing up for your arrival. She's done the same thing every day for the past 7 months, and every day you've come in the door just in time to catch her act. This time, however, you're going to enter the house through a different door. You'll only have to do this enough times to confuse the heck out of her so that you catch her a bit off guard when she hears the car door slam. Now she's not so sure where you're going to come in, so she has to think rather than react with her automatic, going-bonkers routine.

2. Cool your heels. Delay entering through a door for a few minutes. When you do enter the house, wait until your pet is quiet so that you are not unintentionally reinforcing his barking or jumping on the door. Only when he is calm should you enter the house.

3. Get out of the way. Ignore her by walking directly to the kitchen table, or the bedroom, or other locations that will get you out of the entryway. If you stand still in your foyer, your cat will be happy to tangle through your legs so that your only choices are to fall flat on your nose or wind up kicking her. And by the way, if your cat has established a pattern of zooming off the walls and over the end table, knocking off your picture each time you walk in the door, why not get rid of the picture or place it face-down on the table, at least for now. If you enter with grocery bags, you may find that your dog's not only excited to see you but also thrilled that you brought her such tasty treats, which she will be more than happy to help you clean up if—in her excitement—she can bump them from your arms. Keep moving; however, you may wish to shuffle your feet as you progress into the house so that you don't step on someone's feet.

4. Distract a pee-er. If you have an excitement urinater, usually a dog (but sometimes cats will back up against the wall and spray baseboards), he'll probably be wagging his tail and maybe even his whole rear, and he probably won't be a jumper. Most of these dogs remain fairly close to the ground, although they may move their front legs up and down in excitement as they proceed to pee. If you follow step 3, you may well have a trail of pee from where you entered to where you wind up in the house. Your strategy with a pee-er will be different. As soon as you enter a door (preferably a different door from time to time), squat down and toss a treat to his left or right so that he becomes interested in food rather than your greeting. Don't look him in the eyes right away because eye contact is arousing to dogs. More excitement is not a good thing. As he's eating his tasty treat, you may get up and proceed to the kitchen, bedroom, or some other room because it is quite difficult for a dog to eat and pee at the same time!

5. Go right back out. On arrival home, minimize the Big Greeting you two get into by taking your dog outside immediately to relieve herself. Following that, you may wish to get rid of her energy by taking her for a walk, throwing a ball and allowing her to run around in the backyard, or inviting her to be with you as you return to the house to complete your arrival ritual with hugs and kisses and sharing the events of your day. She's been alone all day, and now she needs some social contact and reassurance.

6. Do not leave again! Ever! If you absolutely must leave, consider taking him with you or using one of the following suggestions for keeping him busy in your absence. Many separation-anxious dogs learn to cope with their anxiety so that it remains below threshold—below the point at which they can no longer contain themselves and act out. But if they've managed to be good all day, and you leave them again in the evening, for some pets, that's the last straw—it's too much. And they'll let you know it! If you must leave again, even for a short trip to the store, and you can't take him along, be sure to give him some good exercise, and review steps 1 through 7.

In general, arrival-related accidents can be reduced by decreasing your dog's or cat's excitement, regardless of whether you have a mauler, a pee-er, or a zoomer. By confusing your pet regarding which entrance you will use, by not waiting for dog jumpers or cat figure-eighters to ambush you at the door, by changing the way you present yourself so that you're less exciting, and by eliciting an eating motivation that is incompatible with excreting, you should be able to see a significant reduction in your pet's arousal from week to week.

DR. WRIGHT'S INSIGHTS

Human nature being what it is, as much as pet owners hate the results of arrival ela-tion—the scratched arms, wet carpets, barking, knocking over of knickknacks—they sometimes find it difficult to deliberately take the steps that will greatly lessen the dog's or cat's frenzy when they arrive home. I mean, isn't unconditional love the one reason you have a pet? If your spouse just grunts when you walk in the door, you might really appreciate your pet's enthusiasm. So don't undermine your efforts to stave off separa-tion-related problems by encouraging a crazed cat or dog when you walk in the door. Your pet will still love you. Be strong!

🐾 Signs of Separation Anxiety in Dogs

Because so many varied symptoms are associated with separation, it is often difficult for an owner to diagnose them correctly. The dog has not forgotten his housetraining, he pees because he's anxious. He is not just spoiled, he barks all day in the backyard because he's anxious. He doesn't knock over the 10-gallon potted plant and push it down the stairs because of exuberant play, he pulls it over and attacks it because he's distraught and anxious. And the list goes on.

Here's a list of the most frequent signs of separation anxiety in dogs, as re-ported to applied and veterinary animal behaviorists. You can probably add a few of your own if your dog has separation issues!

- Digging holes in sofa and chair cushions
- Scratching floors and doors
- Chewing holes in windowsills, furniture, and bedding
- Pulling down drapes, curtains, and Venetian blinds
- Knocking over potted plants
- Howling, barking, whining, or any other kind of prolonged distress vo-calization
- Excessive licking of his coat, scratching, or biting parts of his body
- Panting, salivating, or foaming at the mouth
- Extreme lethargy or depression
- Refusal to eat
- Inappropriate urination, defecation, and diarrhea
- Aggressively blocking the exit from the house or biting when you try to

leave. (Only if this is the *only* time he is aggressive can you blame separation anxiety.)

🐾 Don'ts for Separation-Anxious Dogs

Here are seven things to avoid if your dog is showing symptoms of separation anxiety:

1. DON'T crate him, especially if he's not used to being crated, or he's had a bad day in his crate. If he already has separation anxiety, crating him may decrease the damage to your household, but it's likely to damage him, to say nothing of the crate. Imagine the distress your dog experiences, trying to cope with separation anxiety while being trapped in a crate. Some dogs lose it, have a full-blown panic attack, and wind up breaking their teeth and nails attempting to dig or chew their way out. Some dogs manage to get their heads wedged between the wires, but they remain trapped like that until their owner returns home. I'm sure that this scenario is not what you had in mind! Curiously, I find it necessary to state that the idea is not to build a stronger crate.

2. DON'T make her do obedience exercises or make similar requirements of her that involve restraint prior to leaving instead of giving her a lot of exercise.

3. DON'T shut him in the bathroom, a closet, or another small room, especially if he begins distress vocalizations, scratching at the door, or other indications of discomfort that don't stop or decrease after a minute or two. Although some of his distress is due to your absence, he may also be anxious about feeling trapped or not being able to see where you're coming or leaving from, especially if he considers it his job to be the sentry in the house. Some dogs are likely to settle right in as long as they can see down the driveway to the street or the sidewalk that leads to the door they exit and enter from. Older dogs who may have abandonment issues, such as those adopted from shelters, may not have pleasant memories of being "trapped." ("Am I being abandoned all over again?")

4. DON'T punish her. Punishment is likely to increase the dog's distress and anxiety and to make acting out in your absence more rather than less likely.

5. DON'T automatically decide to get him a puppy to spend the day with. Here are some of the cons: (a) He'll become jealous of the puppy, and his anxiety will increase; (b) if his anxiety is caused by being apart from you and your leaving the house, the puppy will be irrelevant to him, and he'll still experience distress; (c) the puppy may learn to do unruly things in the household, just like his big brother. Here are the pros: (a) The dog may be amused by the

puppy long enough to break the pattern of destruction associated with your leaving; (b) the dog may actually form a secure and stable bond with the puppy, and your departure and absence will become less important to him, but it's a crap shoot. If it works out, that's terrific. If it backfires, now you have *two* problems on your hands, and twice as much work to do with the applied animal behaviorist you decide you needed after all.

6. DON'T ignore signs of separation-anxiety disorders in your older dog who has never shown the problem before. When dogs begin losing their sensory abilities, they sometimes become more dependent on their owners and more anxious about being left alone.

7. DON'T ask your veterinarian for a drug that will solve this problem as if by waving a magic wand. Psychoactive drugs are tools therapists use in a treatment program, not a solution in and of themselves. I will talk about drugs in more detail in "The Truth about Dogs and Drugs" on page 340.

🐾 How to Keep Your Pet Calm When You Leave

Now let's look at some do's! Try these six ways to lessen your pet's anxiety:

1. DO give her plenty of exercise before leaving. A tired dog has less energy to fuel her anxiety.

2. DO change the order of the things that you do prior to leaving. Continue mixing up your predeparture events, including when you take him outside and when you feed him, so that he is less able to predict your imminent departure. A confused dog is less likely to build anxiety as he strings together the things you do in the house before you leave, each one leading to the next, each one building more and more anxiety.

3. DO eliminate from your preleaving routine as many events as possible that your dog relies on to predict your departure. If turning off the TV or radio is what she waits for before racing to block your exit at the front door, use a timer that turns off the appliance some time after you're gone. If putting on your shoes raises her blood pressure, leave them in the garage or in your car, so that she doesn't see you put them on. If she becomes alert when you jingle the keys in your purse, take off the key you're going to use and put it in your pocket where it won't jingle. The fewer events you string together that predict leaving, the less anxiety that can build.

4. DO try to involve him in a nonsocial activity prior to leaving. If he eats well alone, feed him or toss a dog biscuit, rawhide, or chew toy over his head and away from you so that he has to move away from you to get it. Inducing a

nonsocial motivation in a location away from you may decrease his attention and concentration on you.

5. If you have more than one door available for leaving, DO use them randomly from day to day. Even if the car is easily accessible through the hallway door leading to the garage, you can decrease your dog's investment in lying in wait there by breaking the chain of prediction and using a different door.

6. DO sneak out. Don't draw attention to yourself or otherwise increase his arousal and excitement prior to your leaving. You need to give up the big good-byes and the meaningful talks (I know that you feel guilty about leaving him). Instead, leave unobtrusively. Sneak out. If you do this while he's involved in some other activity, like self-play, eating, resting, or watching TV in the bedroom, so much the better. A calm dog involved in an "important task" is a dog who is less interested in your leaving and is, therefore, less anxious. Don't feel sneaky and guilty about this, either. It's for his own good!

Cat Scratch Fever

Destructiveness in cats can take many forms, but with fewer owners willing to declaw their pets, the biggest problem animal behaviorists run into today is damage due to scratching. After litterbox problems, destructiveness is the second most common cat problem in the country.

Borrowing from our model of always offering the animal a substitute behavior or opportunity to get rid of an important need, the idea for cats is *not* to stop them from scratching, which is a basic need and source of security for them. Instead, you need to stop your cat from scratching on the inappropriate surface that she likes—such as the oriental carpet or your maple bedframe—and substitute a location that is preferable to you and also okay with your cat.

Scratching is a highly motivated behavior, one that the cat is quite prepared to display for more than one reason. Certainly, not all scratching is part of a pattern of destructiveness motivated by separation anxiety or separation-related problems. But when you come home and find the new sofa ruined, you probably won't care for the moment whether Tiger was anxious, bored, or just feeling the urge. You're not going to stop the cat from scratching—he's going to do it anyway—so it's important to enable him to scratch on the right kinds of surfaces as part of his daily routine, even if you successfully remove the stress and anxiety producers in Tiger's life.

Your cat is likely to want to scratch as part of a muscle-toning stretch when he wakes from one of his many catnaps—rather like a dog's play-bow. He may

Cats naturally stretch out and scratch when they wake up from their naps (which is often!). Give them a scratching post that's tall enough for them to stretch out on and covered in rope or another material they can really sink their claws into.

scratch to exercise the digits of his claws, to help remove old claw sheaths from his front paws, or to mark a vertical or horizontal location with scratch marks (a visual sign) and his scent (an olfactory marker). You know where he scratches that you wish he wouldn't. Here are some low-stress tips to help change his mind.

🐾 Make the Object Less Attractive to Him

Here are three easy things you can do to make a formerly desirable scratching site positively yucky to your cat. They may not be too attractive, but once your cat gets the idea, you can return the room to normal. I'm going to use the sofa—a frequent target—as my example, but these tips will work no matter what he's scratching.

1. Cover the sofa. Use double-sided sticky tape or Sticky Paws (a transparent tape designed specifically for this problem). Or try bubble wrap, which will feel weird to your cat and pop disconcertingly when your cat tries to sharpen his claws on it.

2. Put an obstacle in his way. Block access to the location with a side table, plant, or magazine holder so that he can't get to it.

3. Move the sofa to a different room. Or, switch its location to a different part of the room, in case he doesn't really care about the sofa but has developed a location preference, and the sofa just happens to be right next to his bed, within stretching distance.

🐾 Focus on the Location

If the location of the sofa, and not the sofa itself, is attractive to your cat, here are six ways to divert his scratching onto something more suitable.

1. Bait and switch. Place a scratching post next to or near the sofa arm. The post can be a kitty condo completely wrapped in sisal rope (the best option) or carpet, one with a vertical log or two as part of it, a commercially available vertical scratching post, or even a log with bark on the outside. Make sure that your cat can completely stretch out her body on the post, with her front legs extended.

2. Try masking. If your cat rips up carpeting by stretching out his front legs and pulling up the wool pile on your favorite rug, mask the area with a material he can't get his claws into (like plastic carpet runner) and place a cardboard-type scratching strip (available at pet stores) next to or on the damaged area. (Remember that you can take up the plastic when he has switched to the cardboard strip.)

3. Make sure the new stuff is cool. Cats prefer materials they can get their claws into, including vertically ribbed carpet strips, horizontally ribbed rope (the direction of ribbing is less important than their ability to stretch their digits and "get into" the substrate), and wood logs complete with bark (after all, that's the natural substrate they would be using if they were still living outdoors).

4. Use the same material as the sofa arm for a scratching decoy. This works well, but only if your cat has not established a location preference. If she has, as soon as you remove the double-sided sticky tape, you risk her returning to the sofa arm. You can still have some success using the sofa material if she has established a location preference by changing the location of the sofa. (She's thinking, "I really like this material for scratching, but only if it's right next to my bed.")

5. Move the new scratching post. After your cat has developed a preference for your clever option, try moving it perhaps a foot or two a week to a location that you prefer. Be careful, however, because he will quickly show you his location preference for scratching. For some cats, 3 or 4 feet is as far as they'll go from the original target, if location is an issue. For other cats, wherever you place that darned post, they will seek it out and use it. Let's hope yours is one of them.

See if it works. When your cat has established her scratching preference in the new location, you can try gradually to take off the masking material from the old location. Remove a third of it every 3 or 4 days so that after less than 2 weeks you'll have your sofa back for sitting on, and she'll be happily into her pattern of scratching something else in another location.

6. Use a remote device. Should this scenario not play out for you, there is always the remote citronella device by Animal Behavior Systems, which activates whenever he gets within a couple of feet of the sofa. But if you use this method, you risk his choosing never to curl up on your lap when you're sitting on the sofa (a fate worse than death!). There's always a risk with punishment.

🐾 But What If He's After the Sofa?

If your cat is hooked on the sofa itself, and not its location, there's still hope. The solution is to give him his own furniture. Donate the scratched sofa arm

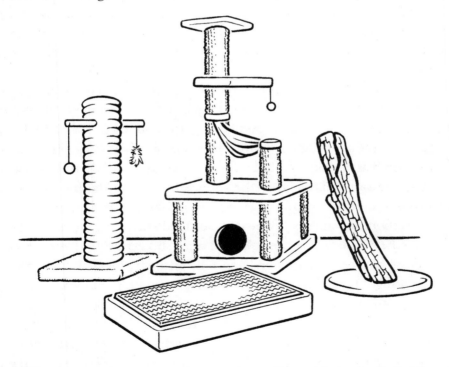

You can choose from a variety of materials for your cat's scratching post. Just make sure that the material has lots of texture so that your cat can connect with it. It has to feel good to her! Some kitty condos have legs that are covered in different materials so that your cat can choose her favorite.

DR. WRIGHT'S INSIGHTS

It's important for you to try to displace your cat's scratching without punishing her. Punishment may be regarded as the initiation of social play or may totally freak her out. She will find another location to scratch, and you will be faced with the process all over again on a newly discovered piece of furniture. You also must resist the temptation to show your cat where you want her to scratch by taking her front paws and running them down the post. It may work for showing your 2-year-old how to pet the kitty, but your cat is likely to regard this "hands-on" stuff as a form of punishment and avoid that new place like the plague, or she may decide that your arm is the new "preferred location"!

to him for scratching and place it someplace where it won't be as unsightly, provided that he hasn't also established a location preference (and you're ready to part with it!). You would hate to buy a new sofa, only to discover that he has a location preference and not a material substrate preference.

We've donated a red chair to our cat, where Domino is free to stretch out and scratch. "Her" chair is on the screened-in porch, and cushions on the chair mask the scratched area. You might call this giving up, but we prefer to call it our private little compromise! We could have tried a log next to the chair to attempt to displace her scratching onto a different location but decided that it would be less unsightly having one rattan chair scratched from behind than a log with bark that we'd have to clean up on a regular basis.

Handling Fears and Phobias

Sometimes a dog will develop separation-related problems if he was traumatized by a scary event that occurred in your absence. Unfortunately, one bad scare from a thunderstorm can turn into a full-fledged phobia the next time there's lightning and you're not around to comfort him. I've had one pair of clients whose dog was so fearful that the owners actually chose to take turns staying home from work when thunderstorms were in the forecast! Otherwise, they would find a trembling dog in a torn-up condo. Now, there's a poor pooch and poor owners, too.

So if your dog is suddenly, inexplicably fearful, think back to the day it began and try to recall if there were storms, scary firecrackers, or any kind of unusual happenings that day. Then take it from there. By the way, the biggest day of the year for me and other animal behaviorists is July 5, without fail. You can't always shield your dog from thunderstorms when you're away, but you can bring him inside the house and put him in a comfortable room with music playing to mask any popping noises outdoors when you go out to watch the fireworks on the Fourth of July. This is just a word to the wise!

The Truth about Dogs and Drugs

We're always looking for an easy fix, but dosing your dog with drugs to resolve behavior problems is not one of them. Prescription psychotropic drugs for dogs have created a firestorm of interest in recent years. (A psychotropic drug affects a dog's or cat's behavior. It must be prescribed by a licensed veterinarian, usually only after a valid behavioral history has been taken, a tentative behavior problem diagnosis made, and a treatment plan developed for the dog or cat by a competent behaviorist.) In the late 1990s, Novartis Animal Health received FDA approval for marketing Clomicalm, a tasty form of the active ingredient clomipramine, for treatment of separation anxiety in dogs.

The manufacturer recommends, as do I, that it be used in conjunction with a behavior-modification program. Although the drug is designed to reduce the dog's anxiety, it is not designed to treat the behavior problem. Rather, Clomicalm provides the owner with the opportunity to teach the dog positive behaviors. After the dog has learned them, the proper course of action is normally to wean the dog gradually from the drug over the course of a number of weeks. The length of the weaning period varies, depending on how well the dog adapts his learning (or generalizes his learning) to the new feeling of his non-drug-induced emotional state.

Some dogs need to remain on the drug to maintain good behavior, but at this writing no studies have looked at the effectiveness and clinical safety of Clomicalm for long-term use in dogs (longer than 3 months). The effectiveness of the drug has been assessed in 8-week studies. They found that compared to separation-anxious dogs who received only a behavior-modification program, separation-anxious dogs placed on Clomicalm *and* a behavior-modification program resolved their behavior problem more quickly, with most gains observed in the first week of treatment.

However, there is no evidence that more dogs resolved their misbehaviors with Clomicalm plus behavior modification than with just behavior modification. The jury is definitely still out on the "magic pill" approach. Consult your vet for further information about Clomicalm, the appropriate doses, and the potential side effects.

Thunderstorm Phobias FAQ

Here are three of the thunderstorm-related questions I'm most often asked by my clients. Maybe you've been wondering, too.

Q: Does moving to a new house sometimes reduce a dog's fear of storms?

A: Yes, sometimes it does because the context in which he experienced the "upset" is different (he experiences the storm in a different setting or location that is not conditioned to the fear associated with the storm). But if he was severely traumatized by a bolt of lightning and a clap of thunder, he'll probably start shaking as soon as he hears the first rumble or sees the first flash of lightning miles away. If he's really put off by lightning flashes, consider desensitizing him with a strobe light. Bounce the light off the walls in a corner so that it approximates the real thing.

Q: Should I put him outside in the backyard where he has his doghouse if he was traumatized indoors and a storm is imminent?

A: No, it's likely to make his fear worse. Although the context would be different, most of the sights and sounds of a storm are amplified outdoors. One of my clients tried this with her dog, who was chained in front of his doghouse. But she had forgotten that the roof of the doghouse was sheet metal, and that the sound of the rain on the roof made the storm sound way worse than it was. The owner finally found her dog—leash, stake, and all—burrowed-in beneath the house, the only location that muffled the storm sounds and provided a tight physical surrounding for him to wait out the storm.

Q: Should I shut him into his crate so that he can have his "den" to protect him during a storm?

A: No, if you close the door, it's not like any den in nature. You're more likely to add to his anxiety. If you trap him inside, he's likely to damage himself trying to escape. However, allowing him access to his crate is a good idea. Provide him with a secure, comfortable blanket in his crate, and leave the door open.

Only one other drug, Anipryl (active ingredient, selegiline hydrochloride), has been FDA approved for use in dogs to treat clinical signs of pituitary-dependent hyperadrenocorticism (to manage symptoms of Cushing's disease) and problem behaviors associated with canine cognitive dysfunction (see chapter 9 for more on this). The use of *all other drugs* in dog and cat behavioral medicine is "off-label" use. Although they have been prescribed and used for behavior problems for years, their use involves more uncertainty and risk than FDA-approved drugs. This includes the active ingredients in drugs like Prozac, Valium, and Buspar.

How Much Freedom at Home?

Many of my clients want to allow their dogs and cats as much freedom as possible in their homes. After all, they are part of the family, and you don't make your spouse or kids remain in their rooms all day! Still other clients believe that their dogs and cats should have access only to the kitchen—and maybe the den if their feet are clean—but never the living room or other parts of the house that could show paw prints or hair shed on the sofa. The amount of household restriction is rightfully the homeowner's call. As long as your dog or cat has a high quality of life or, at the very least, isn't experiencing any discomfort due to a restrictive arrangement, it's up to you.

Remember that you are your pet's advocate. You must represent his needs in your decision, because he's not Marmaduke, so he can't verbalize his opinion. He is probably experiencing discomfort from the living arrangement if you see any of the following signs: excessive panting (he's wet from panting when you get home); hair loss, wounds, or bald spots from excessive licking; whining, crying, howling, or barking (turn on your cassette recorder or videocam to check for this while you're away); or damage to the room.

These symptoms may also be signs of separation anxiety. To determine what is bothering your pet, phone your veterinarian for an opinion or referral to an animal behaviorist. Sometimes owners restrict their pets' access to the house *because* they are acting out due to separation anxiety, whereas other homeowners act to prevent problems—they just don't want to take a chance that their pets *might* damage something.

At any rate, if you're both comfortable with the living arrangements, that's terrific. And if one or the other of you is having a problem with too much or too little access, try to come to a compromise. Ultimately, if you're like my clients, your decision will be based on how well-behaved your dog or cat is in your absence.

🐾 Room to Roam at Home

When your pet is at home and you're not, you have some decisions to make about how much freedom versus how much confinement you'll both be comfortable with. There are certainly different kinds of options for leaving your non-separation-anxious dog and cat at home during your absence if you live in a rural, suburban, or urban setting.

Dogs living in rural settings may have access to the outdoors where they can occupy themselves with many different kinds of jobs, which make your leaving seem rather trivial. They can chase away critters from the porch, keep an eye on livestock, dig up roots (which creates huge holes somewhere in the lower 40), ambush crows and other birds that flock to steal recently planted seeds, and so on.

The same is true of cats. The time-honored tradition of ridding the home area of pests is still a job cats and their owners value in many rural households, despite our efforts to keep cats indoors 24-7 for the protection of birds and to save the cats themselves from disease and injury. Resting on the front porch swing or sitting on the fencepost scoping out other cats and critters becomes an important part of some cats' daily routine. As a result, their owners' absence from the household seems relatively unimportant. The drawback, of course, is that the life expectancy of cats who live outdoors in this way and spend little time indoors is significantly less than that of strictly indoor cats. In fact, in some shelters today, the trend is to not adopt out cats to families who do not plan to keep their new pet indoors.

Suburban dogs and cats are likely to spend most of their day indoors, but they may have access to a doggie or kitty door leading to a fenced-in area outside, or at least to the deck or patio. They too can occupy their time and thoughts with nature's outdoor activities, either by directly interacting with them or by watching them from the back of a couch, the top of the kitty condo next to the sliding glass doors to the deck, or a picture window.

Rufus and Muffy have fewer options if they're living in an urban setting. The yards are smaller, more people are crowded into a tiny area, and crime is a real problem. (So much for the doggy door, where Rufus can go outdoors and risk getting stolen, and a burglar can crawl indoors and make himself at home.) If you live in a high-rise apartment or condominium, even the view is not as exciting. Sure, there are pigeons lighting on the windowsill to excite Muffy. But the wonderful people-friendly views overlooking the city, lake or ocean, or park are less exciting and dog- or cat-friendly than if you lived at ground zero and had at least visual access to squirrels, cats, dogs, and people passing by.

❧ Making Home Life More Exciting

To help Muffy and Rufus avoid boredom while they're home alone, consider arranging for the private viewing of an entertaining videotape for your dog

Dr. Wright's
✉ E-Mail

Dear Dr. Wright:

I would like to know why dogs are attracted to TV. My collie watched TV, and whenever an animal barked, he would growl.

Thank you,

Jimmy

Dear Jimmy:

Dogs respond to things that satisfy their curiosity, and many of the whines, whimpers, and roars that come from animals on the screen are enough to prick up the ears of most dogs and cause them to focus their attention on the video screen. When your dog's eyes are on the TV, you can watch his nose twitch in response to what he sees. (Remember that dogs are much better at investigating with their noses than we are!)

The two-dimensional critters on the screen look enough like something he's seen before to warrant a bark or a growl. Do you think that he growls because he perceives a dog in his room, or a prey animal that needs chasing? It is not unusual for dogs to respond to familiar sounds on TV as though they were real. My clients frequently reveal episodes of their dogs running past the TV to the front door when a doorbell chimes on the screen.

or cat to occupy his time during the day. How about making available dog or cat toys that when batted or "nosed" about the living room floor in the right way release a food treat? Or how about a gadget that shoots a ball across the floor at random times of the day and that contains treats that can spill out? The moving ball not only activates your dog's or cat's predatory response but also provides her with the opportunity for participating in self-play with the ball, just like the Buster Cube, and is less harsh on your baseboards.

✿ What about the Great Outdoors?

Some owners feel more comfortable leaving their dogs outside the house in their absence. For them, a fenced-in backyard is an option if they're lucky enough to have the financial resources and enough property to have one.

Isn't an electric fence cheaper? It may be, although you do have the cost of burying it. An electric fence is buried underground. To keep the dog in the yard, a small box is attached to his collar to remind him when he gets too close to the edge of his turf. However, the fence will not keep intruders out of the yard. And if a squirrel, bunny, cat, or other dog enters his territory, your dog may pursue it, electric fence or no electric fence. A hotly debated issue is whether the shock is, in fact, painful to the dog. One company's consultant described the dog's yelp when "stimulated" as similar to a "surprise" response, rather than pain.

There have been several court cases involving dog bites allegedly caused by dogs who were shocked by their collar and then decided to attack either a bystander in the yard, someone walking by in the street, or another dog. But many dog owners swear by the underground fence. Although several of my clients have reported that their dogs shoot through the thing if there's really something neat on the other side (one client said her dog tightens up every muscle in his body and just runs through the shock!), most users seem pleased with the underground fence. They're happy and more confident that their dogs will remain at home.

One company advertises its underground fence with the slogan "Never Walk Your Dog Again." Unfortunately, that's what some people will buy. So let me state unequivocally that the underground fence should not be used to substitute for walks and other interactions with your dog!

If children are actively playing in the fenced-in backyard and the parents don't feel comfortable allowing the dog to have free access to the children, they may opt either to tether the dog or to provide the dog with a smaller chain-link enclosure built somewhere in the backyard. For either option, the dog should have access to a nice doghouse and water. So far, so good? Well, it has potential.

Almost any kind of arrangement for housing your dog outdoors for short periods of time is fine, as long as it's humane. ("Humane" in this case means that the dog has access to shelter, shade, food and water, toys, and space to move around and lie down.) Of course, tying a dog to a stake with a 5-foot chain attached to his choke collar is not a good idea. Neither is tying a dog to any kind of chain or restraint for longer periods than 8 hours a day. I will explain the consequences of this type of owner behavior in chapter 15.

New Millennium Alternatives to *Home Alone*

The best thing for your household companion—just as it is for a child—would be to have a stay-at-home mom and/or dad. You might be able to arrange to work from a home office part-time or to start your own business, but these options are just not possible for most people. This doesn't mean, though, that your dog or cat has to sit home alone all day. Dog and cat owners have a number of other choices in the new millennium, as people focus more and more on the quality of life for their pets. If your dog or cat doesn't have separation problems (knock on wood!), he or she will probably spend the better part of the day sleeping while you're away. But some pets will spend hours looking out the window, waiting for your car or your footsteps on the sidewalk outside. Whatever their style, it's nice to provide your pets with some variety if you can.

Feel free to mix and match these options for what to do with Bowser while you're at work. (But first, you should nip in the bud any early signs of separation problems rather than wait until they become full-blown!) I had a client in Atlanta who drops his dog at doggie daycare 2 days a week, has a pet-sitter 2 days a week, and leaves the dog with his sister at a friend's (who works at home) the remaining day. Each morning, the dog's and owner's predeparture ritual results in one of a number of attractive alternatives. The dog loves the variety, and the early signs of separation are no longer a problem. Although not something most pet owners would do, it works for them. See which of these seven suggestions are worth trying for you and your dog.

1. Take him to doggie daycare, where he can play with people and other dogs.
2. Scope out a certified pet-sitter to play with her once or twice a day.
3. Form a neighborhood pet co-op so that he can romp at a friend's house.

4. Take her to work with you. Many employers now permit this.

5. Make arrangements with your breeder to take him for occasional "alumni" days.

6. Consider driving her around the block in the car before you leave so that she feels as though she's been on a trip outside the territory with you and can now relax.

7. Send him to canine day camp instead of a kennel while you're on vacation.

Here are a couple of caveats about sleep-over camps and doggie daycare. These new options for pets need to be examined much as you would if you were leaving a child at a daycare center. Look the place over and make sure that it meets your standards for cleanliness and attention to the animals. These places range from rather spartan or simple "babysitting" arrangements in a modest home to resort-type places that engage and pamper your pet to the hilt. If you find that your dog or cat is going to be spending the greater part of the day in a crate, he would probably be just as well off at home. Or if the people are not interacting with the animals except in a "15-minutes-a-day play session," I would think twice.

You may find a waiting list to get in, and the facility can afford to be choosy, too. Your dog or cat will have to pass a screening herself based on inoculations and behavior; these places won't (and shouldn't) accept campers who are likely to cause trouble or spend their time howling for Mommy. They aren't for everyone, and they can be pricey; however, if the facility and the people appeal to you, go for it! Don't worry: When it's time to go home, your dog and cat will still like you best. I promise!

Chapter Fifteen

Keeping Kids Safe around Pets—
And Pets Safe around Kids

Just as with teaching the pets themselves, it's best to teach youngsters how to meet or greet dogs and cats when the children are young, before they become old enough to have bad habits that they must undo. I know about these bad habits from personal experience. I have been invited to do talks for kids in different grades in elementary schools regarding how to "read" a dog, how to treat and greet a dog, and the do's and don'ts for responding to a strange dog (or the neighbor's dog or even their own dog, if she makes them feel uneasy or afraid, or if she looks grouchy and wants to be left alone). I even go over what to do around a threatening dog so that they can reduce the risk of dog bite.

It's been my experience working with children and a "live" dog or two that first- and especially second-grade children are the most open and interested, and the easiest to work with. If we can instill in these young children humane, respectful, and safe ways to interact with their pets, it is time well-spent.

By contrast, we need to work at changing the opinions and beliefs of some parents, especially when their parents—and their parents before them—may have passed on ways of treating dogs and cats that have little place in today's household. Today, the dog and cat are treated as beloved companions, not owned objects to use or abuse as one sees fit. But there's hope even for these unenlightened parents. Children go home and instruct their parents about what they have learned in school. And it seems that parents are more open to listening to their children's point of view and watching their children's nurturing ways of treating their dogs and cats, or even their furry stuffed animals, than to taking advice from a professional.

Unfortunately, the window of opportunity for training children is small. By the time these kids get to third and especially fourth grade, they know "all about" dogs. Many boys at this stage seem to need to show how "macho" they are in front of the girls. Despite the honored speaker's opinions and suggestions, these boys are certain that they could outrun a vicious dog (bad idea!), or they revel in talking about how they were able to run faster than a dog who wound up biting them. ("See my scars"—wow!)

However, as enlightened parents and pet owners, you can help reshape your children's attitudes—and ensure their safety—no matter how old they are. This chapter will help you understand what you need to know and do for your own and your children's safety. Some of the material may be disturbing—but it's not half as upsetting as an actual attack upon a child, which should and many times could have been prevented.

Why Do Pets Bite?

Dogs and cats don't come with signs that say "Nice" or "Nasty." And a pet or neighborhood animal can be nice one day and apparently turn nasty "for no reason" the next. Many factors influence why dogs and cats bite. Here are some of them: the breed, genes, rearing, health, veterinary medical conditions that may be painful or irritating, sex, reproductive status, size, and age.

Then we need to consider the same number of variables about the "victim," and the motivational context that brings the two together—in other words, why Sassy bites Johnny. Movement, eye contact, lack of familiarity—they don't seem very good "reasons" to us, but to an aroused animal with just the right combination of factors, the attack seems perfectly reasonable.

Some of those reasons involve:

- Territory (entering or leaving)
- Food, toys, or other items or locations within the territory that can be possessed and valued by the dog or cat
- Other pets living in the household
- People living in the household
- The dog or cat taking on the role of protecting others
- Access to his (the pet's) "sibling" or "parent" (owner)
- Situations inducing fear or pain
- Awakening or startling the pet
- Punishment or the threat of punishment
- Play
- Competition over some desired resource other than food or toys, such as attention from Mom or Dad
- Violation of the pet's "rules" associated with status, as in the dog not giving a person permission to move quickly, to stare at him, hug him, or to push him off the sofa.

So now you know why the phrase "But the dog bit me for no reason!" doesn't really resonate for an animal behaviorist! With so many different factors to consider in determining the cause or causes of aggression, you can see how difficult the correct diagnosis of a bite can be. For this reason, serious instances of or propensities for aggression should be handled by professionals and not by well-intentioned but ill-equipped self-proclaimed experts or your

Safety Information for Kids

Many humane societies have developed educational programs for kids, as have several large animal-control departments, so that children have the opportunity to learn how to recognize the dog who is likely to bite, and what to do when they run across an aggressive animal. They also learn to regard animal-control officers as people who can not only help dogs and cats who are being mistreated but also help remove dangerous dogs from children's neighborhoods.

   One of the best videos on the subject, <u>Dogs, Cats, and Kids: Learning to Be Safe with Animals,</u> is narrated by my colleague Dr. Wayne Hunthausen, and is targeted at elementary school children. It shows the danger signs of aggressive dogs and cats, what to do when a child sees an aggressive dog or cat, and more important, what <u>not</u> to do.

neighbors, who sometimes freely offer their own personal menu for how to correct Fido's biting or fighting or Fluffy's scratching or latching-on and biting.

🐾 Who's Riskiest and Who's Most at Risk

We know that intact (unneutered), 1- to 3-year-old male dogs who have been isolated on chains in the backyard 24-7 are dogs who are ready to bite. If they have skin allergies, or are large, well-muscled dogs such as German shepherds, Rottweilers, pit bulls, or even Chow Chows, the risk factors for dog bite increase even more.

We also know that male humans, especially young boys from 5 to 9 years old, are bitten more frequently than any other sex/age group combination. They are probably bitten more often because they take greater risks and are more active and assertive (eliciting chase from dogs or exaggerating their movement in a threatening way for the dog). Boys in this age range are less responsive to "let's stop playing now" signals from tired or defensive dogs (see chapter 4 for more on this). And they may be more prone to prolonged eye contact with dogs than other people.

Most bites take place in the late afternoon and evening in hot rather than cold weather. That's probably because it's the time when children come home from school and dogs are let out of the house, or are allowed to play, and there is a higher likelihood of dog-person contact. Most victims of reported dog bites are the neighbor's kids, followed by someone in the dog's family. Only about 10 to 15 percent are individuals unknown to the dog (strangers).

Dog bites are less likely to be reported to animal control if the dog is a member of the family. From information I have gathered from my own clients and from colleagues who have talked with other clients, only a handful of people whose own dogs have bitten them or a member of the family report the bite to animal control or the health department. Thus bites to family members is grossly underreported. Only the most severe bites are reported. Likewise, reports of dog bites from neighbors' small dogs are less likely than dog bites from medium-size and large dogs. It's a social-psychological phenomenon: Only wimps would report being bitten by the toy poodle next door, but it's quite reasonable to report a bite from a Rottweiler or pit bull, even if the bite wound is not as severe as that from the smaller dog.

Besides, people usually try to get along with their neighbors. And growling, snapping, and even biting by a neighbor's dog may be overlooked—especially if the kids are friends of the neighbor's kids—until the dog is clearly dangerous. Then the owners may finally recognize that it's time to consult their vet, who can refer them to a professional trainer or applied animal behaviorist. So we can expect that many more bites actually happen to family members and neighbors than are reported to the Centers for Disease Control and Prevention (CDC), humane societies, or animal-control offices.

🐾 Beware of the Attack Cat!

If a housecat has ever killed anyone, I am not aware of it. But cats can inflict painful bites and oh-so-awful scratches, which are much more prone to infection than dog bites are.

The cats who are reported to bite or scratch people are mostly adult females, but their reproductive status is unknown because a large percentage of the bites are delivered by unowned or roaming cats, whom no one identifies following a bite report filed by animal control. The victims are mostly younger adult women, and the bites most frequently occur in the early afternoon and

Wanna Get Bitten? Try This!

We behaviorists know that staring at a dog, approaching it straight on, bending over it, and reaching down for its head are risky behaviors for anyone. Taking a dog's toy or food, reaching toward puppies with their mom present prior to weaning, hitting dogs, or intruding into their territory or home are foolish things to do if you don't want to be bitten. And standing at the end of a dog's driveway or an adjoining driveway and waiting for a school bus to arrive, pushing a family member during play as some young boys are prone to do, or even hopping on the dog's back to take a ride are situations that place children at risk for a dog bite. Although "teasing" is rarely considered to be a primary cause of biting, it may lead to biting because you rarely know when the teased dog may have had enough, and his irritability sets the condition for him to bite the closest moving child.

Usually, these are dogs, victims, and situations where bites are often described by clients as accidental. For example, "He didn't really mean to do it," or "It was just play." These scenarios don't even come close to those involving trained attack dogs or guard dogs whose digs have somehow been violated and who have been trained to attack whoever wanders into their area—whether the person can read, is old enough to read, or lives in the same house with the owner of the attack-trained dog.

DR. WRIGHT'S INSIGHTS

As much as movement incites dogs to action, loud voices alone can drive a cat up the wall. "Be gentle, walk softly, and speak quietly" is always the best policy with cats. And forget about the big stick!

early evening following lunch or dinner. The incidents may involve feeding the strays or taking out the garbage where outdoor cats hang around.

Scratches and bites by cats are not commonly reported, though, if they involve one's own cat. So the reported cat-bite scenario is probably biased because outdoor cats may carry disease. Seeking help and reporting a scratch or bite from a potentially diseased animal is "okay" socially (there's that social psychology again), but "I know my cat's not diseased, so I'm going to treat the wound myself."

Indoor cats commonly bite in many of the same contexts or situations that dogs bite. But with cats, status doesn't seem to be as big an issue—at least in bites to people, kids included. Although little investigation has taken place regarding dominant cats and subordinate people, it looks like a simpler explanation for cat bites involves inappropriate or rough play. Or the cat becomes too emotionally aroused to inhibit his behavior, or had never learned to inhibit biting or to retract his claws during rough play when he was a kitten. (See chapter 6 for more on socializing cats.)

Likewise, it may be that the child has not been taught an appropriate way to stroke and play with cats. When he approaches to touch or even be kind to a cat, the cat may react fiercely because of a previous bad encounter with him or even another child. Princess may not be willing to "see if" Johnny or another child is going to do something nice, but instead runs off, or latches onto Susie's arm or face or scratches the kid's hand if she's foolish enough to reach under the couch to try to drag Princess out.

Remember from chapter 5 that cats remain aroused for up to 2 hours following emotion-laden or traumatic events. And they're really set off by other cats' urine (or pheromones), so even if the child is not directly responsible for producing those odors, the closest moving individual to the odor is guilty! Perhaps urine or cat hair is brought in on someone's feet or clothes. For example, the next-door neighbor kid will bring his cat's odors with him when he comes over to play with your kid.

Sometimes cats will growl if children get too close to their food, and they may hiss or bite if the children don't back off. Their signals should be respected! Sometimes cats respond by biting if you pet them too many times. There is some evidence that "dominant" cats are allowed to lick (groom) other cats, usually on the face or head, but if the subordinate cat returns the favor, it is swatted or chased off. We don't know enough about the way cats regard people in their relationships. It may be that if Princess grooms and licks you, that's okay, but if you return the favor (grooming, at least!), you're treated like the subordinate cat who wasn't given permission and are reprimanded for it. It's an interesting hypothesis, anyway.

Somehow, although a biting, scratching cat may be a health risk, some people believe that turning the cat over to animal control isn't an okay thing to do. And that's a shame because it's actually the responsibility of the cat's owner to keep the cat indoors and treat any signs of aggression. And it's also the owner's responsibility to see that his female cats are spayed so that they don't give birth to unowned outdoor cats who may become a health threat to others.

Porcupine or Pet? Check Out Those Hackles!

By this point in the book, you're well aware of the communication signals and warning signs of aggression, based on a dog's or cat's body position, tail, teeth, and eyes. You're also tuned into the snarl, hiss, spit, or bark that lets you know that you're heading for trouble! You should review these signs with your kids and make sure that they can tell the difference between a friendly wag and a warning swish.

Here's one more simple way for a child to see that a dog or cat is about to lose it. No matter what's bugging the animal, raised hackles are a sure sign that he's teed off. Dogs and cats raise their hackles—the hair on the back of the neck—when they become aroused, either by surprise or when they become fearful or angry. Sometimes, their fur literally stands on end all over their bodies. Both dogs and cats raise their hackles in the same situations: those involving a choice between "fight," "flight," and "freeze." "Freeze" is a third popular strategy among many cats who become very large-looking, with fur on all parts of their body standing on end (the classic Halloween cat). Dogs, on the other hand, don't usually try this trick (you've never heard of a Halloween dog, have you?). Instead, they commonly raise the fur from the back of their ears down their backs to the tips of their tails.

Although dogs and cats in this aroused state look "bigger than life," and it

When you see a cat in the classic Halloween shape, with his body arched and fur on end, he's most likely afraid. Don't try to console or provoke him! Just stand still and speak quietly until he calms down.

may be useful for them to look scary to an opponent, neither animal has the ability to "will" themselves to raise their hackles or the fur all over their bodies. They don't think to themselves, "Hmmm. I think I'll raise the hairs on my body so I'll look bigger than I really am!" Instead, the Halloween cat and raised hackles are a result of their sympathetic nervous system automatically turning on the "erectile hairs button." After the dog or cat perceives himself to be in a "fight-or-flight" situation, each little hair follicle reacts to the message sent by the brain, and they all stand up. It's similar to the reaction some people have when they've been insulted or embarrassed in public: Their faces flush, and they feel the little hairs stand up on the back of their neck—it's part of their nervous system's preparation for fighting back or getting the heck out of there!

In either case, the best thing for a kid to do when faced with a Halloween cat or a dog with hairs standing on end is to stand still, speak in a soothing voice, and smile—without looking the pet in the eye. Eventually, the pet will calm down and move off.

Stranger Danger, Canine-Style

Dog bites can be very serious, even life-threatening—especially to children. But children can effectively defuse the situation when they are confronted by a strange dog. Make sure that your kids remember these three "street-smart" tips. If your child sees a dog approaching him and he feels afraid or uncomfortable, he should react as follows:

1. Stop moving and stand still.
2. Don't look at the dog; look away. This is a hard one, but important—eye contact now can be like waving a red flag.
3. Do not reach to pet the dog or let your arms dangle at your sides. Stick your hands under your armpits so the dog has nothing to nibble or bite.

Facing a kid like this, most dogs are quite content to sniff a leg or hip. And because the child is not arousing the dog by looking into his eyes or by running or kicking or pushing the dog with his hand, the dog will likely lose interest and move away. By remaining motionless and passive, the child also gives an older caregiver or parent the time and opportunity to intervene if necessary.

🐾 What's a Parent to Do?

Most stray dogs—that is, truly unowned animals—are generally too fearful to approach children and especially adults and are likely to run in the opposite direction. But dogs who are owned and are roaming free are susceptible to the same kinds of mood elevations as our own dogs. So if you need to intervene to protect your child from attack, you should attempt to clap, say, "Good boy, good boy!" in a playful way, attempt to find things to make the dog happy, like "Where's your ball?" or "You want a bone?" You might also try "Wanna go for ride?" or even "How's that good boy?!" These words and phrases and voice intonations make the parent or big sister more interesting than the child and motivate the dog to move in their direction.

Other options in one's arsenal of phrases most likely to control the dog's behavior and move him away from a child include commands like "Sit!" (many dogs are familiar with this command, and will in fact sit if told to do so), followed by "Come here, boy! Go-o-o-od bo-o-oy." Or even "Go home! Go on! Go home!" The interesting thing about these three different kinds of verbal control is that some dogs respond better to demands that they go home, others respond better to instructions to behave in a way they know gives them some control and typically leads to praise, and still others respond best to "jollying up."

🐾 If His Bite Is Worse Than His Bark

Should the child find himself bitten by a dog or, worse, by more than one dog, the safest thing to do is try to remain standing motionless, or if brought down, to protect his neck and face with both arms wrapped around them while trying not to move or scream. I know these instructions sound difficult for a child under attack. But kids can and should practice this crucial move until it is second nature so that they're ready to bring it into play should the situation arise.

Have the kids role-play the situation. In one elementary school I worked with, I asked second-graders to pair off—one played the dog, and the other played the child. When one barking/snarling second grader approached the other (no touching was allowed by the "dog"), the "victim" practiced standing still, cupping his hands under his arms, and looking away from "the dog." Then I suggested to the dog that "this isn't a very interesting boy/girl, I think I'll go back home." (Thus, teaching both children the concept and the outcome of correct behavior.) Then, of course, they switched roles. Next, I had the kids practice with two child "dogs" and one child lying in a supine position on the ground with his arms protecting his neck and face. The "dogs" were instructed to nudge his shoulder and head, as a dog might do. The children under "attack" are taught not to move or call out—no matter what. Learning to "play dead" can spare their lives.

Unfortunately, some dogs either have been trained to attack people or lose their inhibitions to bite people when extremely aroused, and about 17 to 20 people a year—mostly children—fall victim to killer canines. I have examined several of them for court cases designed to assess responsibility. In one horrible incident, three pit bulls ran up to and threatened a 4-year-old boy. The child's 7-year-old cousin had learned in school how to respond to a vicious dog. She correctly told the boy to "fall on the ground and freeze," which they both did, just before the dogs got to them. The dogs stopped their charge. However, when one of the dogs nudged the 4-year-old's shoulder with his nose, the boy tried to push the dog away with the arm that was touched. Within just a few seconds, the 4-year-old was dead.

The 7-year-old girl, in the meantime, got up and raced to the side door of a neighbor, who had heard and seen the commotion. The girl rushed inside the house just as one of the dogs leapt and crashed into the glass door—which, fortunately, didn't break. The dogs were soon caught by a well-trained animal control officer and were eventually put to death. The owner served a 5-year prison term for negligent homicide.

Case after case of severe attacks on children indicate the same thing: Dogs that bite increase their attack when the child (or adult!) kicks or struggles.

They bite whatever moves—and eventually release their grip when the child (or adult) remains still. It's worth role-playing/practicing, especially if you live or visit an area where there are no leash laws or the leash laws are not enforced.

Safety Strategies for Teens and Parents

Teens and adults whose jobs bring them into contact with dogs on a daily basis can employ other strategies. Most of the gadgets that use sound are not likely to work. Even pepper spray is not particularly effective (or humane) for dogs who are bent on biting. And if the wind is blowing toward the user, he can blind himself. Some sprayed dogs become even more angry, and with eyes burning, latch onto the user and shake him. Tasers or electric shock devices can be effective (do you consider them humane?), as long as the dog doesn't get hold of the user's arm and make him accidentally shock himself. Sometimes the sound of the tasers activating (before any contact is made) is enough to frighten some dogs. And a new citronella spray named Dog Stop seems to give dogs pause without overstimulation.

Opening an umbrella in front of a charging dog was formerly a favorite technique of some letter carriers. But as it turns out, the umbrella may not be as effective open as it is closed. One instructor, teaching a seminar, opened his umbrella, but the charging dog came up underneath it—resulting in 20 stitches in the trainer's hand.

If a dog is lunging at you in spite of your best efforts to make yourself uninteresting, and he will not "jolly up," use a small closed umbrella. (Animal-control officers use a "bite stick," which looks like a closed umbrella.) Take your umbrella and hold it parallel to the ground, with hands on both ends and the length of the umbrella between you and the dog. In this way, if he does lunge to bite, he'll bite the umbrella, not your arm. The idea is not to hit him with the stick, as movement makes the attack worse (and if you hurt him a little bit, you're liable to anger him more), but to make whatever you're holding

DR. WRIGHT'S INSIGHTS

It's always best to remain motionless to stem off any attack. But if attack is imminent, and you have no stick or umbrella to hold in front of you, try offering the dog the sleeve of your shirt or jacket by taking one arm out and dangling it in front of you for the dog to bite.

available for him to bite. Then don't let go, but back out of his territory or back up to your car and get in, or allow others to draw his attention away from you and give you some help.

Although there is no panacea or guarantee against being bitten, the bite-stick approach is an effective option to consider. Most certified animal-control officers have several options available to them should this nightmare occur. And you can bet that they've thought about what to do and how to do it, and have role-played their options so that, if the situation presents itself, they don't panic and wind up in the hospital, severely bitten.

Raising a Kid-Proof Family Pet

If you have kids, it's even more important to make sure that any pets you bring into the family have steady, calm temperaments. Whatever you do, *never invite a dangerous pet to join your family.* Here is a word to the wise: Top-notch responsible breeders or shelters won't let you adopt a dog or cat that they know is not safe around children. Think twice before you bring in a stray animal off the street or buy one through an ad in the newspaper or on the Internet.

Here are some of the guidelines one nationally respected humane organization uses to help its volunteer workers decide not to kennel a dog for adoption:

- The dog is an adult animal who is known to have bitten someone.
- The dog is an animal who can't be restrained, examined, or petted even after 24 hours because of offensive or defensive aggression.
- The dog is highly aggressive to other types of animals in many situations.

This humane organization warns that animals with a history of aggression to children should never be adopted out to families with children, and it prohibits kenneling any dog who is a wolf hybrid. (This list is especially frustrating to me and other animal behaviorists because we know that many of the problems that will make these throw-away dogs unadoptable could have been prevented or reduced by caring owners or resolved with our help.) At any rate, use common sense, and never use your kids as guinea pigs, thinking you'll be able to "straighten out" that aggressive dog or nasty cat because she is just so cute, or you feel sorry for her.

🐾 Mine! Kids, Pets, and Possessive Aggression

Through exposure to, and repeated correction ("Wrong!" or "Off!") for even sniffing the baby's toys, your dog should know even before the baby arrives

Dr. Wright's
E-Mail

Dear Dr. Wright:

I have a dog who is part sheepdog mixture and part coyote. I'm wondering if domesticated animals grow more prone to attack as they age? Do they grow more protective of the owner and more likely to be aggressive to guests?

Thanks,

Shane

Dear Shane:

If your dog is really part coyote, its "wild genes" are likely to make it less social with age (as far as accepting people is concerned!). The reason we domesticated the dog was to increase tame behaviors and decrease aggressive behavior. Although we haven't been totally successful—by a long shot—we're better off sticking to domestic genes in our selection of pets than wild genes. The wild genes have been selected in animals who have to worry about where their next meal is coming from, which mate is available and most likely to bear young, and who the competition is for the mate—none of which is a concern for the common companion dog.

So some of the aggression toward visitors may be the wild genes expressing themselves. Your dog may also have

that there are his toys and there are Baby's toys. It helps a lot if the dog toys are obviously different from the baby toys. (Baby's are furry or soft-rubbery, and the dog's are hard-rubbery or plastic like a Buster Cube or bone-shaped like a Nylabone.) Many dogs learn to respect those differences, usually with learned a role as "protector," and you have allowed him to function in that role—maybe it makes you feel more secure. Not only do dogs get more aggressive with age, but older dogs also have more experience with testing limits and being aggressive: running to the door, barking, and growling, which becomes a pattern. And this pattern, if not stopped, becomes a job!

Dear Dr. Wright:

Our family adopted a Siamese cat, Suki, who is beautiful, but who hisses and spits if my kids try to pick her up. Why would she do this?

B. J.

Dear B. J.:

First of all, check to make sure that your kids aren't hurting Suki when they pick her up. They should be gentle and support her under the front legs and behind the rear legs. But I suggest that they leave her alone for the time being rather than risk a bad scratching. It sounds as though your cat may have missed some important human contact when she was young. Typically, cats are most open to accepting "others" (cats *or* people) when they're somewhere between 2 weeks to 2 months old. After that, they become more fearful of whatever they haven't been familiarized with, and it seems to get worse as they age. Ask your children to stick to petting when Suki approaches them. Some cats, sadly, just aren't "picker-uppers"!

a lot of Mom's and Dad's help, but sometimes it's necessary to just get rid of a baby toy and anything else like it if the dog is much too vested in it for retraining to be successful. Few parents want a teether to be put back in Baby's mouth after the dog drools on it, anyway!

Nothing about life with a dog and kids is perfect. Consider what can happen when Baby becomes a toddler or preschooler. Your toddler finds it helpful to clean up the house by picking things up and putting them in a pile, but your dog takes offense at her removing everything he took such pleasure in depositing about the house—mostly his items that should not (from his perspective, anyway!) be messed with. Now you have a problem and a potentially dangerous situation.

Possessive aggression occurs when a dog growls, snaps, or bites if someone comes too close to an item he's possessing or to one of his belongings, even if he's the only one who's aware that he has defined the item as "his"! Some dogs seem to love to "own" paper, especially toilet paper, and they seem to take great pleasure in letting people know that it's theirs. I can only speculate that, for example, the first time the cocker spaniel unraveled the toilet paper and paraded it around the living room, he received a lot of attention, laughter, and comments from his family, which reinforced the "winning" behavior for the future.

Oddly enough, the dog may become so compulsively possessive of the toilet paper that some people might call him "anal retentive," in a Freudian sense, although those people would probably not include me! People generally handle this problem by shutting the bathroom door and by keeping wadded-up computer paper off the floor and in the trash can (or in containers with lids) where it belongs. Obviously, the same goes for clean or dirty diapers! But what happens when your toddler spots the toilet paper next to the cocker spaniel, and you've walked across the room for a moment and are not there to supervise directly who gets what?

Teaching Raffles the command "Off!" or, better yet, "Drop it!" and "Leave it!" (if the item is nearby, but not in his mouth yet) is very useful for anyone who has dogs and children. Any one of these commands may quickly defuse a potentially dangerous situation.

But you've still got the toilet-paper problem (or whatever he's being possessive about) to tackle. Here's what to do: The idea is to find something the dog likes even more than toilet paper. He gets this prize, in a training session first, *but only after he drops the toilet paper.* Give him praise and more praise. This technique uses the same idea as the "mood word" list I talked about in chapter 5. Tennis balls, plastic balls with reduced-fat peanut butter or light cream cheese in the holes, or even rawhides are on most dogs' "favorites" lists, and they're all higher on the list (more interesting) than toilet paper.

The earlier you teach your dog to drop it, the easier it will be for him to drop new items that you don't want him to have. Although the training is always done by offering your dog the prize when he drops whatever is in his mouth, it almost becomes reflexive for some dogs to release the item when they hear the words "Drop it!" and to look to the child to present a prize, getting him out of a dangerous situation. Now the child can more safely pick up the toilet paper. It's also a wonderful idea to begin teaching your toddler the rule "Don't approach Raffles when he has something in his mouth," or "No, that's Raffles', leave it." Yes, kids respond to such concepts, too. If you can teach both your toddler and the spaniel to "leave it," won't life be much easier? They'll both walk away from the TP and "leave it" for you to roll back up.

❧ Three Steps to Toddler-Proofing the Food Bowl

Your child may think that turn-about is fair play. So another aspect of possessive aggression deals with food—the dog's this time. This food adds even more risk to kids! You can and should teach small children not to touch Fang's things, *especially* food. But these three steps will help you short-circuit your dog's developing a stingy attitude when it comes to her food. Again, it's best to start this training when she's a puppy.

Step 1

Take a bowl—preferably not her food bowl, which she may already "own"—into the family room or living room (someplace where the dog doesn't eat), and sit down with some hard dog food in your hand shortly before her dinner time.

DR. WRIGHT'S INSIGHTS

Here's a common situation fraught with peril to kids. Toddlers and preschoolers love to march around with stuff to eat—ants on a log (celery with peanut butter and raisins), wedges of cheese, slices of baloney—usually sashaying by at doggy's eye level. This is too much temptation for even the best-trained dog who wouldn't think of grabbing food from the table. The grabby dog may scare the child, nip him by mistake, or invite an unthinking swat from the hungry tot.

  The key to short-circuiting your pooch's "I want" behavior is to say in a low-pitched, firm, monotonous voice, "Don't even think about it." This should stop him in his tracks. Then reward his stopping with a treat so that he'll learn to expect something even more tasty if he ignores the toddler's gooey cookie.

Toss or place one nugget in the bowl for her to eat, which she will, and let her finish. Do this again and again. Repeat. Ditto.

This set of repetitions teaches her that you—or rather, your hands—reaching into the bowl results in food that she can eat. Hand reaching to food bowl predicts food. It's a good hand. Now have someone else in the family do the same thing, and then help your toddler copy you so that little hands as well as big hands in a bowl predict food.

Step 2

Take a nugget and place it in the bowl as before, but before your dog can eat it, place your hand in the bowl again and take out the nugget. Your dog should appear confused. It's important not to tease the dog at this point, and not to wait more than a second or two before you do the next part. As you're giving the dog a Liv-a-Snap with your other hand (or any healthy treat that's "better" than the nugget of food), say, "Good treat!" or something else in a happy, fun way. Now repeat this step. Do it again. Ditto.

What she learns is that when one hand goes into the bowl, even if it takes something out that was hers, she gets something even better than what was in there to start with. Now that's a good hand! Have the rest of the family repeat this step, and help your toddler do it, too.

Step 3

Move the bowl into the location where you want to feed your dog (the location is important), and practice Step 2 there a few times. Use that bowl as her food bowl from now on. Fifi's new bowl stands for fun food games, and not serious possession issues that can lead to biting little hands!

🐾 Nibbling (People) Is a No-No!

Part of your child's daily routine with your dog will involve social play: you, her, and probably a Frisbee, ball, or other toy. From their pet's puppyhood on, people who play with their dogs have their hands or fingers nibbled on or bitten. They don't connect this with aggressive behavior, even though it is.

Nibbling can lead to things like lunging and biting to control the object (you or your kid!). So you must stop this bad habit before it starts. From the beginning, it can be a simple discrimination-learning task. The dog needs to discriminate between your hands, which are never bitten or mouthed, and an appropriate object—such as a toy, rawhide, or rope—

that can be held by your hands and mouthed or bitten.

The best way to accomplish this task is to remove your hand or fingers as a stimulus and make the ball or other object you want him to mouth or chew available an inch or two in front of the dog's mouth. As he's mouthing the ball, stroke him under the chin or on the side of the head or down his back with your hands. Thus, he connects your hands with making him feel good. He can then connect anything other than your hands with biting and controlling things. Never use your hands for hitting him, slapping him, or doing anything other than making him feel good.

An open palm, yours or someone else's, should never indicate anything to your dog other than comfort and pleasure. An open palm should never be the source of pain or even discomfort; hands should never be used to hold his mouth shut, slap him, tap his nose, hit him—none of these punishments!!! It may take a few weeks for him to learn that fingers are not okay to nibble on, but you'll be able see him get better from week to week.

Make Sure That the Kids Learn the Rules

You and the kids and others who play with him must know and agree to the rules, too. You can acknowledge with your children that it sometimes feels really good to have little puppies nibble your hands or fingers. But add that this is the time to restrain yourself and make those digits unavailable while offering the dog something else to chew on. So whether he's cutting teeth, trying to control a toy, attempting some rough play with you the way he would with another dog, or working out any other scenario, enable him to be a good, safe dog by learning that it's never okay to put his mouth on or around people's hands or fingers. You'll be glad you did when the neighbor kids come over to visit and stick out their hands inappropriately for the dog to sniff. (They obviously didn't read chapter 4 on how to communicate with a dog!)

Now your pooch will probably not think, "Since hands are okay to place in my mouth, and they sometimes mean that I'm going to get slapped or hit or hurt or pulled on, I think I'll try to get this hand before it touches me and just hold onto it." And the child, as a result, is less likely to jerk her hand away and suffer a tear wound because she thinks the dog is trying to bite her. In any case, teaching dogs this simple discrimination task—early and consistently—will result in a safer, happier pet and family because you have nipped your dog's biting in the bud.

DR. WRIGHT'S INSIGHTS

In a culture that increasingly endorses respect for living creatures, certain behaviors should be cause for alarm. Deliberate, repetitive cruelty to dogs, cats, or any other kind of pet or animal by a school-aged child can be a predictor of serious psychological and social problems down the road. You must intervene to protect the animal and seek professional help for the child; cruelty to animals can be a predictor of cruelty to people as the child gets bigger.

🐾 Don't Pull That Tail!

Whether it's a cat or dog, stray or family pet, reaching for a tail is a no-no. But it's really difficult for a toddler to ignore a tail because, at that height, one is usually waving in her face! The solution comes more under the heading of "kid training" than dog training or applied animal behavior because the dogs and cats can't stop being dogs and cats, and it's not nice to remove (dock) their tails. So a rule and approach like, "Don't pull the dog's tail, it will hurt him. How would you like someone to pull your tail?" might be a good place to start. (Praise compliance—it works for kids, too!)

As they get older, you can help children look at the dog's or cat's feelings and reactions through their pet's own eyes. Ask leading questions: "How do you think it makes Fluffy feel when you pick her up by the neck?" And be proactive: "Prince's tongue is hanging out and he's panting. What do you think we should do for him?" Some kids will be naturally sensitive and empathetic toward pets; others will be less so. Any chance you have to impress upon your children that dogs and cats are not possessions or objects, but important members of the family, will be time well-spent. Children who love animals usually grow up to be loving adults, respectful of all the life forms around them. I feel very sorry for grownups who don't like dogs or cats and teach their children to fear or dislike them. Think of all the fun and loving they miss!

🐾 Help Your Kids Become Leaders

Teach your children to take a leadership role with your dog and cat, and make sure that they know how to correct the pet without challenging him. Correcting a pet is not the same thing as hitting her when she misbehaves or yelling at her. It is not telling her she's a bad dog hours after the incident so that she doesn't know what's wrong but does know that you don't seem to like her anymore. The object is not to make the pet afraid of the child but to have the child understand what to do to get the kind of behavior he wants from the dog or cat. (With cats, make sure that your child understands the limited amount of "obedience" that can reasonably be expected!)

Don't allow young children to decide upon lengthy "time outs" or drag the pet into her crate if they become frustrated. Instead, encourage them to call Mom or Dad if they need help. This will help prevent having *two* out-of-control little animals in the house!

Do help your child learn the simple training commands like Sit, Stay, Come, Down, Off, and Drop it, and to give commands by encouraging the right behavior more than punishing the wrong behavior. You can emphasize that the dog can follow only one command at a time. If the dog fails to respond to the command, let the child know that he is not to react with anger. He must simply say the dog's name and calmly repeat the command. Then make sure that he reinforces the dog's good behavior with a treat, kind word, or stroke, which will make the two of them better pals. Most dogs are very forgiving of children's frustrations and inconsistencies, but some aren't.

How *Not* to Create a "Head-of-Household" Dog

Dogs who have no reason to view humans as leaders sometimes decide to take over, and in my practice I have found these to be the most likely to cause injury

Kids and Dogs Together

Review this list with your kids so that they know what's safe to do with the dog and what's dangerous. You'll avoid biting and aggression from your pet and give your children and pet the best chance of having a great time together!

RISKY ACTIVITY	SAFE ACTIVITY
Nibbling fingers	Nibbling treats
Rewarding barking	Rewarding retrieving
Keep away	Throwing balls for pup to chase
Backyard wrestling with Bowser	Doggie massage
"Horsie" ride on dog	Cart or skateboard pulling
Teaching "attack" moves	(only after vet's okay)
	Teaching agility

to children and other family members. (The German shepherd who inspired the title of our book *The Dog Who Would Be King* was one such dangerous tyrant!) If you don't want a pet who thinks he's the boss, follow these rules:

1. Don't take a pup from his mother and littermates until he's about 2 months old, or he won't know the rules.

2. Pass up wrestle-on-the-floor games that seem like World Wrestling Federation at home; instead, choose balls, a Frisbee, or agility training, so that playtime will be a win-win game instead of a power struggle.

3. Handle the puppy a lot, but forget the roughhousing and hurting with hands. These games may teach the dog to be aggressive.

4. Give him his space. Don't let the kids pounce on him when he's eating, just waking up, or any other time!

5. Take your dog to puppy preschool and training classes, and follow through with his education. Be in charge, without hitting or yelling.

6. Enable him to win (succeed at what he's doing), and praise him as though the behavior were his idea.

7. Reward him with a kind word, stroke, or dog treat when he "lets" you be the boss.

8. Discuss with all family members—even the kids—the need to take a leadership role. (See page 366 for tips on teaching leadership to kids.)

9. Don't allow your dog to mouth your hand or fingers, even if he's gentle. This could escalate into mouthing your arm and become a sneaky way of controlling you. And even a "gentle" hold could be disastrous for a frightened 2-year-old.

10. Ignore your dog's nips, stares, or barks. Giving in to them will encourage her to control you. Instead, give her an alternative activity that's acceptable to both of you.

11. Don't get sucked into petting the dog so that he'll stop bugging you. If he starts to *demand* petting, don't do it! Remove your hand, sit on it—or get up and walk away.

12. Figure out if she needs something or just wants it—it may fulfill her needs, but she doesn't need something just because she wants it. How freely you give of yourself is up to you, not your dog.

If your pet is already biting or giving lots of warning growls, it may be too late to prevent a serious behavior problem. It won't just go away, and it's not a do-it-yourselfer. Consult a certified applied animal behaviorist before someone gets hurt.

🐾 Put an End to Chasing

To train a dog not to chase kids, have a child known to the dog (a family member or neighbor child) walk quickly while the dog is on a flex-line or a long leash. When the dog leans forward to start running after the kid, give him a "Wrong!" or "Uh! Uh!" followed by a prompt from the leash to stop. Then say, "Good boy!" Don't jerk him or try to yank his neck off—remember that the leash and collar (or head halter) are communication devices! Have the kid walk faster and then start to run. After this exercise, ask the kid to throw a ball or something the dog *can* chase, while the dog is still on his long leash. When the dog does not chase the ball, praise his compliance. Repeat the exercise in a different location, and use a less well-known kid, so that the dog generalizes the rule "It's okay to chase 'things' but not any child, anywhere."

🐾 Chained-Up Dogs + Kids = Disaster!

Dogs restrained on a chain for more than 8 hours a day are significantly more likely to be aggressive to people—especially young children—than dogs who are restrained by a fenced enclosure for several reasons. He could be protecting his territory, his food or water bowl, his doghouse, or the area he lies in. It's hard for a dog to be restrained from being with people, and he might not even know how to act around them. He may be afraid or frustrated. And he may have—quite understandably—gotten into a bad habit of straining at the end of his chain, making lunging forward a frequent daily activity. He may be trying to possess something he considers "his"—even a stick will do—and if a little kid reaches for it, the kid will likely get nailed. Finally, the tied-up dog may have no other way to use his pent-up energy than to use his teeth, jaws, and claws to rip up whatever is moving within reach.

Sometimes dogs who have been relegated to the backyard chain have behavior problems that keep them from being welcome in the owner's home—destruction, aggression to family members or other pets, or, as in the case of Pal in chapter 2, the threat of fleas and ticks being brought into the house. Now that there are behavior therapy options for pet owners whose dogs begin to show unacceptable problem behavior in the household, fewer dogs will wind up unhappily isolated in the backyard on a chain, wondering why their family is living life without them.

The very popular new underground electric fences—while giving the dog his freedom up to the perimeter of the property—can present their own set of difficulties for families. The biggest threat, it seems, is a false sense of security for the kid or adult outside the electrified area. Some dogs learn to steel themselves and ignore the mild shock from the box on their collar as they race across that invisible line in hot pursuit of a rabbit—or a jogger or a preschooler. (These types frequently lose their nerve after the chase is over and find themselves "trapped" outside the fence, unwilling to take another hit!)

This chapter should get you through those hectic family years. Before you know it, you'll be waving goodbye to the kids and looking forward to some time alone with your ever-faithful companion. That category may or may not include a spouse or housemate, but it will surely include your ever-loving pooch and kitty!

The Empty Nest:
When Your Pets *Are* Your Kids

Okay, so you're an empty nester. The kids have gone—graduated from high school or college, moved off to parts unknown (or down the street), and left you with a quieter home to rattle around in. What are you going to do with all that extra time? Perhaps it's time to retire or start a new business, travel more, go out more often, take up a hobby or sport, and enjoy your freedom.

So to some extent you have made plans for your extra time, but there's always a point in time where being home with your pet—your old friend—feels best. And because you have no more kids to worry about (okay, just because they're adults now doesn't mean you don't worry—most of us still do), you begin to have concerns about how to care for your pet as you both get older.

We've already covered the geriatric pet and how to meet her needs in chapter 9. But this is different. The older dog or cat still had a whole family to live with, and everyone was still available to stroke, watch, protect, and make

371

your pet feel important in the family. Now that everyone but you and your spouse is gone, the day-to-day environment is different.

And now that your schedule and activities have changed, what about the quality of life of your dog and cat? With all that commotion and furniture gone (kids always take the best stuff!), what is your cat or dog to do with all that room? All that time? From her point of view, there are fewer people around to help meet her psychosocial needs, there are fewer comfy locations to nap on— your son's ever-present pile of dirty clothes or your daughter's pillow—and you begin to think about her loneliness. What is a pet's place in the older home?

Suddenly Single or Separated

Downsizing isn't always something you've planned. Perhaps your spouse or significant other has died, and you're mourning your loss. Some pets spend great amounts of time at the side of an elderly individual, and the bond between them seems stronger than ever. In chapter 9, we covered how to grieve over the loss of a pet. Now the tables are turned, and both you and your pet must deal with mourning the loss of a loved one. And of course, divorce can have the same effect on your pet as the death of a family member. Now that it's "just the two of you," each is likely to become more important to the other, and you'll have greater expectations of one another than when your spouse was there.

For many people, aging leads to relocating to a smaller home or condominium, or even a retirement village that allows pets. I will discuss moving-related issues later in this chapter. There will be new places to walk. There may be some new rules and regulations to follow regarding what your pet can and cannot do. You may have to finally get used to walking your pet on a leash, with poop-scooper in hand. Maybe it's time to hire someone or make other arrangements to exercise your dog and to take her outside. You need to consider the life stages of both you and your dog now!

Small Dogs, Large Problems

Many empty-nesters acquire a pet after years without one, when they find themselves alone and lonely. And too often, in my opinion, seniors choose a toy or other small breed because that is what is permitted in their new digs or what seems the easiest to handle in their big empty home. Large animals just seem to present more "work" and just might be harder to move (true—if a shove is the only way you know!). Or maybe they have always had small breeds. Although this line of reasoning may work out just fine regarding cats, small dogs can spell big trouble for older or elderly owners.

Fortunately, older pet owners today have plenty of help at hand. Perhaps it's time to consider a doggie diaper for your aging dog or a doggie litterbox indoors. (Can you stand the thought? Will the retirement village board allow it?) There are now aids like the electric, self-scooping litterbox for your cat, so you don't have to bend over and scoop as often. When the passage of time forces owners to slow down, it's great to be able to have some help from the latest gadgets designed to lighten your load. That's what I recommended for one retired couple who found themselves with a dog perhaps better suited to a pair of energetic Yuppies.

🐾 A Useful Deterrent

Sarah and Benjamin Blakely were pretty helpless in the face of the frisky apricot whirlwind who tore through their lavish suburban "empty nest." At this stage in their lives, a snoozing cat or laid-back Labrador would have suited their situation better. But they thought that a small dog—a 1-year-old toy poodle named Taffy—would keep them company and be "no trouble" in their retirement years. And after all, they did have a housekeeping couple to do most of the work—or so they thought.

The Blakelys were quite elderly, so my options were limited. The dog desperately needed obedience training, and I suggested that one of the staff take her to puppy kindergarten or a basic obedience class. This idea was not received well by the household help—they wanted nothing to do with the rambunctious little dog. I could readily understand why. The dog chewed everything, and if anyone tried to correct her, she bit or nipped. After cleaning up after her and taking her for the required walks, the housekeepers virtually ignored little Taffy.

Mrs. Blakely was a bit forgetful, she moved slowly, and she just couldn't cope with the adorable little bundle of energy she had brought into her home. She didn't have a clue what to do with the puppy, although she confided to me that she had occasionally swatted at the little dog with a feather duster borrowed from the maid. "But Taffy's too fast for me," she said, shaking her head.

The dog had a habit of pouncing on moving objects like feet. This posed a rather large problem for the frail Mr. Blakely, who sat in a rocking chair in the living room most of the time. He had a cane sitting next to his chair.

Whenever he would get up and start inching across the floor, here came the little dog, barking and rushing at his feet.

Of course, Mrs. Blakely was quite concerned that her husband would trip and fall and break his hip, and she had absolutely no control over the little dynamo. Because of the limitations imposed on this couple by their age and infirmities, I suggested that they use a hand-held sonic device until I could recommend a competent trainer who would make house calls for a crash course in good manners.

In the meantime, this electronic device would begin to give Mrs. Blakely some important control over Taffy, assuming that she used it correctly. I instructed her to point the small hand-held device at Taffy as the dog approached to nip her husband's feet, and say, "Wrong!" while pressing the button for 1 or 2 seconds. The SDG emitted a high, ultrasonic tone that the poodle could hear, but that people could not. It was not painful, but it definitely stopped Taffy in her tracks.

As Taffy stopped, the owner was to praise her and throw her a toy to pounce on instead of Benjamin's L. L. Bean slippers. Then Mrs. Blakely was to praise Taffy for "getting" the toy. Soon my client was able to use the word "Wrong!" without the sonic device, and have the dog respond appropriately. Thanks to technology, Taffy was now ready for professional training, and she was well-enough behaved for the Blakelys and visiting grandchildren to enjoy.

🐾 More Tips for Breaking Bad Habits

Another elderly client of mine was being driven nutty by a 5-pound ball of fire named Chica. That Taco Bell Chiquita has nothing on this little Chihuahua. Mickey Burgos was disabled, walked with a cane, and needed a calm, obedient dog that she could control. Instead, she had Chica. And even though the dog was small enough to pick up and move off the couch, her strong will and high energy made her a bad choice for this one-person household. The owner would have been better off with a 60-pound creampuff of a Lab or a golden retriever. She could not move the bigger dog with her foot, true, but these gentle breeds would obey her commands without question. (In fact, she would have been better off with an obedient dog of any size, for that matter!)

It was somewhat ironic that both dog and owner had locomotion difficulties—Mrs. Burgos from crippling arthritis, and little Chica from previously snapping her front leg bones as she flew off the back of the sofa in a burst of spitfire energy. I defy you to find a chocolate Lab who will do that! At any rate, for performing the service of lifting the little dog off and on the couch, its owner was rewarded with growls, nips, and other unpleasantness.

Not only that, but—the reason I was called to help—the dog kept visitors virtual prisoners in the house, biting at their feet if they dared to get up and move toward the door. As a result, when they were able to make it to the door, it was vamoose! They went away and were loathe to return until their friend gained the upper hand over the pint-sized tyrant.

"I knew I could never shove a big, lazy dog around the room if he was blocking my way," moaned my auburn-haired client. "So I picked out this little one." She scowled at the saucy Chihuahua, who was eyeing my ankles, waiting for me to make a move. I sat very still.

"Andale, andale!" My client swept her arms around, looking keenly at Chica. Chica looked around, drew herself up to her full 7 inches off the floor, and barked back in protest. She was not going to leave my feet, not when I might get up any minute. She had to stop me from leaving!

"Sometimes I give the naughty dog a little poke with my cane," Mrs. Burgos announced. That explained why the little dog seemed to be even more ornery when her owner brought the offending instrument out of the closet and attempted to go up and down stairs. The dog would bark, rush at the cane, and generally make it hazardous for Mrs. Burgos to be on her feet at all.

The situation was downright dangerous—certainly one that could not have been anticipated when the vivacious woman picked out a little dog she thought that she could "handle." At the same time, with the dog's fragile legs at risk as well, I had to admit that the duo was somehow quirkily well-suited for one another!

It was my theory that Chica was unhappy when her mistress's attention was diverted away from her, even for the amount of time it took to walk a guest to the door. By making a pest of herself, Chica—like the kid who disrupts class at school—was seeking attention of any sort. Negative attention was better than no attention. And by barking and nipping at those feet and ankles, perhaps she was actually driving those troublesome people away (never to come back, in some cases). Chica certainly had power over that household! Wow!

Actually, for all the havoc she wrought, Chica was well-trained in obedience and could be counted upon to do her "potty" business in a cat box in one of the bathrooms. In this way, her disabled owner did not have to worry about walking her very often, and in that regard the toy breed was an appropriate

Dr. Wright's
E-Mail

Dear Dr. Wright:

I have a 5-year-old neutered, declawed male cat who has begun to bite me a few times a month. I am confined to a wheelchair and do not want to give McGriff away or have him destroyed, but my caretakers say I should consider that. He had been mostly an outdoor cat with a large territory and was gregarious. He came home very late one evening in December and has been very timid and unhappy about going out. McGriff is very important to me, and my local vet has not been much help. Please help me if you can!

Lillian

Dear Lillian:

Try displacing his attacks onto something that you can toss away from you and he'll chase and bite rather than continuing to allow him to bite you! If he has too much energy or this is aggressive play, the worst-case scenario is that you'll have to keep rolled-up balls of aluminum foil in your pockets all day and collect them every evening—my apologies to your caretakers! Also, if you can involve your cat in several brief (2- to 3-minute) play sessions throughout the day, especially just before those times when he's likely to get feisty, you should see a decrease in his biting. Look for progress from week to week, not day to day, as behavior patterns take time to change. Good luck!

one for the situation. But as is often the case with obedience-trained dogs distracted by behavior problems, the motivation to sit and stay more or less disappeared in the face of tempting ankles and feet.

"I think Chica has figured out that when she nips your friend, she gets some extra attention from you, am I right?" I asked Mrs. Burgos with a smile. She put her hand to her cheek.

"Oh, Dr. Wright, you have no idea!" she said, shaking her head. "I am so embarrassed, I must scold this dog and pick her up and hold her until the people leave!"

Just as I suspected, the dog was loving every minute of it.

I gave Mrs. Burgos some techniques for keeping Chica's attention away from departing guests and on something more desirable. A rubber mouse, a ball with treats inside, and an old sock were Chica's favorite toys. When one of these was thrown across her field of vision just as a departing guest arose to go, it took only a couple of weeks to do wonders for everyone involved.

As a backup, Mrs. Burgos, like the Blakelys, invested in an ultrasonic device—a whistle—which she could use to give a sharp warning signal to the dog if it looked as though she was ready to misbehave. Chica was never again permitted on the tops of furniture. Luring her down with treats at first and then firm commands ended that worrisome habit and preserved her fragile legs in the bargain.

Although the little pooch still craved lots of attention, she soon tolerated guests without assaulting them, her owner's social life improved, and everything calmed down very nicely. So even though we had a mismatch between breed and owner, again we could institute some techniques that allowed the pet to stay in the family.

So I was surprised to receive another call a few weeks after everything was under control. I had labored under the misconception that Mrs. Burgos understood that her dog was a difficult one for her to have chosen, in spite of its seemingly convenient size. She had enough confirmation of that right on her own ankle! So I was surprised, and I might say dismayed, by what she had to tell me now.

"Dr. Wright," she crowed. "Guess what? I've got another dog! So now Chica can pay attention to her *and* to me, and not get into so much trouble!"

I was floored. I wished she had consulted me. Well, I sure hoped it was a golden or a Lab. "Is it a nice dog?" I asked, burning with curiosity.

"Oh, yes, she's very nice, but they are both under my feet now," she laughed nervously, "But so far, so good! The new puppy, she's a little Yorkie. I named her Tequila."

Oh, boy. I gave her the advice she was looking for—how to keep the dogs out of each other's food bowls—and told her farewell. Not goodbye: I was pretty sure that I would be hearing from her again in the not-too-distant future.

Check Out Your Pet's Short-Term Coping Strategies

Moving, divorcing, sending kids off to college, mourning the death of a spouse, growing older, and other significant life changes are certainly stressful to pets. But just as with people, there is diversity in your pet's ability to cope: Some will do better than others. You can get an idea about your cat's or dog's strategy for dealing with different stressful situations by paying attention to what she does in response to short-term changes in her routine.

Cats and dogs reveal through their behaviors what strategies they are likely to use to reduce the anxiety or distress of the moment. Sometimes owners report that their cat's or dog's personality seems to have changed now that someone is absent from their daily lives. This response is not unusual because many dogs and cats not only form different relationships with each individual in the family but also often take on different roles when one person is present and the other absent.

For example, when I leave town, our standard poodle, Roo-Roo, takes on the responsibility for letting my wife, Angie, know when there is someone or something outside. Roo spends more time during the day and most of the night watching the driveway from the front foyer. She barks when she sees or smells or hears something, not that Angie wants Roo to keep her awake at night barking (and, yes, we're convinced that she sometimes just barks to hear herself talk). Angie describes Roo's penchant for behaving like this as "Roo-Roo is that 'other dog' again" when I call from out of town.

Roo-Roo typically spends her night either on our bed or around the corner from our bed and away from the door on the bathroom floor, where it's cool but where there is no possibility of protecting us from the night. So there are some aspects of Roo that only "come out" when Angie's alone—and one could use this example to suggest that she has a mutiple-personality disorder, that she becomes a "different dog" when I'm not there. Another possibility is that she always has "had it in her," but she just suppresses that behavior, or gives up the responsibility for that behavior, when I'm present.

I should tell you that Roo-Roo does the same thing when Angie travels and

I'm home alone. Thus, there is an alternative explanation for Roo's apparent personality change. Roo-Roo seems to have learned to look out for whichever one of us isn't home yet by staying at the front window, and because she can explore more of the world from the front door and window than she can from the bedroom, she barks more.

Regardless of the actual explanation for Roo's behavior change, the point here is that "different" dog and cat behaviors are in part a function of who is in the household at the time. And further, when significant life changes present themselves, take care to observe what effects those changes have on your pet, what kinds of behaviors she shows, and how often she displays those behaviors because of someone's absence.

I see cats and dogs who have been unable to handle a short-term, yet significant, change in their lives (like vacations without them, relatives visiting, or even cleaning up after a storm or construction to the kitty-condo room after a hurricane). As a result, these pets exhibit behaviors like obsessive-compulsive disorder, which—you will recall—involves doing something over and over and over again, like licking a paw until it's raw or chewing on crate wire until her gums bleed. But most cats and dogs discover that they can use strategies or ways to "wait it out" until all the nonsense that's disrupting their lives is over. It's almost like they're willing to switch gears until things return to normal.

🐾 Sleeping and Other Strategies

For some cats and dogs, switching gears is less of a change in what they do, and more of a change in how long they use a coping behavior like sleeping. Although prolonged sleeping is one sign of depression, cats who sleep longer during the day in the presence of significant disruption are not necessarily depressed. Longer sleep periods can result from prolonged exercise, lack of sufficient food, lack of sleep the night before, and other causes, including what seems to be the cat's attempt not to face the stressful situation, rather than to overcome the obstacle that is causing the stress. Thus, cats and dogs may sleep longer during stressful periods—say 16 hours rather than 12 in a 24-hour day. Let sleeping dogs and cats lie.

If you recognize that your dog or cat sleeps too much when visitors stay over and so on, then they'll be likely to use that strategy when you move or your family changes in a significant way. Other strategies designed to avoid stressful life changes include spending more time in dark locations, like under

(continued on page 382)

FAQ: When a Pet Grieves

If I ever doubt that dogs and cats mourn the absence of a loved one, be it human or otherwise, I think of the first real pet depression case I ever saw, many years ago in South Carolina. A little beagle named Snoopy saw his mother get hit by a bus and die. He became very depressed, and about 2 weeks later the family called me for help.

What does a dog do when his mother dies in front of his eyes?
I think, first of all, that Snoopy had to have been traumatized by this, but his reaction could be shared by any dog or cat who finds himself suddenly without a close companion. After the accident, he went up and sniffed his mother, and when he went home, he sniffed around the house. Then he kind of moped around and lay down in the corner. He went on to lose his appetite, his housetraining, and any interest in going outside.

Did the family go out and get a new puppy to cheer Snoopy up?
No, they (correctly) didn't think another dog would help at that point, especially since Snoopy didn't seem in any condition to put up with a frisky new puppy. Many people assume that another dog is a good remedy. But for dogs who experience serious loss—emotional loss—the cause of the problem is loss of attachment to a specific other, not any old other! Consider the replacement to be similar to getting another husband to "replace" one who dies . . . hmmmmm. Doesn't really work, does it?

What was the family doing wrong?
The family meant well. They were terribly worried and also sad about the death of Snoopy's mother. They tried to help him. First, I asked them to stop going to Snoopy to offer him petting, food, and other bio-psycho-social needs. This behavior was reinforcing his inactivity, even though, of course, it wasn't reinforcing the depression, since you can't reinforce an emotion.

What do you mean by that?
I can't increase your depression by saying "Good girl" every time you say you're depressed. I can't increase your anger or any emotion unintentionally by praising

it. But the family could certainly reinforce or reward the inactivity by doing things for Snoopy or not praising him for any effort.

What's wrong with petting and extra TLC?
There is nothing wrong with it, of course! The extra petting helps to build the pup's relationship with the family and decreases the void left by the dependence on the dead mom. It doesn't "magically" reduce depression.

Well, what did you tell them to do?
I asked them to watch their pup closely. Even if the dog only picked up his head, they were to say, "Good Snoopy!" The idea was to change the emotion-laden behavior so that he would want to resume his activities. The family had to make him do the biological things first, such as going outside to pee and poop, essentially starting housetraining all over again, with lots of praise. They had to lead him to the food bowl and give him some really tempting foods. They called Snoopy to them for treats and petting so that he would play an active role rather than receive everything passively.

Did it work?
Yes! From week to week, Snoopy ate a little more, resumed his housetraining, and even started to play again. The outcome was really good.

Wouldn't a dog get over his grief eventually, anyway?
Some might, but the veterinarian, the owners, and I agreed that, without intervention, Snoopy probably would have died. He was getting too weak to eat and seemed not to care about living—what we call failure to thrive. It was one of those situations where the dog looked like he might have died of a broken heart. Had we known about the effects of Valium as an appetite stimulant, we might have tried that, but this was at a time (1982) when behavioral medicine was still in its infancy. Today, we would use a combination of medication and behavioral solutions to help a mourning dog like Snoopy.

the couch or bed or in the closet; bolting out the door, essentially escaping the problem or the source of the problem (Granddad's need to pick me up and wrestle with me); or waiting in one location until things return to normal, like spending all day on the bed, or continuing to look out the front window from the foyer 24-7 until Dad comes home.

🐾 How Dare You Change Things?!

Some dogs and cats seem to respond to change by trying everything in their power to return things to the way they were. These cats and dogs have a tendency to act out, to vocalize, and to become irritable or even aggressive. "How dare you not feed me at 6:00 P.M.? If Dad were here, I'd be fed by now. Can't you hear me clicking my nails on the kitchen floor, jumping up and down? I've been doing this act for at least 10 minutes now. What's it gonna take? Try this!!" (And the dog growls and snaps, or the cat latches onto your thigh.) And the beat goes on.

Unless you change and redirect this type of pet's activity at once, you're likely to wind up in a confrontational relationship with her where none existed before. Remember that she's reacting like this because some need isn't being met. She expects something from you that she is not getting. She's frustrated. And some dogs and cats respond to frustration not by giving up, not by denying that this is happening to them, not by avoiding the source of the change by hiding in the closet, but by becoming emotionally aroused and even aggressive. Review the ways to change emotional behavior that I discussed in chapter 5—it may come in handy here.

🐾 The Mourning Vigil

Sometimes pets show great savvy in realizing that Dad isn't coming home. A cat or dog no longer walks into his bedroom to stay by his bedside or to lie with him following his departure. Or the dog or cat becomes dependent on the remaining person in the family, much as when Mom or Dad was gone for short trips in the past. Now the remaining person has taken up the slack for the other's absence. I don't know what enables some cats and dogs to "know," and others to continue to wait for weeks or even months for things to return to normal.

Are we likely to attribute such human characteristics to the pet that he still misses his human dad, he's still mourning his loss, and it appears as though he'll wait there for him for the rest of his life? I suspect it's easiest to assume

DR. WRIGHT'S INSIGHTS

One of the changes that some of us will need to consider is that our pet will outlive not only our other pets but us, too. Rather than making this inevitability add to our daily anxiety, fears, and uncertainties about the future, why not deal with it now? Making plans for your dog or cat when you are gone (who will take her, what arrangements will have to be made, when should she go, and where will she live?) will enable you to rest assured that she will not be mistreated, abandoned, or (horrors!) euthanized after you're gone.

that Fergus and Fluffy merely begin a short-term coping strategy that becomes a habit, and unless the remaining person in the family disrupts the pattern and redirects the dog or cat into more healthy kinds of activity, it just continues.

Most pets are likely to try these short-term strategies for coping with loss and change first. What happens next and when it happens vary greatly in pets. Whether they are mourning the loss of a human, canine, or feline companion, some dogs and cats give up, become depressed, and fail to thrive. This was the story of Snoopy, the South Carolina beagle.

Prolonged stress that a dog or cat does not deal with effectively can have really devastating psychological, behavioral, and physiological consequences. Of course, some traumatic changes can't be predicted. But whenever possible, consider effective strategies for your pet's changes in life before things deteriorate this far. As with most inevitable major life events, spending a little time now planning ahead for such changes is preferable to spending more time later, when you're least prepared to deal with it.

As you can see from Snoopy's case, the remaining person or people in the family have to take the lead in providing new options for their pet's daily routine. Sometimes the daily routine will be quite similar to the one your pet had with all family members (or pets) present, but sometimes it will be a different routine altogether.

The Hazards of Mobility

Moving is a fact of life for people—and especially empty-nesters. When I first started doing house calls in the early 1980s, about 30 percent of my clients had recently moved, and there was something about moving that contributed to their dog's or cat's behaving badly. Now I have an additional 10 percent calling in preparation for moving. They know that moving is going to present a potential problem for their dog or cat, and, as in human medicine, they want to design and take part in a behavior program that will prevent the occurrence of behavioral problems down the road.

In no other single act is there likely to be as big a challenge to a dog's or cat's ability to deal with change than in moving. The three components of stress—novelty, unpredictability, and uncontrollability—are all present to some extent in moving. And the dog or cat has to take along the baggage of figuring out how to live with a baby in the old household or without Dad or Mom all of a sudden. Changes in their routine must be made yet again.

Some dogs and cats get better at it, as in, "I figured out how to adapt successfully to the baby in the family. Now I get a chance to adapt to another new routine in this neat place that has three sets of stairs and two places my family comes and leaves from, not just one like in our apartment, and a lot of new odors to figure out, both indoors and out. I'm excited!"

On the other hand, some cats and dogs loathe changing what they have to do yet again: "Too much stimulation! Too much uncertainty! Where's the closet? And where's my other owner?"

❧ New Home, Old Odor

New homes are riskier than the old homes, especially for cats, who may be seen hovering over several carpet locations and displaying the flehmen response (opening his mouth, pulling back his lips, curling his upper lip so his canine teeth are visible, and licking the roof of his mouth to sample the odor he's "tasted" from the carpet). He's probably taking in some kind of deposit from a cat or other critter who recently left the home. It's a pretty good sign that in the near future, your cat will deposit his own pheromones on the carpet, either from just being too aroused or from marking his new territory. It could also be some combination of stress and trying to do what's right—trying to

DR. WRIGHT'S INSIGHTS

Dogs and especially cats tend to bond to the location they are used to. They tend to organize their lives around not only people and the family's activities but also the odors, the squirrel who lives up in the tree, the dog next door, and the pattern of the sun on the carpet. So you have to be a little patient; your pet needs to get used to the new place gradually, just as he would a new person in the family.

figure out which urine smell is supposed to signal the appropriate location of his new litterbox. (Wait about a week, and then get rid of the old wall-to-wall carpeting. Trust me, it's worth the additional expense.)

In general, for both dogs and cats, establishing a routine right off the bat is the best solution for preventing a pet's breakdown. The more familiar the dog or cat is with what to expect and in what order from day to day, the less likely he is to freak out. Calling to your dog or cat in the morning to establish yourself as the leader of all good things to follow in the day is a good way to start. Treating your dog or cat as though he was just introduced into your family as a kitten or puppy will probably reduce the number of misbehaviors in your new household. (See chapter 6 for more on introductions.)

You can take your first big step toward preventing chaos by shaping your dog's or cat's behavior throughout the day for the first few days, just as you did in preparation for a new baby, and before that when you first brought your pet home. Planning ahead by deciding who in the family will do what with Fergus or Fluffy, when during the day (also, how many times), and where in the house or outside, will keep you and your companion animal on the right track in those important first few days. The avoidance of a traumatic experience now, and the development of a new pattern—a daily routine that meets your needs and your pet's needs—will keep you from seeing people like me in the weeks to come.

🐾 Preparations for Moving

Many people prefer to board their dog or cat the day before moving actually takes place and bring them into the new household the next day, or as soon after the move as possible, so that they don't spend too much time at kennel. This option works particularly well for local moves, where driving to pick up or drop off the dog or cat doesn't take up most of a day. One factor to consider in deciding whether and how long to leave your dog or cat at the kennel is your pet's ability to handle the predeparture and arrival aspects of your move comfortably. It depends a lot on her and what you think that she can handle. So let's look at her characteristics and how they'll affect your move.

Your Pet's Characteristics
Among the important pet characteristics to consider are personality, size, age, and reaction to strangers. Is she likely to react badly to seeing boxes packed with her stuff—biting the packers or fleeing before she gets packed or

stepped on by people who are helping with the move? Will she bolt out the opened front door in an attempt to escape from all the commotion, never to be seen again?

As far as personality traits are concerned, calm, laid-back dogs and cats may be less at risk for getting into trouble than excitable, reactive pets. If you don't have time to take care of and console your pet during the move, happy animals will be better off than overly aggressive or fearful animals or pets who run to you as a source of comfort during times of stress and uncertainty.

Size is also a factor to consider. All cats and small dogs can get underfoot and be hurt during all the commotion, and very large dogs can hurt you, both directly and indirectly! How will the laid-back large pooch be affected? Consider the dog who decides to lie in doorways and passages during the move—he can create a liability for yourself and others. Have you ever considered that you or a friend of yours might trip over your sleepy bloodhound while carrying a box full of your priceless porcelain items to the truck? And once your brother's stumbled over her, will she lick or bite the guy?

Age can be a factor to consider in how long to leave your pet at the kennel. Adult animals who are more secure and confident of their place in the family probably need less time in the kennel than others. Puppies and kittens are likely to make pests of themselves, believing that all those boxes surely need to be jumped in or hidden in, and that all the wrapping paper for the breakables was probably meant to be scattered all over the place. And boy, now that you've moved that leather chair from the corner of the living room, it smells like a great place to pee! Older dogs and cats can also be a problem, mostly for themselves. Geriatric pets may experience anxiety but not show it. They may try to stay out of the way or play it safe and just not move. Invariably they wind up getting hurt by movers or others who don't understand their typical lethargic tendencies and inability to move quickly when someone or something is bearing down on them.

Unless you plan to do the moving all by yourself, you need to consider not only your dog's or cat's reaction to strangers who may have volunteered or been hired to help you move but also the remote possibility that someone could actually not like your dog or cat, and not really care if he happens to get in the way. It is better to be safe than sorry.

For some pets, the longer they stay at the kennel, the better. They may be as upset with the unpacking as they are with packing, and there will be open doors in the new house to escape from. It's best to wait until most of your items have been unpacked and situated in the new house before bringing a pet like this home. Then, revisit chapter 6 on how to introduce your new pet to your home.

Remember that, although neither you nor the furniture is new to your pet, everything else about the new house (which, of course, she can experience through all of her senses) is different, and she needs to become familiar with the newness of it all. Play this one by ear. Do not force her to become acquainted with all the rooms at one time. She needs time to explore the different floor textures in the house (some will be more slippery than others), the neighbors (some will be scarier than others), the stair steps (especially if she hasn't encountered stairs yet), and other new parts of your house that you can master much better than she can.

You know that you're going to be there permanently, but she doesn't, and she may be looking for a way to leave. Work at a speed she's comfortable with (revisit chapter 4 to become familiar with signs of discomfort and fear). And although it's okay to coax her with a friendly, upbeat voice—or even a treat— to explore throughout the house, it's not a good idea to pick her up and carry her all over the place. If she struggles to get loose, you risk the possibility of injury and trauma (to her and yourself), which she'll now connect with some part of the new house or even with you.

Your Own Characteristics

Finally, consider your own personality characteristics and temperament in deciding what you would prefer to do with your pet during the move. It's okay to consider your own nerves sometimes, and not to place your concern for Fergus or Fluffy first, the way you frequently do. Remember that, even if she doesn't have a great experience at the kennel, dogs and cats are resilient, and what they learn at the kennel is often left there. After you have your companion animal back with you, in your new residence, you'll feel better about being able to devote some quality time to her. You can help her become acquainted with your new household and take pleasure in doing it together.

🐾 Options for Moving Day

Kenneling is not always an option for people who move, for a variety of reasons. Sometimes a dog or cat has had an unpleasant experience at a facility, and you just don't want to risk his having another. Or you forgot to get him a shot for kennel cough, or his rabies or other shots aren't up-to-date and you can't find a kennel willing to board him. Or you're moving on a holiday, and because you've always gotten a pet sitter to stay in your home during previous absences, you had no idea that you need to make reservations 6 weeks in advance. All the kennels are full, you're moving from New York City to Atlanta, and you must take your dog or cat with you. Depending on your circumstances (and whether he's up-to-date on his shots), you may wish to consider one of the following options:

1. Have him visit a friend (preferably yours, unless his friend's owner gives you permission!).

2. Have her visit your neighbor. When you're ready to drive off, you can just walk next door and get her.

3. Have him visit his breeder. Many breeders are thrilled to see an old friend, and vice versa.

4. Have her visit the pet sitter. Some pet sitters care for pets in their own homes.

5. Have him visit a daycare facility or day camp. (Be sure that he qualifies, and call ahead.)

DR. WRIGHT'S INSIGHTS

Find a new veterinarian before you need one!

It's a good idea to make sure that your dog or cat has had his shots well before moving because you'll have more options open to you for boarding him, and you're probably going to have to find a good veterinarian to use in your new location. You will want to take some time researching the possibilities before or as soon as you arrive. If you should need to see a veterinarian on arrival at your destination, it's better to avoid making your selection a crap shoot.

You may wish to ask your present veterinarian to recommend a DVM or VMD in your new location. You can also check your options with friends living in the area to which you're moving. You could consult the American Veterinary Medical Association (AVMA) to identify vets who are Board Certified in a specialty area, should your pet require one, or consult the American Hospital Association (AHA) for a list of which veterinary hospitals meet the stringent criteria for membership as an AHA animal hospital. Planning ahead will help you avoid a crisis later on.

🐾 What about the Trip?

Most dogs travel well in a car if they are used to riding with you to places that are fun. Angie and I frequently take our dogs for a romp to different locations where we have friends or family. When we say, "Who wants to go?" we all rumble down the stairs that lead to the garage and everyone jumps in the SUV to go on the next adventure. With four dogs, there is not much else we can take in the SUV, but we knew that when we made our last move, and things worked out well.

What we hadn't thought much about was that we would also have to take our cat, Domino, and she had never made a trip in the car with all four dogs before. Luckily, we didn't have to explore that option. Domino rode to the new house with Angie, in her car. Do think ahead!

Talk to your veterinarian about giving your dog or cat something to make the trip easier, if you're not certain how your pet will react in the car on a long trip, or if you must drive your dog or cat to the airport and crate her for the plane ride to your destination. Although sedating a pet may sound like a good idea, it may keep a crated pet from being able to control himself while being jostled about, and he may actually experience more anxiety as a result. Selecting an anti-anxiety drug may be a better choice. Be sure to discuss the possibility with your veterinarian well before the trip so that you can cover all the

Soft, luggagelike travel bags are an alternative to rigid carriers. The soft carrier's material is easier on your pet's bones, which makes for a more comfortable ride, and it's easier to store after you've reached your destination.

bases and decide what will be best for your dog or cat. Most vets recommend crating a dog or cat for long trips, for safety's sake, and the use of drugs is becoming much less frequent.

Finally, should you need to crate your dog and fly him, be sure that you know what the rules and regulations are for crating with the airport and carrier you select. Some airlines permit small dogs and cats to travel on board with you. Of course, before you leave, you should let your dog or cat get acquainted with the crate or "travel bag" they'll make the journey in, regardless of the mode of transportation. You don't want to get off to a bad start with an animal who decides to go nuts when you need him to cooperate most. (See chapter 12 for how to acclimate your pet to a carrier or crate.)

A 12-Step Program for Curing the New-House Blues

Here's the program I recommend to dog owners when they're planning a move. It will really help to relieve your pet's anxiety and make every aspect of the move as smooth and painless as possible. Try it!

1. If possible, take your dog to the new home a few times before you actually move so that she becomes familiar with the lay of the land. Make each trip fun. Perimeter-train her by walking around the borders of the yard together each time you go.

2. If no one is living in the home, walk her through the house, room by room, with you by her side. Make it fun—talk to her and pretend you're having an adventure together.

3. No matter whether your move-in day is only a few days after you close on the house or you must wait until the previous owners have moved out completely and the house has been cleaned, it's still a good idea to acquaint your dog with the new house and yard. Be sure to enter you new home (territory) together, spend a bit of time together, and leave the house together several times. She'll soon get used to the whole endeavor and it won't seem so scary once you move in for good. Just imagine what a good job she thinks you're doing, both of you checking out a new den site, before moving the whole family!

4. Meet your neighbors, if possible, ahead of time—or at least find out if they have dogs. It's always a plus to know if the guy living two doors down has three pit bulls and a Rottweiler-Chow cross (all intact males) before you venture past his home with your sweet little sheltie. If the neighbor on your left keeps his dog outside and a fence separates your property, recall that the

best way to help dogs meet each other is to do it "off territory" first. At the very least, think about how you'd like it to go so that Gretel doesn't experience a trauma in her new home right off the bat.

5. After you've moved in, repeat Steps 1 and 2 so that she smells the new digs along with the "old" furniture. If you've moved a great distance, settle in after the move and begin the introduction to the new house in the morning, when everyone is more fresh and ready to go. This should help your dog associate the perimeter and confines of the new residence with "her stuff."

6. From the beginning, arrange the motivationally important things in their "proper" locations in the house, and walk with your dog to those places so that she knows where to find her food, water, and bed or rest area. Place her things (toys, bear, chews, Kong) around her bed so that she gets the idea. When she settles down, sit with her there for a minute, stroking her so that she know it's the right place to relax: It's her place, and you approve. If you'd prefer to allow your dog to select her place in the house, so be it. But after she has established her preference, it will be difficult to change. So if she looks like she's about to plop down at the base of the stairs and take a snooze, make sure that you encourage her to move to the side.

7. When you want to teach her where to go to eliminate, walk with her toward the kitchen door to the backyard (or whichever exit you'll want to use), say, "Outside?" (or your own cue word), and take her to the location where you'll want her to relieve herself. Stand there for a few minutes, not looking at her (eye contact is provocative to a dog, and arousal tightens sphincter muscles, making it more difficult to eliminate), until she has a chance to sniff (preferably in the groundcover, off the grass) and mark the area. You may wish to place an old stool there so that she'll know more readily where she's supposed to go. If you're in the city, take her out the door and in a direction you'll want to use for eliminating as a regular routine. Praise compliance, and before taking her back inside, walk her or play with her for a minute or two. If she's still not sure of herself and prefers to head back indoors immediately, allow her to do that—don't force her to "enjoy" the yard or the city sidewalk right away.

8. Begin walking her to the car for rides and including her in your daily activities as much as possible. The idea is to let her know that you're not going to abandon her and to acquaint her with the physical aspects of her new home.

9. Increase the distance of your walks only to the areas you want her to equate with her "home range"—the total area she'll traverse on walks while

DR. WRIGHT'S INSIGHTS

Older cats or cats in a new household sometimes become disoriented and just stand in the dark and vocalize . . . "Yo-o-o-o-ow." It's loud enough to awaken the owners, and they seem to be as confused about what's going on as the cat is. At these times, involving your pet in something he knows how to do and is familiar with is often helpful. Give him a treat in his bowl, take him to bed with you, or restrict his access to the location where he was vocalizing. These steps should help calm and comfort your bewildered yowler.

living there. Whereas the perimeter training is to acquaint her with her yard (or territory), the walks outside her territory allow her to explore and become acquainted with who else is in the neighborhood—kids, dogs, cats. First-time meetings with neighbors and their dogs should be planned, fun, social events, where everyone leaves happy. The leadership and control you exert over the quality of these first experiences will spill over into other potentially scary new situations. As a result, your dog will feel increased trust in you and confidence in herself—she'll believe that she knows the "right thing" to do.

10. After she's familiar with the different rooms and levels of the new home, practice helping her relax in her areas (places like her bed) when you're not there. (Did you have to teach her how to climb up and down steps? It's easiest if you start at the last step down, then the second step down, and so on, until she can come down all the steps at once. Use the same approach for learning how to go up the steps.) Practice giving her Sits or Sit-Stays or just say, "Uh, uh!" if she needs help settling down some distance away from you so that she doesn't become too dependent on you. When she's settled in, or relaxed enough to lie down (try ignoring her, especially any unnecessary pleas for attention), walk to another room and return before she has a chance to get up and follow you. Or give her a Sit-Stay, and do the same quick-leave and return, followed by praise for not following you there. The idea is for her to be able to relax in the new house, with you out of sight (for longer and longer periods of time). You don't want her to take on the responsibility of keeping track of your whereabouts at all times.

11. Begin practicing graduated departures, as in the treatment program for separation anxiety. (See chapter 14 for how to reduce separation anxiety.) If she appears to be getting too anxious about being left alone, consider options

for where to leave her during the day (see chapter 14) until you can desensitize her to being left alone in her new home.

12. Pay attention to her needs, and try to anticipate ways to fulfill her needs in your new home. Plan locations and times for playing with toys and for social play, hugs and licks, walks or other exercise, and other nonbiological, yet important, psychosocial needs. Soon, you'll be into your new patterns and rituals, functioning successfully again as one happy family.

Moving As a New Lease on Life

If you are moving because of a downsizing of your family, treat the new household setting as an opportunity to restructure your pet's relationship with you. You can enjoy new activities with him, or the same activities at different times and more or less often. In this way, you can constructively wean your pet from his old routine into a new living arrangement with you as his primary, perhaps only, attachment figure. Keep track of and use the kinds of incentives and environmental enrichments he has valued in the past to reshape his daily routine within the confines of what you're able and willing to provide for. Just do the best you can. Most cats and dogs are resilient and will respond to you if you communicate what you want them to do. You'll give your pet and the change in lifestyle your sincerest and best effort. And, if you need a break now and then, consider the options for time off in chapter 14.

Whether you are downsizing or moving with the family intact, you can consider the move a chance to change any problematical behaviors your pet may have gotten in the habit of enjoying at the old place. "Empty nesters" Becky and Richard Gardner used their move of just two miles to accomplish a difficult task: successfully turning their mostly "outdoor" cat into a safer "indoor" pet.

Opal was an 8-year-old female who liked to prowl around outside from 4:30 in the morning until dinnertime. In the summer, she reversed her schedule, staying outside from about 8:00 in the evening until daybreak.

The Gardners were concerned that, with such a short move, the cat would attempt to find her way back to the home she had lived in all her life. That home would now be across the Chattahoochee River and a four-lane highway! I convinced them that even with a successful move, the cat would be far more likely to live a long, healthy life if they could curtail her time outside. They agreed that it was worth a try.

At the old house, the cat was used to eating and lounging around a large, low, wraparound deck and then sneaking off to pursue rabbits or mice and visit some neighbors and a cat friend as she pleased. When I accompanied the Gardners to their new home, I saw that it too had a deck—but this one was on the second floor.

The new setup was perfect. Although Opal was used to scampering off a deck, I was fairly certain that she would not be foolish enough to launch herself into the air for a 15-foot drop to the yard. But just in case, I asked Becky and Richard to buy a cat harness for the initial exposure to the new deck, which would become her only "outdoor" territory from now on. I also had them install a cat door so that Opal could come and go as she pleased after they were sure that she wouldn't try to escape. By the day before moving day, we had our plan ready to go.

❧ The "Outdoor-Indoor Switcheroo" Moving Plan

Here's what I had the Gardners do with their cat, Opal. If you're trying to accustom an outdoor or indoor-outdoor cat to life in the great indoors, you can follow the same program.

1. Keep Opal in her normal routine until moving day because she has never been boarded. She can be outdoors; the packing and moving action will be indoors.

2. Set up her room in the new house so that she has all the biologically important things in their place: food and water in a different part of the room from the litterbox; bed away from the food and water and the litterbox.

3. On the day before moving, if she behaves as if she's upset with all the changes going on in the house, consider giving her something to calm her nerves for that evening and then take her to the new house in the morning before the movers arrive. This strategy would remove her from the anxiety-provoking situation as soon as possible, placing her in a less stressful environment in the new house, under the influence of an anti-anxiety medication. The disadvantage of this option is that there will still be the commotion of the movers moving in, her daily routine will be disrupted, and we risk her experiencing trauma in the new house as her first impression. The use of drugs does not eliminate stress—drugs are a tool but not a solution in this case. (Fortunately, Opal proceeded to jump into bed with the Gardners in their bedroom as usual. The status quo was calming for all concerned. It's not unusual for people to take apart their bed the next morning

to avoid a motel stay, and it worked in our favor this time, for Opal's sake.)

4. On the morning of moving day, let Opal outside after Richard showers and dresses, as usual, and when the cat shows up at the neighbor's deck about half an hour later, ask them to call you so you can proceed with the move.

5. After the moving van completes the move, return to the old house as if nothing had happened, about 6 P.M. Drive in the driveway as if returning from work, open the garage doors with the car opener, and put the food bowl out on the back deck for Opal, but do not feed her. When she meows for her dinner, let her in the house.

6. Pick up Opal, put her in the cat carrier she uses to travel in the car, and take her to the new house.

7. At the new home, introduce Opal to her new room. Sit passively on the floor to provide the cat with both people and objects from the old house that remind her of the safe, familiar aspects of home. Offer Opal her dinner and some treats, so that she can associate being in the new location with good things.

8. Decide whether the anti-anxiety drug is called for. If Opal isn't freaking out, forget it. If you wish, sleep with Opal in "her" room that first night. Your fabulous master suite can wait!

9. The next day, set up a routine that involves play, allowing the cat to explore one section of the house that can be closed off from the other parts by shutting the hallway door. This will decrease her need to immediately reestablish her preference for the outdoors.

10. During the next week, gradually introduce Opal to the use of the harness, so that you can take her out on the deck to begin to explore her new "outdoor" area.

The transition to the new house was successful, although it took several weeks to get Opal used to spending longer periods of time in the house. We learned that, in this battle of wills, it was more prudent to put a kitty door in the door to the deck than to try to change completely the habits of an 8-year-old cat who didn't want social contact in the morning or at other times when she was "outside." She must have decided years ago that this was her time to explore and sneak about, and the owners only interfered with that time. So we compromised by allowing Opal access to the deck for short visits in the morning and again in the afternoon. The owners locked the kitty door at other times and ignored Opal's vocalizations and attempts to get out there in the next 2 weeks.

(continued on page 398)

DR. WRIGHT'S
C A S E B O O K

Duchess and Scully

Cheryl West and her two 7-year-old Boston terriers, Duchess and Scully, had recently moved from Tennessee to Atlanta. Newly single, Cheryl was glad to find a house with a security system; she had never lived alone before.

Duchess had always done strange little things, unlike her littermate sister, Scully, who was just about the perfect pet. Duchess liked to eat Scully's food, she urinated in her own food bowl if it was empty, and she really took to chewing, which her owner attributed to her generally nervous personality. As a result, Duchess was crated during the day, while her good sister had the run of the house. Of course, Scully chose to lie next to Duchess in her crate most of the day, as any good dog would. As thanks, Duchess would promptly attack Scully when let out of her crate!

This was not a problem for Cheryl. But the first time she left the new house and activated the security system, which had a motion detector, she forgot that one dog moved about during the day. The security system worked, all right: There was a security team at her front door to prove it! After the embarrassment and the apologies, Cheryl went to see the dogs and to release Duchess from her crate. Apparently Scully was able to mask the noise of the alarm by crawling in the closet, and she was okay. But poor Duchess was shaking from her black nose to the white tip of her tail, and she was soaking wet from panting, being unable to escape from her crate and the loud, blaring horn that would have sent any potential burglar back out the window.

So accidentally, disaster raised its ugly head. We had a severe case of not only separation anxiety (Duchess would no longer tolerate being left alone in the house) but also what can best be described as "home-a-phobia." Now that Cheryl had moved into a nice, safe suburban home, her dog hated it and preferred the outdoors to being anywhere near her crate, even in the hallway leading to her bedroom, where the crate was kept. And for a Boston terrier, it's quite unusual to prefer the out-of-doors to the comforts of home.

Well, we successfully treated the dog for separation anxiety in the presence of an anti-anxiety drug, and Cheryl also had to change security systems. Poor Duchess became conditioned to the sound of the "beeps" used to set the alarm system, probably because they were novel and distinctive sounds that she had only recently become acquainted with in her new surroundings. (Recall from chapter 3 that dogs learn to do things most easily if the words used or sounds heard are new and easily distinguished from other sounds.) Thus, poor Duchess learned that when she heard the beep, beep, beep, beep, of the keys Cheryl used in setting the code for turning on the security alarm prior to leaving, it was time to set off a full-blown panic attack.

Who would have thought of that?! Sometimes you just get lucky when you move, and things work out. In Cheryl's case, it just wasn't going to be so. Fortunately, the setting for her new security system was different from the beep, beep of the old one. It sounded more like a ding, ding, ding, and she practiced tossing Duchess a treat each time she set the alarm system. Cheryl did this several times a day, but didn't leave. You get the idea. Duchess seemed to be content lying on the second-floor bannistered hallway overlooking the large picture window where she could watch for Cheryl coming and leaving. The crate was a thing of the past. Comfort and being able to see out seemed to help immeasurably, as did the antianxiety medication I asked the vet to prescribe. Eventually the medication was reduced to a maintenance dose.

Duchess still continued to pee in her empty food bowl and occasionally to take out her frustrations on innocent Scully. But, for Cheryl, that meant things were back to normal, and with Duchess now uncrated, their new life as three single "girls" was off to a much less alarming start!

By 6 weeks after the move, Opal had a new routine firmly established. She made no attempts to find ways off the deck, and the Gardners considered the transition to have gone well. Retired from hunting, Opal had now become much more of a lap-sitter and a more playful companion, and Becky and Richard looked forward to many more happy years on the deck of their new home with their bird-watching feline friend.

If Your Pet Gets Lost

Another client of mine wasn't so lucky. His dog, an adolescent mixed-breed hound named Jasper, was introduced to his new home by being hooked up to a run in the backyard, which adjoined a beautiful golf course and country club. Shortly after the move, the dog's collar was found on the ground, still attached to the long line.

It's funny how, when you don't adopt a pup from a humane society or shelter, it's easy to forget things like neutering and obedience training. (Jasper had been a stray.) And there's another funny thing about hound dogs: They have a tendency to take off helter-skelter after a rabbit without thinking or seeming to look up long enough to realize they've run through uncharted territory for 10 minutes or so, winding up somewhere in God's country. And oh, how exciting those new smells are to a reproductively intact boy dog who has just come into sexual maturity!

Well, two agonizing weeks later, the lucky family found their burr-covered, skinny puppy lying exhausted on the back step. The dog had no introduction to the boundaries of his turf. He hadn't been neutered, so there was ample motivation for him to rid himself of his shackles. The owner was very fortunate that his dog decided to find his way back home. Jasper went to the vet the next day for neutering, and I gave his dad a crash course in how to introduce his dog to his new turf properly. Instead of just tying Jasper outside, my client spent some time walking him around the grounds of the golf course so that Jasper could figure out how to get home from various spots, should he ever escape again.

A missing or lost dog or cat is a horrible situation for both owners and pets. Loose pets are unfortunately prime targets of unscrupulous "nappers" and are also at great risk of being hit by cars, taken in by abusive people, being attacked by another animal, or dying of starvation. Sometimes dogs continue going in the wrong direction and become irretrievable. But sometimes they run into a kindly dog lover who decides to do them a favor and keep them.

Animals are at high risk for becoming lost in the first days and weeks after a move. But if it does happen, the sooner you initiate a search, the more likely you are to be successful. Jasper's owner went door to door, put up signs, and called the local police. You can do more.

🐾 Best Bets for Retrieving a Lost Dog

Many of these steps are just as effective in finding lost cats. Don't give up until you've tried them!

1. Check daily with animal control, all the animal shelters, and the vet clinics in your county and in the adjoining county. If you've had your pet (or your pet's ear) "micro-chipped" for identification, the search is less subject to error at the department of animal control or the Humane Society, and chances are they'll call you before you have a chance to call them!

2. Take the other dog in your household for a walk in the direction you think the lost dog took, in the hopes that he'll pick up the other dog's scent, or the other dog will pick up your scents. Dogs' urine trails are deposited on vertical objects on walks, and other dogs use them to figure out who's been there, how long ago, and where they came from and went. That's why you see dogs with their noses to the ground, following other dogs' trails—hopefully back home!

3. Repeat Step 2 at different times of the day and evening. If the lost dog is searching, he may be able to pick up a fresh trail if you provide him with one more frequently than just once on the day he became lost.

4. Place an ad in the paper in the pets column. Sometimes people look there when they find a dog, but for some reason are less willing to call animal control.

5. Drive to places that you have taken your dog, like the veterinarian, a shopping center where you purchase groceries, and other locations your dog is likely to recognize by scent, sight, or hearing. One client whose dog frequently escaped from the yard realized that Rufus invariably wound up at the veterinarian's office some mile and a half down the road. The vets knew him and enjoyed his company until his owners came to pick him up.

6. Consult a pet-rescue service in your community. They have communication networks that help with recently lost dogs.

Chapter
Seventeen

Dr. Wright's Guide
to Relationship Building

In this chapter, I want to help you put some extra zing in your pet's life. With dogs and cats becoming more like "family" every day, owners are really getting creative about including their pets in human activities—or inventing games, sports, and pastimes just for dogs and cats. And everything you do together helps to create a really special relationship that goes way beyond feeding, walking, and just hanging out.

First, though, I want to make sure that you understand the basics so that you can go on to add some icing to that feline or canine cake. If you start these activities in puppy- and kittenhood, after the little ones get to know their routine and can obey some basic commands, you and your pet will have the foundation for a great relationship. Teaching your brand-new cat or dog the rules of the house is much easier than teaching "old dogs new tricks." But don't let your pup start out on the wrong foot: Too much confinement or a not-quite-right relationship with you can spoil the basics, let alone the extras. So you must address those things first.

Hate That Crate?

If your new pup is miserable in his crate, here are eight ways to give him more freedom without driving either of you nuts.

1. Use baby gates and confinement in the kitchen, a heated and cooled sunroom, or your favorite puppy play area.

2. Supply paper or litterbox for elimination. Eight-week-old pups will probably need to "go" every 2 hours or so because their bladders are still small. They'll especially need to eliminate after napping, eating (15 to 30 minutes after meals and snacks like dog biscuits), and exercise or play. House-soiling is more likely to be urination than defecation because puppies can hold poop longer, until you finally let them out. Be aware that some large breeds can take up to 6 months to finally learn not to soil in the house, and so can some toy breeds (less than 20 pounds), who may seem bent on surprising you with new inappropriate places to pee.

3. Furnish a bed or fluffy towel to nap on (and not eliminate on!), placed away from the papers on the floor or the litterbox.

4. Provide a location for toys, which have three different textures and shapes to keep your pup interested (rope, hard rubber Kong, and Nylabone, for example). It's fine to place toys on or around his bed.

5. Keep cabinets and closets closed, and things that could go bump in the night (glassware, china) away from the pup and inaccessible. Move stored chemicals from the bottom cupboards. In other words, "puppy-proof" his area. Throughout the house, put shoes and slippers away if you plan to wear them again!

6. Consider expanding the confinement area as the pup becomes an adolescent, but watch for destructive chewing.

7. Engage in a good play session before and after confinement, no matter how large the confinement area.

8. If you require more guidance, try the Suzanne Hetts videotape on crate training. (See Recommended Reading and Viewing on page 426.)

Dr. Wright's

✉ E-Mail

Dear Dr. Wright:

We just adopted a 6-year-old neutered male sheltie named Andy. Andy immediately became attached to our 7-year-old daughter and is okay with my wife. But he seems afraid of me—he will not walk past me and even runs away from me. This is a real problem because I am going to be the primary caregiver. How can we allay these fears?

Thanks,

Rick

Dear Rick:

Shelties sometimes have a problem with fear, especially if they're not yet comfortable in their new home. The fear may be revealed as a "personality trait," in which case the dog is fearful of all people, or it could be selective fear, in which anything "unfamiliar" is scary. This often resolves itself as the pet settles in. Neither of these types of fear seems to fit your situation, though, because Andy became attached to your child right away.

There's a third cause of fear, based on an unfortunate experience he's had with someone in his past. If something about you reminds Andy of that nasty experience, his fear is not likely to go away, unless you desensitize your sheltie to whatever "parts of you" remind him of the source of that painful event.

So now you've got to put on your detective hat. How many ways does Andy "experience you"? He uses all of his senses—he sees you, hears you, smells you, and so forth. Do you stand for a collection of separate but related parts that he classifies as "men"? Or maybe he's just afraid of your hat, your deep voice, your beard, or even your height!

Ask your wife to wear your baseball cap. Does the dog shy away from her or treat her any differently? Have her put on your t-shirt. Maybe it's your cologne! After you've identified what Andy is afraid of, try to eliminate it. Rather than trying to desensitize him to your sunglasses, it's better to just take them off around him.

Obviously, if he's afraid of "men," we have a problem. We can't eliminate you! So we'll have to try desensitization. Let's say that Andy cowers when you (1) look at him, (2) bend over to pet him, and (3) say, "Hey Andy!" in your nice deep voice.

If his idea of "scary men" consists of those three actions, stop doing them together at first, and then add small instances of each behavior until he can handle them all together. Try stopping about 10 feet from him, squat down to his level (decreasing your size), look off to his left (decreasing the arousing eye contact), and say in a gentle voice, "Good Andy." As he approaches to greet you, praise him. If he knows how to sit, you may choose to have him sit when you make your initial approach. Before long, he should be responding much better to you, just as he does to your wife and daughter.

Spice Up Your Pet's Life

Everything on the right track now? Then it's time to think about quality of life and aim for some enrichment for you and your pets. Organizing your dog's or cat's day to consist of predictable, controllable events is likely to make him less stressed out, but be sure not to bore him with the same old things day in and day out. Vary the kinds of things he does to keep him interested and challenged from day to day. Stimulate his curiosity and problem-solving skills by trying some of these "advanced" (but easy) games.

🐾 Fun and Games for Pets

For most pet owners, game-playing doesn't go much beyond "fetch," or a half-hearted toss of the catnip mouse now and then. But keeping your dog or cat happy and stimulated can be easy and fun—for both of you! Here are some of my favorite dog and cat games.

Interspecies Hide 'n' Seek

During his play periods, try a version of hide 'n' seek with a treat or yourself as the "thing" worth finding. My wife, Angie, started a game of hide 'n' seek years ago with her rescue dog, Peanut. And despite Peanut's hearing loss and poor eyesight, they continue to play (nothing like a good nose to help you find who's hiding!). The game usually starts when both come in from outside. When Peanut walks by, Angie scurries off into another room to hide on the other side of the bed, or in a closet, or in the shower. She started off with easy hiding places, and made it more interesting for Pea by choosing progressively more difficult spots when our terrier got the hang of it. When Pea finds Angie, there's nothing but smiles all around, and then a familiar bout of hugs and kisses.

Where's the Cookie?

For some dogs, a dog biscuit (my dogs love Woof-a-ronies) or a homemade cookie can be substituted for a human during hide 'n' seek. In our case, we keep a couple of our dogs outside and then bring them in and say, "Where's that cookie??"—which means Angie and I have hidden four or five treats in various places in the family room. The two dogs that compete best are Charlie, our Lab, and little Lucy, the mixed Chihuahua. They know it's their game, and they race around with smiley faces, sniffing under the couch or chair, behind the privacy screen, beneath an overturned rubber bowl, and so on. Afterward, they seem satisfied with their ability to uncover the goods, and, of course, both think they've won. Of course, we congratulate both girls and tell them how they are surely tracking-dog prodigies!

Toss-a-Teddy

Toss-a-teddy is another great playtime exercise that both Lucy and Charlie seem to love. It's the old game of fetch, but it's done with the dog's favorite bear toy, which may never be relinquished for the next round because she gets into self-play with it! (Lucy's is actually a small furry kitten, and Charlie's is the almost indestructible Vermont-style bear—no eyes, nose, or mouth to swallow!) Play it with one dog or two, but be sure that you have two players who find play more important than "winning"—you wouldn't want to create competitive aggression where it didn't exist before!

We say, "Charlie, toss-a-teddy?" I hold her bear so that she can see it for a second (I'm usually on the family room couch or chair). I know she's ready when her ears are perked, her tail goes up, and she tenses, ready to get to it first before the other pets. Then I toss it over their heads so they have to run to get it. If you get it back to toss again a few times, that's terrific. Be prepared to watch a fascinating game of toss-a-teddy, with each dog carrying it proudly, and then tossing it for the other pooch to get, or in Lucy's case, tossing it just far enough so she baits Charlie with it, but takes it back on a whim. It's great fun and an excellent energy release. Some of our friends tell me that they play this game with their dogs in their living rooms. Other people, no doubt, will prefer the outdoors, but remember that teddies that stray outside may wind up dirty or even lost or buried. If that happens, I guess you can resort to a game of "buried treasure"!

Hide 'n' Ambush

Hide 'n' ambush is for cats only! Try placing a nice woven cotton mat in front of the bathtub or other stalking place your cat loves. Then watch her as she discovers a purrrfect match: soft cotton material she can scrunch up with her paws and an ideal hideaway from which to plan her next zoom-zoom sneak attack. Watch as she plops on her side and attacks the mat's corners, scratches the middle with all fours, kicks and bites the imaginary critters that are trapped inside, and—as she becomes wrapped up in the mat (literally)—peeks out with wide eyes. But be careful not to move—don't help a good cat go bad!—a walking human is great "prey" to ambush from the confines of a soft, warm rug!

Add-a-Ball

For a variant on hide 'n' ambush, play it on the safe side and add a ball or her favorite small toy on top of the rug. Now when she needs to unleash her attack on something that moves, she can bat, grasp, and bite the ball. Besides, she didn't give that darn thing permission to be on her rug anyway!

Toss-a-Tennis-Ball

Toss-a-tennis-ball is obviously a variation of toss-a-teddy, but this game is for cats! Unfortunately, "retrieve" isn't in most feline vocabularies, so you'll probably have to spend more time going after the ball. Just think of it as good exercise.

Tasty or Tempting Toys

Finally, playtime can involve different kinds of treat-filled balls—variations on the Buster Cube. After your dog gets the hang of getting treats from one kind of food-dropping toy, switch to another. The variety of textures, sizes, materials, and ease with which he can obtain a treat will keep him interested and occupy his playtime while you're there watching. Sometimes a dog can become so engrossed in the activity that he may not notice you've slipped off to do something else (a hint for separation-anxious dogs, whose self-play motivation makes separation less traumatic).

The Chase and Pull is a toy that most dogs get into, whether they're terriers or herders. It looks like a cat dangler but consists of a hairy, bearlike toy on a rope that squeaks when your dogs bite it. You can participate in your dog's play because the rope is attached to a flexible shaft, which you can hold onto or toss. Luckily, most dogs seem more interested in pouncing on it than on the neighbors' irritating kid.

Ain't Got Game? Try a Clicker

Dogs who just can't get the hang of a game but seem to want to participate anyway, may benefit from some clicker training. As he approaches a new treat-giving ball, click-treat him. When he touches it with his nose, click-treat; moves it with his nose, click-treat; until the first treat falls out, at which time

Quick Fix for Stick Eaters

Some dogs—especially retrievers, who like to have things in their mouths—would much rather sit down and munch on that stick than retrieve it. Here's how I tell people to squelch that kind of behavior in their dogs. If he hasn't developed too strong a habit, you may be able to teach him to know the difference between sticks and a couple of items he **is** allowed to chew on. Play with him daily using the same rope or rawhide bone so that he'll be able to smell and taste the differ-ence between these and sticks.

If your dog begins to go for a stick, interrupt his progress by saying, "Ah! Ah! Ah!" or "Wrong!" and call him back, praising him as he returns. Offer him the "good" item instead. The idea is to teach him that rope or bone means play—any-thing else means correction. This task is easier than trying to teach dogs that it's okay to pick up something they love to eat, but not eat it! (Try that yourself with a candy bar!)

he should be on his own! (See Recommended Reading and Viewing on page 426 for books and videotapes on clicker training your dog or cat.)

Match the Game to the Breed

Try these variations of what to do during play with purebred dogs so that pre-dictable playtime remains interesting and challenging. Some dogs seem better prepared to enjoy some kinds of games than other games, and breed type may play a role. For example, scent hounds like bloodhounds and bassets seem to enjoy tracking games or olfactory hide 'n' seek games, where a well-used tennis ball, mostly empty peanut butter jar, or even a stuffed Kong is hidden somewhere in the fenced backyard. They seem genetically prepared to sniff and track, as opposed to visually surveying the landscape for a clue, or to look and track, which some sight-hounds (Afghans, for example) may prefer.

My sister Judi's golden retriever, Scout, used to find all the tennis balls her son, John, hit into the myrtle. Scout was on duty as a ball boy when his tail started twirling around in a circle like a helicopter blade. Of course, after Scout found the ball, it was too soggy for John to hit against the garage door—so the basket of "Scout's balls" was always filled to the brim!

Retrievers retrieve, herding breeds herd (or at least run after things and try to move them somewhere, or keep them from moving), and terriers dash,

How to Get Your Cat to Buy Into Walking on a Leash

Leash-training your cat is not easy. She may want to go out, but walking attached to you certainly isn't on her agenda. However, it can be done, and both of you can learn to enjoy it. Here's how:

1. Start leash-training when your cat is young. Put a harness on your kitten.
2. Get the right equipment for the job: a cat harness that fits. Don't even think about using a heavy dog leash, please! A collar might slip over her head as she resists, or hurt her neck, which is more delicate than a dog's. So don't attach the leash to her collar!
3. When Midnight is relaxing on your lap or the floor, approach her with the harness. Stroke her gently as you slip it on.
4. Let her wear the harness around the house for a few minutes, building up to a half hour. She may try to scrape it off at first, but that should stop as she finds more interesting things to do. (This is a good time for treats, feeding, or playing.)
5. Remove the harness gently.
6. When she seems used to wearing the harness, attach the light cat leash to it.
7. Walk around the house for a little bit, going where <u>she</u> wants to go, as she gets used to the added leash.

pounce, and shake things. (Please don't let your terrier use a toddler as the "thing"!) Shape your games accordingly, and you'll have one very happy dog.

🐾 Same Old Walk? No Way!

"Ho-hum, not the same old walk again!" Remember that you're not the *only* one who's bored! Try some exercise alternatives, and you and your dog will both be happier.

A New Leash on Life

For dogs who are still into sniffing wherever they go (and which dogs aren't?!), try walks in new and different places. If you're sure that your dog is under voice control when you walk on a street or sidewalk, invest in one of those extra-long retractable leashes so that he can follow his nose. That's bound to

8. Walk around the house where <u>you</u> want to go.
9. When she's comfortable, move it on outside and enjoy yourselves! (If she obviously isn't comfortable with the harness after several tries, both of you might be better off if you hang up the leash and forget about it. Never force her to give up her independent ways if she doesn't do so willingly.)

Your cat will feel more comfortable—and you won't have to worry about her escaping—if you put her in a halter and attach the leash to that, not to her collar.

be much more stimulating than the same restricted walk down the same sidewalk, in the same direction, day in and day out. And if you have access to a free-run area in your community (or in your neighbor's backyard—better his than yours), try some off-leash activities with dog friends.

Just be sure that you know who the players are, and that the fence gate is closed before you let them go at it. If it looks like your Jack Russell is getting in over his head with his German shepherd friend (or he's making you feel uncomfortable by digging up your neighbor's new peach tree), practice calling him. He'll probably be glad you did, and you'll build the trust he's looking for in a secure human-pet relationship. (Can't you just see him daring the host's German shepherd to try to get him now that he has you to hide behind?) So next time you call him, he'll be more likely to come. He has fun. You have fun. And your neighbor gets to clean up the poop!

Pick Up the Pace

Many dogs enjoy accompanying their owners on a daily jog or run. Just remember that your pooch has no place to sweat except through his footpads, so
if you are mopping your brow, he's probably starting to pant and get dehy-
drated. Don't push your dog beyond his capabilities—most pets would rather
drop from exhaustion than stop running beside their best friend. If your dog
is older, has painful joints, or is too small to keep up, try some "interval
training" when you exercise with her—a little walking, a little jogging—and
save the serious stuff for the gym.

Let's Dance! The Freestyling Fad

Maybe your dog is in great shape and finds walking or running a bit too
tame. You and he might enjoy the latest craze sweeping the dog world:
freestyle! And, why not "dance" with your dog? It's a relatively new canine
sport and may be an option for those whose dogs (or owners) can't do agility
training. In agility, the owner must run with the dog through the course,
so, obviously, you both have to be fit. Additionally, large dogs in their for-
mative months may injure themselves in negotiating the course (it may place
too much stress on tender hips, for example).

Freestyle can be done with any size and age of dog. Further, my clients who
have tried it say that it is as much fun for them as it is for their dogs. Freestyle
essentially involves learning how to "dance" with your dog to music, in a
planned performance. It offers a rare opportunity to bond with your dog and
challenge his interests! It's way up there in generating doggy excitement and
smiles. Making freestyle a part of your social routine with Marcus won't disappoint either of you. There are two divisions in the WCFO (World Canine
Freestyle Organization, Inc.): Heelwork to Music and Musical Freestyle. More
information is available at their Web site (www.woofs.org/wcfo) or at the Canine Freestyle Federation, Inc., Web site (www.canine-freestyle.org).

Like Team Sports? Flyball's Your Game!

Another dog sport that is sweeping the country—and the world—is flyball. In
this lightning-fast competition, teams of four energetic canines each race the
clock to run (and I *do* mean run!) down a 51-foot lane set with four hurdles, and
release a tennis ball from a box at the end. Then the first dog races back to the
beginning, and the second dog takes off. The first team to finish with no mistakes wins. This sport welcomes dogs of all breeds and sizes—in fact the little

guy is especially prized because the height of his team's hurdles is set at 4 inches below his shoulder (from 8 to 16 inches). Top dogs can run this course in less than 5 seconds, a speed usually seen in dogs only when Mom leaves a steak unguarded for a moment on the countertop!

Catwalks

I don't know about disco or relay races for felines, but you can certainly take your cat for a walk—if you have plenty of patience and resolve to take it slowly. Whereas dogs often burst into paroxysms of joy at the sight of their leashes, cats are a different story. You may have been able to get your cat to wear a collar—even one with a bell and ID tags attached—but felines tend to go hide under the bed when you pick up a leash and summon them for a nice walk. That leash looks more like an instrument of torture to a freedom-loving feline! Still, it's one of the only safe ways for indoor cats to experience the outdoors.

But you can't just clip a leash onto a cat's collar and start walking down the block. Chances are good that the cat will stage a sit-down strike right there and then, or should I say a "slide on my tummy if you decide to pull me" protest demonstration. The area around her neck is rather sensitive, so you don't want to be pulling on it. Here are a couple of tips: Try a harness—many cats get used to walks wearing a harness, which gives you quite a bit of control and her a lot of freedom. Another safe way to help her experience the great outdoors is to try putting a stool inside a mobile dog crate and then placing this mobile playpen outdoors on the patio beneath a tree or on the deck in a shady spot with you. She can explore the outdoors with all of her senses, and she's sharing it with you.

Dogs, Cats, and Other Pets

Often, a dog or cat joins a household with (or is joined by) a myriad of birds, fish, reptiles, or smaller mammals that elementary-school-age kids tend to collect. Like Beanie Babies and baseball cards, this is usually a passing fad, but it can make for some pretty interesting encounters. "The more the merrier" is a very giving attitude when it comes to pets, but you need to give some thought to the food chain, in addition to teaching Johnny to love and appreciate animals.

First of all, wild or protected animals should not be kept at home. This includes iguanas, deskunked skunks, box turtles, wolf-dogs, and snakes you find in the backyard. (One of my sister, Judi's, children put a snake in an aquarium, and she promptly had 18 babies, which would all have died in captivity.)

(continued on page 414)

DR. WRIGHT'S
CASEBOOK

Brandy and Stolie

Playdates work for dogs just as they do for kids—but don't try them with cats, unless you enjoy spitting, growling, swatting, and general havoc. Stephanie, a single 20-year-old college student, lived with her two dogs— Brandy, a mixed adult female Lab whom Stephanie had rescued from the street when Brandy was already pregnant, and Stolie, Brandy's even more mixed 2-year-old daughter. The three loved their time together—studying in the evening, sleeping late in the mornings, and partying hard with friends and pizza on the weekends.

It was at one of those parties that Stephanie met Dave and Rhonda, a couple who shared a rented home as she did and kept an only dog named Critter. Critter was a neutered 10-month-old boxer mix with a pronounced underbite. (His owners were the only ones who didn't regard Critter as a bit facially challenged.) He had not been around too many other dogs and was not yet obedience trained. He was driving Dave and Rhonda crazy, using them as "exercise equipment" each day when they returned from school. Because they were all three students in one of my psychology classes, we shared conversations about our dogs in the hallways between classes, or whenever we happened to be walking together on our way somewhere on campus.

It turns out that Stephie was getting into a rut with her dogs. They were a great comfort to one another, but Stephie thought that their lives could be so much more than just sitting at home all day, going out to eliminate a couple more times before bedtime, and then repeating the performance the next day. It wasn't the holistic, well-balanced life she had imagined for her dogs. But her college life was pretty full of things that needed to be done, and she was only one person. So she defaulted to "no time to walk tonight" or "that's all the time for play I have right now" when her daily schedule became packed with more immediate concerns.

One Monday afternoon on the way to the parking lot, we were talking about what was in store for Stephie and her dogs. David and Rhonda were just escaping from their classes, exiting from the same building, just to our left. They easily joined in our conversation: That's what we always did when we happened on one another, we talked about dogs. This time, we got an update on Critter. There was nothing new on that front, either. Critter was still full of energy, and like Stephanie, David and Rhonda talked about what they would be walking into when they greeted their lonely dog as they arrived home.

After hearing David and Rhonda describe how Critter was driving them nuts (they were hoping against hope that I would give them the magic judgment that would make their lives "normal" again), and admit that they had gotten into a bad habit of yelling at the frisky pooch all the time, the solution came to all of us at once. Neither "family" was happy or healthy, and both had a similar problem. I suggested that they consider a "playdate" for Critter—let's see if Brandy and Stolie think Critter is as interesting as his owners insist he is, and let's get the two females on the road to cavorting with a male dog who would love to have them visit.

The three students quickly worked out the details. Stephanie would take her two female dogs over to visit and play with Critter at David's and Rhonda's house. They agreed on a playdate three days a week after school. (Tuesday and Thursday were out because Stephie had a late class those days.) On the weekend, Stephanie and the dogs had plenty of "environmental enrichment" with friends, salsa, pizza, and CDs.

Here's an etiquette tip to make sure that you and your dog are always welcome for your playdates. If you're the visiting owner, provide snacks for all the dogs (and people) on the playdate. Check with the host or hostess about any dietary restrictions (theirs and their dog's!). Just as with a kid's playdate, be sure to offer to help clean up when you and the dogs leave, or you might find the welcome mat pulled out from under you! It is also a good idea to have a friendly chat about damages and vet bills before you turn those pooches loose at someone else's house.

As a general rule, you need to protect any small pet from cats and dogs. Cats will definitely stalk, lift the lids from cages, and otherwise find a way to terrorize your child's chick, mouse, or even hamster. Try amusing the cat with a video made for cats so that the daily default doesn't become another animal. (For more information, see page 423.)

My sister Judi had the horrible experience of having her cat, Kitty, knock over the cage of our mother's beloved canary, Hi-Fi, leaving nary a feather as evidence. You can't blame the cat or dog for these mishaps or change their predatory natures (with a few exceptions as noted later), so just make sure that there is no access to other, helpless pets if your kids like to have a menagerie.

🐾 Making Friends

Someone asked me recently what to do about their chinchilla and their terrier. After I stopped shuddering at this combination, I thought about the owner's concerns—the dog was "beside herself," sitting outside the door of the chinchilla's room and crying all day. The terrier had even lost interest in eating. The family unfortunately hadn't thought ahead and anticipated these reactions—after all, terriers were bred to root out rodents and other ground-dwelling animals and destroy them. The poor dog was only doing her job.

As I told the owner, there have been numerous cases of "nurture" over-riding "nature"—examples of natural mortal enemies who have become constant companions. For example, cats have accepted and successfully reared mice and rabbits along with their own kittens on more than one occasion. In fact, maternal cats groom their kittens, baby bunnies, and mice equally often without incident. Kittens have also been reared with puppies without incident. I actually knew a retired greyhound who befriended and reared a very young squirrel! But what all these examples had in common was a "youngster." One or both species need to be young for this to work.

In this case, here's the program I suggested to the owner of the terrier and chinchilla. Try it if you have a dog and a small, rodentlike pet.

1. Exercise the dog before the introduction. Tired is good!

2. In a quiet, relaxing room in the house, have the dog lie down on the rug. Massage her and provide her with a treat or two, while she remains lying calmly on the rug.

3. Have another family member sit on the floor, holding the chinchilla close to her.

4. As the dog begins to show calm interest in the critter, offer her a treat. (Hold the treat so that she must come toward you, not the chinchilla.) Praise compliance.

5. Continue this encounter as long as she's calm.

The idea is for the terrier to connect calm, feel-good behavior with the presence of the chinchilla. Do this gradually, respecting both animals' desire to "leave the scene" if either so desires. And next time, consider getting a goldfish instead—but watch out for the cat!

Adding On

So much for sweating the small stuff, what about getting another cat or dog? This commitment is much bigger, and you truly want the two (or more) animals to be great pals, if possible—not just two animals grudgingly tolerating each other when you can't keep them separated. You probably assume that your pet would like some canine or feline company, but that's not necessarily so.

🐾 Should You Add Another Dog? A Six-Question Quiz

Many adult dogs (and cats) prefer to regulate their own daily patterns of activity, including such things as when and where to rest, watch squirrels or birds outside the back window, bark or glare at the letter carrier, and watch on the landing for the school bus or Mom's and Dad's cars in the driveway. She's been doing it for years. Pets love to be able to predict what's going to happen, when. So before you try to make her life complete by adding a second dog, ask yourself these questions:

1. Is it possible that your dog actually prefers to be alone during the day, but really loves it when you pay attention to her when you get home?

2. Is the decision to get a second dog fair to the second dog? Is it fair to the first dog? If you can't spend enough time with your dog now, how will that work when you're playing with the (more playful) puppy? How will you find the time to socialize the pup and teach him how to be in your family?

3. You know your dog. Is it likely that she will become "jealous" of the attention you pay to her new sibling? Would you be creating a jealous pet where none now exists?

4. Does your female or male like to play with other dogs? Will the two dogs have the freedom to play and run about the house in your absence? Having a sib isn't fun if you're locked in someplace or confined in different rooms.

5. Do you have adequate financial resources to feed, groom, board, and provide medical care for another pet?

6. Is yours the kind of flexible household where it's "the more the merrier"? Or do you feel the need to keep a great deal of control at all times? (That's almost impossible with more than one dog.)

It may sound like I am trying to discourage you from getting another dog. I'm not. We have four dogs, and started out with one, an only dog for 4 years. We love each one, and they all get along wonderfully, despite their very different personalities. I really think the best reason for choosing to adopt another animal—providing that the positives outweigh the negatives—is that *you* want a second—or third or fourth or more!—companion to live with.

These questions are probably a simpler version of what many parents of only children agonize about. You, of course, have an "out" if things don't work the way you envisioned, but if you are spending a lot of time comforting yourself with thoughts like, "Well, if things don't work out, I can always return her to the breeder or humane organization," then you ought to wait until you can summon up a bit more commitment to go with your desire.

❧ How to Keep Tons of Cats—One at a Time

Everyone has heard about those good souls who can't pass up a homeless or unwanted cat. It's usually a "cat lady," and she often ends up in trouble with the law or the homeowners' association or the Board of Health because it is very difficult to take care of dozens of cats without raising some eyebrows concerning sanitation—or sanity!

I don't think people should go overboard—the same questions I posed in "Should You Add Another Dog? A Six-Question Quiz" on page 415 for dogs also apply to cats. Obviously, a cat shouldn't be left to die or be destroyed—anyone who rescues a needy animal deserves heaps of praise. But there is perhaps a better way to "have" lots of cats if you so desire, yet give each that quality of life we are aiming for. Many humane organizations now like to house adoptable pets—especially cats—in private homes while the cats wait to be adopted. Why not consider being a cat foster family? Even though saying good-bye to a cat you have become attached to can be difficult when he is adopted, many more needy kittens and cats are waiting for your love. Not everyone is emotionally suited to this, but a "serial" arrangement instead of multiple cats at once is something serious cat lovers should at least consider.

DR. WRIGHT'S INSIGHTS

Here's how to comfort a "suddenly solo" cat. If your cat is obviously suffering from the loss of his sibling, try to reestablish a healthy, active, ritualized lifestyle by providing him with the same kind of stimulation he enjoyed from your late cat. If he ate with the other cat, try staying with him while he eats. Have a snack yourself. If they groomed one another after breakfast, try brushing him, or at least stroking him, and allow him to lick and rub you in return. If they played after grooming, try initiating play with him. In general, try to satisfy his needs for such motivationally different activities as eating, grooming, playing, and sleeping, and he should begin to initiate and perform them on his own in his own time.

❧ Replacing a Pet

What about replacing a pet who dies? Not too long ago, I heard from a woman whose family had just lost one of a pair of inseparable 12-year-old cats. She was very worried about the suddenly solo pet and wanted to know whether she should go out the next day and "replace" the one who had died. I gave her my thoughts.

Like humans, cats grieve over the death of a lifelong companion. In such cases, they become both emotionally and behaviorally depressed. They mope around the house, vocalize, and generally look lost. As we discussed in chapter 16, they may sleep a lot or stop eating, self-grooming, or using the litterbox. If this describes the remaining cat, he may have discovered that the void created by the absence of his sibling resulted in the loss of things he valued, things that kept him going from day to day, such as thermotactile (warm touch) comfort, social stimulation, and companionship.

Many cats seem to look for their brother or sister for a few to several days following his or her absence, and then they discover that life goes on. I'm acquainted with a rather shy, passive cat who lived with her assertive, take-charge sister for about 8 years, until the latter's untimely death. The shy cat then became more outgoing, sought and received attention more often from family members than she had in the prior 8 years, and even solicited play from her humans on a daily basis—an activity she never attempted in the presence of the other cat.

Regarding the "let's get another cat for him so he won't be so lonely" issue, as is the case with the only dog, probably the best reason for adopting a second cat is that you want a second cat. Although I haven't yet figured out how to read their minds, many felines appear to be perfectly happy spending the day alone—gazing out the window, lying in a sunny spot on the living room rug, and attacking imaginary critters that happen to wander by too closely—especially if they can count on some quality time with their caretakers at the end of the day.

🐾 The More the Merrier—Multiple Pets

Okay, you've decided to add that new dog or cat. (I knew that you couldn't resist; I couldn't either!) Now, any of the many behavior problems we have talked about in this book can arise with the new pet, of course, but you know how to head off or handle just about anything. So you are well-prepared for the additional pet or pets. But you may have more trouble with the relationships your pets forge with each other—and yes, it can be a dog-eat-dog world. If you've picked out "nice" cats and dogs to begin with, you can expect them to learn to get along fairly easily. But you must still be on the lookout for some things because no animal is totally predictable—much less two or three together! Here are some tips for smoothing the waters.

The Secret of Keeping Peace in the Two-Dog Household

I am distressed when I hear about sibling rivalry between two household dogs. Many unintentional bites to people occur when we try to "save" the underdog by sticking an arm or hand into the fray and wind up with a painful puncture wound for our trouble. Quite a few of my clients have turned to me at their wits' end with two dogs whom they love "both the same," except when one is viciously attacking the other and the second is reduced to a pitiful nervous wreck. Then the owners tend not to love the aggressive dog so much!

My first recommendation to readers in this predicament is to consult a certified animal behaviorist or trainer who has experience reducing dog-dog aggression with the use of humane and effective techniques. In the meantime, you may wish to consider a few things.

Let me start by saying that if you are trying to treat two dogs equally, you may be contributing to the problem! Dogs who have problems with one another get along better when we try to *increase* the distance in status between them. Try to treat one dog (the more domineering one) as though she belongs in a dominant role and the other dog as the one occupying a subordinate role. Don't go by their ages, which one you secretly like better, or how big they are. You know your dogs; decide which is the natural leader and which is the less assertive one.

These roles say nothing about their personalities. It's not that we're calling the subordinate dog weak or inferior in any way. You can help the dogs maintain their inequality by offering things (such as food, treats, play, walks, access to desired resting locations, greetings, and other activities) to the top dog first. When you praise the second dog or give her treats, try not to do it in the presence of the first.

Second, make sure that *you* are in charge of both dogs! If the top dog looks like she's about to get ugly, change her mood by saying a word or phrase that elicits a happy mood, and call her to you. This should keep her mind off being "angry" and substitute "happy things." (As you know, she can't be happy and angry at the same time.)

Correct the second dog first and then the top dog, regardless of who is at fault, but don't let the number one dog be a bully. Correct her if she appears to be taking advantage of her status. Praise her for being compliant, and really let her know she's a great girl.

Third, if there are "hot items" that increase the likelihood of a fight (like rawhides or a favorite toy they both like), get rid of them. In fact, whatever you can do to decrease the negative arousal of both dogs is a good thing. If the top dog won't "de-arouse," give her a short (2- to 3-minute) time-out, if she normally responds well to that kind of correction. And never place yourself in a position that puts you at risk for being bitten!

These ground rules, seemingly based on playing favorites, are not going to be easy to follow, especially for parents of children, since treating one as more special than the other has never been a hallmark of good parenting. But trust me! This approach works—and both your pooches will be happier when they know their place. Just don't try this with your kids; you'll never hear the end of "Mom liked you best!"

How to Spot a Top Cat

These same principles apply to cats. Although some cats don't appear to form dominant-subordinate relationships, others do. If you're not sure which is dominant and which is subordinate, here's a tip. The dominant cat often seeks out the higher places in the house to rest on. He's the one at the top of the cat condo or seated on the refrigerator—literally, the top cat. The cat of lesser status is content with the lowly windowsill or cat basket.

If you have more than two cats, provide multiple levels of nesting and resting places. In that way, they can jockey for position and sort themselves out into the proper pecking order. Some humane societies have picked up on this trick, and they swear that it keeps cat fights to a minimum!

🐾 My Take on Multipet Households

I have some final thoughts on whether more than one household pet is the way to go. My own experiences with pets weigh in heavily here, and I write from the perspective of someone who—like the cats and dogs we live with—feels most comfortable when things are rather orderly and predictable. You might thrive in an atmosphere of chaotic excitement, to which your pets will do their best to adjust. Everyone is different.

Over the years, I've found myself sharing a household with a single dog, two cats, one cat and one dog, and one cat and two, three, and now four dogs. Caring for two dogs as opposed to one has its pros and its cons. If you add more than that, you're really asking for trouble!

Here are some observations on the "pro" side:

1. Your present dog will have another of its own kind, a buddy to hang with; for him, the more, the merrier.

2. Since you're getting up in the morning to take out one dog, you might as well take out two!

3. Feeding two dogs is just as easy as feeding one—just separate the two bowls and you've got it.

4. The dogs can get rid of a lot of energy by playing with one another, so you can opt to join them or enjoy the action on the sidelines.

5. Each dog is unique, an individual to get to know in his own right, so you'll have twice as many neat personalities (to say nothing of an occasional "character"!) to love and enjoy.

6. And you get twice the smiles, tail wags, and kisses (licks) when you come home from the grind every day.

The "con" side also has some valid points.

1. Walking two dogs simultaneously is a real trick (unless you're a couple, and you each take responsibility for one dog). If your one dog has a knack for getting you wrapped up in his leash, think about how many knots two dogs can tie you up in! If you add a third dog, we're talking The Mummy! And, yes, walking multiple dogs can be dangerous if something catches their eyes and they start chasing it.

2. Exercising two dogs is a challenge, but handling three dogs is many

times harder than handling just two. Think about it. With two dogs, you have one for each hand. What will you do with three dogs? Looks like you'll have to repeat your routine daily—take out the first two and then start all over with number three.

3. Dogs need shots and other veterinary medical care annually—start adding it up, along with the extra dough you'll have to shell out for food, boarding or a pet sitter, leashes, collars, and other necessities. Add this to your budget, and see if it's reasonable. Be sure to double your penalty if you're considering adding two littermate puppies to your aging dog's homestead.

4. The amount of "doodoo" you'll have to scoop is proportional to the number of dogs you own.

5. More hair will collect on the floor to clog up the heating/air intake register if you forget to sweep for a couple of days. (Poodles and other breeds that don't shed may help here.)

6. Imagine sharing your cozy bed and pillow with multiple slurping, snoring dogs who are psychobiologically drawn to prefer thermotacticle comfort. Now try stretching out.

As you can see, the pros and cons seem to be equally balanced. You alone must decide whether an additional canine is right for you!

Many of the same pros and cons apply for cats as well as dogs. Though possibly a bit more insistent on having their own "space," cats can become bosom buddies, be it curling up together in a contented ball of purr or simply sharing the same home without incident. The idea of having two cats has become so popular, in fact, that two or more is the standard for cat families, compared to dog households.

Does this mean that multiple cats mean more issues? Of course it does! And the more cats you bring into the household, the more likely that issues of marking and spraying will arise at some point. Upsetting the status quo naturally leads to more stress, marking, and other ways cats express their dissatisfaction. So keep that in mind before you agree to adopt the entire litter, okay?

As for cats and dogs together, the old cartoon staple of Fido chasing Boots all over the place rarely holds up in real life. Anyone who owns both species can attest to the fact that one well-placed swipe to the dog's nose usually tells Fido who's boss these days. So go ahead and enjoy your multiples in whatever combinations life brings to your doorstep—just don't go overboard and take on more than you and your human and animal companions can really handle.

Resources

Pet Catalogs

The assortment of pet products of all kinds that you can get through the mail is amazing. Check out the 380 pet-related catalogs available by going to the Buyers Index Web site (www.buyersindex.com), or jump in and order one of the following catalogs.

Drs. Foster and Smith catalog
Doctors Foster & Smith Veterinary Products
2555 Airpark Road, P.O. Box 100
Rhinelander, WI 54501-0100
Phone: (800) 826-7206
Fax: (800) 562-7169
www.drsfostersmith.com

R. C. Steele catalog
1989 Transit Way, Box 910
Brockport, NY 14420-0910
Phone: (800) 872-3773
www.rcsteele.com

Pedigrees catalog
1989 Transit Way, Box 905
Brockport, NY 14420-0905
Phone: (800) 548-4786

Pet-Related Web Sites

Web sites come and go, but here are some of my favorites, still alive and kicking at print time.

100 Top Cat Sites (www.100topcatsites.com)
100 Top Dog Sites (www.100topdogsites.com)
About.com: Cats (http://cats.about.com/pets/cats)
About.com: Dogs (http://dogs.about.com/pets/dogs)
CLAW, the Service Club for Cats Who Care (www.claw.org)
Dog Infomat (www.doginfomat.com)
Dog-Play (www.dog-play.com)
NewPet.com (www.newpet.com)
PurrBalls (www.purrballs.com)
Woofs and Meows (www.woofsandmeows.com)

Cool Stuff for Pets

I recommend all four of these products for easier dog training and happier cats (and owners!).

Head halters or head collars. Snoot Loop (by Peter Borchelt, Ph.D., Animal Behavior Consultants, Inc., 2465 Stuart Street, Brooklyn, NY 11229, 800-339-9505); Gentle Leader (Premier Pet Products, 527 Branchway Road, Richmond, VA 23236, 800-933-5595; also available through your veterinarian and at www.sitstay.com); Halti (Coastal Pet Products, 911 Leadway Avenue, Alliance, OH 44601, 800-367-7387; also available through Drs. Foster and Smith and R. C. Steele catalogs). The advantage of these training collars is that they are not trachea busters. They enable your dog to correct his behavior quickly with the least effort on both your parts, and they are efficient communication tools, minimizing the confrontational aspect of training.

Automatic feeder. For cats who need to be fed on the weekend, at times in your absence, or on a specific feeding schedule for health reasons, the Cat Mate runs on a timer (and battery). You can choose a model that dispenses either two meals (model C20) or five meals (model C50). When it's dinnertime, the lid pops open, exposing a prepared meal for Fluffy. The advantage is that measured amounts of food can be made available at specific times. Also useful for cats and small dogs who need that small after-your-bed-time snack, this feeder can be placed far away from your bedroom door on the other side of your home so that any anticipatory crying for food will be heard only by the Cat Mate. Cat Mate is sold by Pet Mate Ltd. in England but is available by calling Pet Mate's U.S. distributor at (800) 725-4333 or visiting their Web site at www.pet-mate.com to find currently available catalogs.

Cat scratching posts. The Alpine Cat Scratcher (18.25″ × 8.25″ × 10.75″ high) may be the answer for kittens and small cats who enjoy pulling on the pile of rugs (or other horizontal surfaces) and on the arm of your favorite couch or chair (or other vertical surfaces). This little gem gradually slopes up from the floor so that Kitty can sink her claws forward and up into replaceable corrugated cardboard. Her clawing is elevated at about a 45-degree angle, and the package includes catnip for those who partake. Added to this mini-playground is a secret porthole that sports a dangling mouse that your cat can attack once her claws are sharpened! This equipment is available from the Drs. Foster and Smith catalog.

Videos and games for cats? Forget the previews and popcorn. *Kitty Safari—A Video Adventure for Kats* and *Kitty Safari 2—More Video Adventures for Kats* will occupy your feline while you take a break, and wear her out! Both videos are filled with treasures that will interest cats, from loads of bird noises and other wavering sounds to visual movement and footage of critters. They'll induce your cat to pay attention! These videos are available from www.amazon.com.

But my *favorite* is a computer game for cats called *Cyberpounce*, which is available at www.cyberpounce.com. It's interactive—just you and your cat playing together (and besides, I was the animal behavior consultant who contributed to its development!).

For Older Pets

Older dogs and cats often have special needs. Here are some innovative products that will help you help them.

Stair-steps. These steps (25.5" × 16" × 25" high) allow your aging, arthritic cat or small dog to reach the comfort zone of her favorite resting area without your having to lift her. It is carpeted (and comes in different colors), has three steps (one lifts up to store toys in), and has a hide-away tunnel in the base from which your pet can ambush imaginary geriatric mice! Another advantage is that you no longer need to risk that heart-wrenching "Rrrre-e-e-a-ar!" or yelp from her because you accidentally touched her painful zone while lifting her onto your bed. These steps are available from the Drs. Foster and Smith catalog.

PetSTEP folding dog ramp. This ramp (6 feet, folds to 36" × 18" and comes in half-size, too) allows your slow, tender-jointed dog to continue to enjoy car rides without the help of a human to lift him in and out of the vehicle. The 19-pound ramp (when fully extended) hooks over the lip of the back bumper to allow hatch-back access, or over the backseat floor for getting into vans. The surface is ribbed to help reduce slippery footing when wet and holds up to 500 pounds. The ramp is available from the Drs. Foster and Smith catalog or at www.PetsMart.com and www.sitstay.com.

Dog diapers. Doggie diapers for solid and fluid waste are available at www.dog-diaper.com (fax: 419-730-8384), www.PetsMart.com, the R. C. Steel catalog, and the Drs. Foster and Smith catalog. They come in sizes to fit dogs who weigh between 2 pounds and 90 pounds.

Pet Sitters

To find a reliable pet sitter near you, contact your vet, your favorite kennel, your breeder, your breed group, or your trainer. Or contact the following organizations for certified pet sitters in your area.

National Association of Professional Pet Sitters (www.petsitters.com, 800-296-7387)

Pet Sitters International (www.petsit.com, 800-268-7487)

Dog Recreation Areas

Check out these groups for information on dog recreation areas.

TheDogPark.com (www.thedogpark.com)
c/o W3Commerce
125 S. Tremont Street
Oceanside, CA 92054
Phone: (800) 549-3904

Dogpark.com (www.dogpark.com/parkusa.html)
E-mail Vicki Küng, founder, at Vicki@dogpark.com

Cat and Dog Behavior

Here's where to go for professional help with training or behavior problems. (Often, your vet and breeder can refer you to a qualified trainer or applied animal behaviorist as well.)

Association of Pet Dog Trainers. Contact them to search for a trainer in your area (www.apdt.com/trainer.htm).

ASPCA Center for Behavioral Therapy. Consultations are either in the New York City–based clinic, in your home, or over the telephone. E-mail your dog's or cat's behavior problem to companion@aspca.org or write the Center at 424 East 92nd Street, New York, NY 10128.

Certified Applied Animal Behaviorists. For a list, go to the Animal Behavior Society's Web site at www.animalbehavior.org; then click on the "Directory" link under Applied Animal Behavior.

Recommended Reading and Viewing

These are my personal picks for your pet bookshelf.

Dogs, Cats, and Kids

Dunbar, Ian, Ph.D. *Dog Training for Children* (videotape). Berkeley, CA: James & Kenneth, Publishers, 1997.

Hunthausen, Wayne, D.V.M. *Dogs, Cats & Kids: Learning to Be Safe with Animals* (videotape). Chicago: Donald Manelli & Associates, 1997.

An excellent video pitched to reach children and educate families about dog and cat danger signs, what to do when approached by a strange dog, what to do if you are attacked, and more.

Healthy Pets

Ballner, Maryjean. *Cat Massage: A Whiskers-to-Tail Guide to Your Cat's Ultimate Petting Experience!* Gordonsville, VA: Saint Martin's Press, LLC, 1997.

————. *Dog Massage: A Whiskers-to-Tail Guide to Your Dog's Ultimate Petting Experience!* Gordonsville, VA: Saint Martin's Press, LLC, 2001.

Hoffman, Matthew. *The Doctor's Book of Home Remedies for Dogs and Cats: Over 1,000 Solutions to Your Pet's Problems—From Top Vets, Trainers, Breeders, and Other Animal Experts.* Emmaus, PA: Rodale, 1996.

————. *Prevention's Symptom Solver for Dogs & Cats: From Arfs & Arthritis to Whimpers and Worms, An Owner's Cure Finder.* Emmaus, PA: Rodale, 1999.

Shojai, Amy D. *The First Aid Companion for Dogs & Cats.* Emmaus, PA: Rodale, 2001.

————. *New Choices in Natural Healing for Dogs & Cats: Over 1,000 At-Home Remedies for Your Pet's Problems.* Emmaus, PA: Rodale, 1999.

Selecting a Dog Breed by Behavioral Characteristics

Arden, Darlene. *The Irrepressible Toy Dog.* New York City: IDG Books Worldwide, 1997.

Gerstenfeld, Sheldon L. *ASPCA Complete Guide to Dogs: Everything You Need to Know about Choosing and Caring for Your Pet.* San Francisco: Chronicle Books, 1999.

Hart, Benjamin L., D.V.M., and Lynette A Hart. *The Perfect Puppy: How to Choose Your Dog by Its Behavior.* New York City: W. H. Freeman & Co., 1988.

Kilcommons, Brian, and Sarah Wilson. *Paws to Consider: Choosing the Right Dog for You*. New York City: Warner Books, 1999.

Richards, James R. *ASPCA Complete Guide to Cats: Everything You Need to Know about Choosing and Caring for Your Pet*. San Francisco: Chronicle Books, 1999.

Tortora, Daniel F., Ph.D. *The Right Dog for You: Choosing a Breed That Matches Your Personality, Family and Lifestyle*. New York City: Simon & Schuster, 1980.

Rearing a Cat or Dog

Johnson-Bennett, Pam. *Think Like a Cat: How to Raise a Well-Adjusted Cat—Not a Sour Puss*. Newark, NJ: Viking Penguin, 2000.

Moore, Arden. *The Kitten Owner's Manual*. Pownal, VT: Storey Communications, 2001.

Rutherford, Clarice, and David H. Neil. *How to Raise a Puppy You Can Live With*. 3rd ed. Loveland, CO: Alpine Publications, 1999.

Training

Dunbar, Ian, Ph.D. *How to Teach a New Dog Old Tricks: Sirius Puppy Training*. Berkeley, CA: James & Kenneth Publishers, 1996.

Hetts, Suzanne. *Crate Training Your Dog*. (videotape) 2001.

Pryor, Karen. *Don't Shoot the Dog! The New Art of Teaching and Training*. Westminster, MD: Bantam Books, 1999.

———. *Getting Started: Clicker Training for Dogs*. Waltham, MA: Sunshine Books, 1999.

———. *Karen Pryor's 'Don't Shoot the Cat!' Kit*. Waltham, MA: Sunshine Books, 2001.

Tillman, Peggy. *Clicking with Your Dog . . . Step-by-Step in Pictures*. Waltham, MA: Sunshine Books, 2001.

Wilkes, Gary. *Click & Treat® Training Kit*. Available at www.clickandtreat.com, 2344 E. Alpine Ave., Mesa, Arizona, 85204, or (480) 649-9804.

Other Books by John C. Wright & Judi Wright Lashnits

Wright, John C., Ph.D., and Judi Wright Lashnits. *The Dog Who Would Be King: Tales and Surprising Lessons from a Pet Psychologist*. Emmaus, PA: Rodale, 1999.

———. *Is Your Cat Crazy? Solutions from the Casebook of a Cat Therapist*. Edison, NJ: Book Sales, 1998.

Scholarly Books for Further Reference

Bradshaw, John W. S. *The Behaviour of the Domestic Cat*. Cary, NC: Oxford University Press, 1992.

Lagoni, Laurel M., Carolyn Butler, and Suzanne Hetts. *The Human-Animal Bond and Grief*. London: W. B. Saunders Co., 1994.

Lindsay, Steven R. *Handbook of Applied Dog Behavior and Training: Adaptation and Learning*. Vol. 1. Ames, IA: Iowa State University Press, 2000.

This book tells you how to use posture-facilitated relaxation training.

Overall, Karen L., V.M.D., Ph.D. *Clinical Behavioral Medicine for Small Animals*. St. Louis: Mosby—Year Book, 1997.

Serpell, James. *The Domestic Dog: Its Evolution, Behaviour and Interactions with People*. Port Chester, NY: Cambridge University Press, 1996.

Turner, Dennis C., and Patrick Bateson. *The Domestic Cat: The Biology of Its Behaviour*. 2nd ed. Port Chester, NY: Cambridge University Press, 2000.

Voith, Victoria L., and Peter L. Borchelt. *Readings in Companion Animal Behavior*. Trenton, NJ: Veterinary Learning Systems, 1996.

Wilson, Cindy C., and Dennis C. Turner. *Companion Animals in Human Health*. Thousand Oaks, CA: Sage Publications, 1997.

Index

Underscored page references indicate boxed text. **Boldface** references indicate illustrations.

N